What they're saying about Ron Arnold

Arnold warned that unless the environmental movement is brought to heel, "public hysteria is going to destroy industrial civilization."
—*The Washington Post*

Ron Arnold's new book *EcoTerror* reveals the threat to industrial civilzation posed by ecoterrorism. It gives America unprecedented detail about those who try to make policy with sabotage—arson, pipe bombs and criminal trespass. Congress should definitely pay attention to Ron Arnold's vital message and extend legal protection from ecoterrorism to all natural resource workers.
—David Ridenour, National Center for Public Policy Analysis

Arnold established the 'Ecoterror Response Network' to 'compile the first comprehensive list of attacks against Wise Users—and to expose the environmentalist smear campaign to stigmatize the victims.'
—Andrew Rowell, author, *Green Backlash*

The most exasperating thing about Ron Arnold is that he grasps the shortcomings of Big Green environmentalism so much better than "my" colleagues who continue to be seduced by corporate foundation dollars and a self-defeating myth of access with Democratic Party power brokers.
—Michael Donnelly, Friends of the Breitenbush Cascades

Ron Arnold...is gaining increasing national stature and political influence as the arch-druid of the burgeoning movement against environmentalism.
—*The Boston Globe*

Every defender of land rights in America needs to read Ron Arnold's new book *EcoTerror* to see who and what they are up against—violent attacks against their homes, their jobs and their lives.
—Charles S. Cushman, American Land Rights Association

Arnold is now a fixture on the anti-environmental lecture circuit.
—*Greenpeace*

No wonder environmental extremists fear Ron Arnold. While the liberal media turn a blind eye to environmental terrorism and terrorists, Arnold exposes them for what they are: a threat to freedom and the preservation of an environment that includes people. Arnold's *EcoTerror* may read like a Tom Clancy novel, but being the unvarnished truth—with painstaking documentation—it is stranger and scarier than fiction.
—William Perry Pendley, author, ... 'West

ECOTERROR

Other books by Ron Arnold

At the Eye of the Storm: James Watt and the Environmentalists
(Regnery Gateway, Chicago)

Ecology Wars
(Originally *The Environmental Battle,* Vance Publishing, Chicago,
Winner, American Business Press Editorial Achievement Award)

The Grand Prairie Years (historical novel)
(Dodd Mead, New York)

With Alan Gottlieb:

Trashing the Economy

Politically Correct Environment

Edited:

Stealing the National Parks

People of the Tongass

Storm Over Rangelands

The Asbestos Racket

It Takes A Hero

ECOTERROR

THE VIOLENT AGENDA
TO SAVE NATURE

THE WORLD OF THE
UNABOMBER

RON ARNOLD

The Free Enterprise Press
BELLEVUE, WASHINGTON
Distributed by Merril Press

ECOTERROR

First Edition
Published by the Free Enterprise Press

Typeset in Times New Roman by The Free Enterprise Press, a division of the
Center for the Defense of Free Enterprise, 12500 N.E. 10th Place, Bellevue,
Washington 98005. Telephone 206-455-5038. Fax 206-451-3959. E-mail
address: books@cdfe.org. Cover design by Northwoods Studio.

EcoTerror is distributed by Merril Press, P.O. Box 1682, Bellevue, Washington 98009. Additional copies of this book may be ordered from Merril Press
at $16.95 each. Phone 206-454-7009.

We gratefully acknowledge permission to quote from these copyrighted works:
Portions of Chapter Four were reprinted from *Terrorism in America: Pipe Bombs
and Pipe Dreams* by Brent L. Smith by permission of the State University of New
York Press © 1994.
"Cockburn Replies," *The Nation*, by Alexander Cockburn. By permission.
"Unabomber Gores Technology" by Tony Snow © 1995 Creators Syndicate.

LIBRARY OF CONGRESS CATALOGING-IN-PUBLICATION DATA
Arnold, Ron.
    Ecoterror : the violent agenda to save nature : the world of the
Unabomber / Ron Arnold. — 1st ed.
        p.   cm.
    Includes bibliographical references and index.
    ISBN 0-939571-18-8
    1. Terrorism—United States. 2. Bombings—United States. 3.
Deep ecology—United States. 4. Kaczynski, Theodore John, 1942-
5. FC.   I. Title.
HV6432.A76   1997
303.6'25'0973—dc21   97-5377
                                                          CIP

PRINTED IN THE UNITED STATES OF AMERICA

CONTENTS

To Janet Arnold
sine qua non

PREFACE

I resisted demands to write *EcoTerror* for more than five years. My reluctance was not for lack of evidence of crimes committed to save nature, but the contrary: the raw data was so massive no one person could assimilate and analyze it, much less explain it clearly to a skeptical public.

Only the volunteer help of many hands convinced me the task was feasible. Thus, in late 1993, upon completion of *Trashing the Economy*, I agreed with my colleague, Alan Gottlieb, President of the Center for the Defense of Free Enterprise, to design a database of ecoterrorism. It meant computerizing a storeroom full of court documents, newspaper clippings and records of telephone calls from victims. It also meant sending out a call for information about ongoing incidents—the EcoTerror Response Network began receiving messages. The computer database slowly grew.

Then, on April 24, 1995, the project took on unexpected urgency: our friend Gil Murray of the California Forestry Association (CFA) was killed by the Unabomber. In our shock and grief, we suddenly found our Center a possible Unabomber target because of our long friendship with CFA leaders—CFA official Robbie Andersen had served as emcee at our first Wise Use Strategy Conference in 1988. The Unabomber's letters pointed to an environment-related motive for killing his last two victims, and the Center had long been a vanguard challenger of organized environmentalism. Alan Gottlieb said it was time to write the book.

Now the demands to write *EcoTerror* took on a moral dimension: we owed it to Gil Murray to tell the public that the Unabomber was only an isolated symptom of years of hate for industrial civilization that had been incited by the powerful environmental lobby, and that a systematic underground movement of violence against industry threatened the base of our entire society. My deepest gratitude to Donn Zea and the surviving staff of CFA for their extraordinary help in telling their story through dozens of interviews that required they relive the horror of April 24, 1995. To Bill Dennison, special thanks for his many hours reviewing Chapter One. My gratitude to Candy Boak, who spoke through much pain. Gary Gundlach and Rich Bettis gave me in-person interviews and many helpful documents.

Literally hundreds of people volunteered to help with this book. Kathleen Marquardt of Putting People First helped with so many documents and so much time that her name should be on this book, too. Her husband, attorney Bill Wewer, provided essential counsel. Tom McDonnell of the American Sheep Industry Association helped in so many ways I can't count them. Bill Pickell of the Washington Contract Loggers Association made his organization's files available for my inspection and urged his counterparts in many states to do likewise. David Howard of the Alliance for America urged member organizations to cooperate in the research. Dan Byfield of the American Land Foundation gave much-needed encouragement and support. Teresa Platt of the Fishermens Coalition carried out many research tasks. This core group made the book possible.

Perry Pendley of Mountain States Legal Foundation shared ideas from his organization's Ecoterrorism Hotline. Barry Clausen of North American Research opened his controversial files for my inspection and patiently cooperated with many demands for verification. Henry Lamb of the Environmental Conservation Organization shared his extensive files and knowledge with me.

Criminologist Brent L. Smith of the University of Alabama at Birmingham generously agreed to explain terrorism cases to me and granted permission to use materials from his book *Terrorism In America*. Michael Coffman, Ph.D., of Environmental Perspectives, provided valuable help in analyzing data.

I owe a great deal of insight to a number of environmental radicals who agreed to critically review Chapter Five: Michael Donnelly, Tim Hermach, Victor Rozek and Jeff St. Clair. My thanks to Alexander Cockburn for permission to use excerpts from his writings in The Nation.

Law enforcement officers in federal, state and county agencies are acknowledged in the text. I am grateful for their help and their work. To those law enforcement officers who anonymously gave background on particular crimes and criminal tracking methods, my special thanks.

The clerks of many courts were helpful in locating and providing criminal records. I am particularly indebted to the clerks of Josephine County, Oregon, Lewis and Clark County, Montana and the United States District Court - District of Arizona in Phoenix.

The libraries of many newspapers were generous in digging out clipping files too old to be included in their computerized databases, particularly the Tucson Citizen, Arizona Daily Star and Santa Rosa Press Democrat. Many reporters provided background on stories they had covered, and are named in the text. My thanks to them.

My profound thanks to the hundreds of victims of ecoterror crimes who spoke to me on condition of anonymity. Those who refused to speak to me, I understand. May this book give a voice to your silence.

And to my many mentors who are not named here, you know who you are, and you know you have my respect and gratitude.

Any merits of this book belong to these fine people. Any errors of fact or judgment are mine alone.

I have dedicated this book to my wife Janet, who did more than help with research and endlessly review the manuscript. I have never before written a book that gave me nightmares or that required a continual look into the face of hatred or that had so many interviews with people in fear for their lives. Janet's strength of character carried me through unbearable moments. Her bravery infuses these pages.

Ron Arnold
Liberty Park
Bellevue, Washington

Chapter One
THE BOMB

2:52 P.M. APRIL 24, 1995   *Bellevue, Washington*
THE CALL CAME IN ON A BUSY MONDAY AFTERNOON. It was Teresa Platt, executive director of the Fishermen's Coalition in San Diego. Her voice
sounded flat and drained.

"There's been an explosion at the California Forestry Association."

Time froze. Friends worked in that office.

"An explosion?"

"A bomb."

"Was anybody hurt?"

"One person is dead. We don't know who. One person was taken
to the hospital. It's still pretty confused."

"When was it?"

"Half an hour ago. Maybe forty-five minutes."

"Are you sure about this, Teresa?"

"I'm sure. David Howard called because he heard it on the radio.
He asked if I could confirm it, so I called the Association. When nobody
answered, I called the Sacramento Bee. They told me what I just told
you."

"This is hard to believe, Teresa."

"I know, Ron. But it's real. Somebody bombed CFA. And somebody is dead."

THAT MORNING   *Roseville, California*

GILBERT BRENT MURRAY, the handsome, balding 47-year-old former Marine, kissed his schoolteacher wife Connie goodbye. On his way out, he told his two teenage sons they'd better be ready for the big event tonight at Roseville High. He felt the spring morning already turning humid as he went to his car. It would be another hot day. He started the Ford Explorer and backed out of the driveway.[1]

It was a routine 35-minute commute. He'd lived here since 1988 when he first became a lobbyist and the drive had long been automatic: He avoided the Roseville Freeway, taking Baseline Road and State 99 to Interstate 5, then south to Sacramento's urban core and the best available surface route to California Forestry Association's off-street parking lot at 1311 I Street.

Gil Murray pulled into one of the slots marked "Reserved for CFA" next to the front entry. He unfolded himself from the driver's seat and locked his sport utility vehicle. It was not the best part of town. Even though a mere three blocks north of the gleaming white State Capitol, the office was uncomfortably close to a dingy high crime area with its mingled drugs, prostitution, and derelict cars.

The association's building had no entrance on I street. Public access came from the parking lot where Murray walked the last few steps to work. Wrought iron gates like pawn shop bars guarded the brick facade he entered through inset glass doors. The association occupied the rear two thirds of the low building, a good 5,500 square feet of office space. He crossed the entry foyer, opened the heavy double doors into his organization's lobby and stopped in his tracks.

"Good grief, Michelle, your eyes are still a mess," he said in exasperation. "When are you going to call the doctor?"

"I've already taken care of it, honest," the dark-haired young woman replied from behind her reception counter. Her eyes were swollen nearly shut with allergies. "My doctor's appointment is this afternoon."

Murray cocked his head. "It's a good thing! Last week I was ready to call 911 and have them come take you away."

"I bet you would, too," she said, grinning as best she could.

He shook his head as he threaded his way down the corridor, past the open inner office with its file area and computer center, to the kitchen at the back of the building. He selected his favorite mug from the shelf, the one with the motto: CHOCOLATE IS MY LIFE! He filled it with steaming rich liquid from the coffee pot, then slipped quietly into his office.

It was a typical start-the-week morning—swarms of short telephone updates, staff absences to shuffle, members' problems to confront. At mid-morning, he took a quick respite to look with delight at the ultrasound scans of the unborn child of Lisa Tuter, CFA's five-months-pregnant communications manager.

Lunch came and went.

1:58 P.M.

A POSTAL TRUCK ARRIVED with CFA's Monday mail tub. While lugging the container across the parking lot, the mail carrier fumbled and dropped a heavy shoe box-shaped package with too many stamps. He retrieved the wayward parcel from the hot pavement, placed it atop the stacked mail and stepped inside the CFA lobby where he left his burden on the reception counter as usual.

Michelle would normally have distributed the mail, but today the crew had to help themselves.

Association controller Jeanette Grimm, a neatly-dressed fortyish woman of medium height, heard the truck depart and stepped crisply across the corridor into the lobby to pick out the money mail. She found Eleanor Anderson, Gil Murray's secretary, filling in at the receptionist's desk.

"Michelle's not back yet?"

"Just another twenty minutes or so."

Grimm lifted the smoothly-wrapped box to get at her mail. "My gosh, this is a heavy package," she said, needing both hands to lift it. "And it's addressed to Bill Dennison"—former CFA president, retired exactly a year now.

"Who's it from?" Anderson asked.

"It says 'Closet Dimensions, Inc.' Must be a sample of some kind." She turned the package around for Eleanor to see. Anderson, a slim forty-something woman with light hair, leaned forward and took the item in outstretched hands, bracing herself against its six-pound weight. She examined the address label.

Grimm, ever the accountant, added up the stamps while leaning over and reading them upside down. "Look at all the postage they had to pay."

Ten stamps in all: a long neat row at the top with six undenominated "G" Flag stamps worth thirty-two cents each, ending on the right with two purple Eugene O'Neill stamps on which she could not make out the denomination. Beneath them were two colorful $2.90 priority mail stamps depicting a spacecraft. [2]

"And they have our old name on the label," said Anderson, eyebrows arched. "We haven't been Timber Association of California for four years now."

She shook the package, bending close to listen.

"I bet it's one of those folding closet organizers," she said.

Grimm stiffened suddenly. "Eleanor, this is heavy enough to be a bomb."

Everyone in the office had been aghast at the bombing of the Alfred Murrah Federal Building in Oklahoma City only five days earlier. A newspaper spread open on the reception counter bore silent testimony to the nightmare explosion which had killed 168 men, women and children in the worst act of terrorism in the nation's history.

Dr. Robert Taylor, the stocky, studious director of wildlife ecology, appeared from his office and edged past Grimm to pull his mail from the tray.

"What could be a bomb?" he asked.

Grimm pointed to the package. Taylor looked at it for a second and said, "It's probably books." He picked the thing up and hefted it. It felt too heavy and too hard for books, like a metal box, but he made no remark. He put it back and got busy with his mail.

Gil Murray, who had replaced Dennison, emerged from his office after a long telephone session with Donn Zea, the association's vice president of industry affairs, who was working at his Grass Valley home today because of a dental appointment. They were worried that the burgeoning wise use movement fighting government regulations would be falsely linked to the anti-government Oklahoma City bombers. Bill Dennison would have confronted the confronters; Gil Murray was low-key. They would discuss it in the regular weekly staff meeting tomorrow. As he approached the mail tub, Eleanor handed him the package.

"Gil, would you take a look at this?"

He accepted the object.

"Do you think this could this be a bomb?" Anderson asked.

He was mildly taken aback, and scanned the return address carefully. Murray's strong facial lines then creased in a familiar smile, "At least it's not from Oklahoma City. It's from Oakland."

"So," said Anderson, "should we forward it to Bill or not?"

"No," he replied, then reconsidered. "Well, let's see what it is first."

Murray stood before the reception counter and grappled with the puzzling box, looking for ways to penetrate its taut brown wrapping, which wouldn't budge under several layers of rip-resistant nylon tape on the edges.

Jeanette Grimm and Bob Taylor busied themselves with the mail tub.

Grimm finished her sorting and handed the checks to Anderson.

"Here's the deposits for today."

She stepped out of the lobby and to her office door. There stood Lisa Tuter with a question about a project. While they talked, Bob Taylor finished retrieving his mail and said, "If that's a bomb, I'm going back to my office."

"Yeah, right," grinned Murray.

Melinda Terry, the short, busy director of legislative affairs, stepped into the lobby from the ladies' room in time to see Taylor and Grimm leaving the postal tub. She was amused that nobody could wait for Michelle to return and distribute their mail. Melinda told Gil that a piece of legislation was coming up in committee at the Capitol, so she would be gone the rest of the day.

Tuter then stepped forward, saying to the lobby in general, "If it matters to anybody, I have a doctor's appointment this afternoon."

Gil Murray, with his characteristic twinkle, feigned alarm at the striking young woman and said, "What do you mean, you're going to the doctor? Half the office is absent."

Tuter tossed her blond hair with a smile and just patted her tummy.

Eleanor Anderson, seeing Murray still struggling with the wrapping, wordlessly handed him a pair of scissors from the reception desk. At that moment, a call lit up the switchboard and Anderson fielded a request for a Forest Service address. "We're not a government agency. I'll have to go look it up in my Rolodex," she said, putting the caller on hold. She arose and strode preoccupied to her office.

Murray removed the last of the wrapping paper and examined the box before him. He called to Tuter as she vanished down the hallway, "Did you see Michelle this morning?"

"Yes, the poor baby," she said, and stepped into her office.

2:19 P.M.

GIL MURRAY STOOD ALONE IN THE LOBBY. He laid the wooden container on the chest-high reception counter and opened it.

There was a brief flash:

The device expanded in all directions. Four streams of hot metal separated from the spherical blossom of light.

A blast of snipped-in-half wood screws and tiny melted machine parts shot straight down the hallway where Lisa Tuter had stood a moment before, slicing off drywall as it went and snapping into the kitchen wall at the back of the building nearly a hundred feet away.

A vortex jetted straight ahead, piercing the wall behind the reception desk and burying dozens of little projectiles in a row of steel filing cabinets in the next room.

The device's lid lifted on a gaseous ram, shredding the heavy false ceiling, pounding through heating ducts and piercing the roof deck, letting jagged sunlight into its slipstream.

The main thrust of the detonation did not throw Gil Murray back from the counter. It cut through him. It ripped off his face and an arm. It embedded the bulk of his chest in the wall behind him, then bowed the wall into an adjoining storeroom, toppling heavily loaded bookcases on the far side. It riddled the reception area with pieces of his body. What was left fell to the floor. The shattered false ceiling collapsed over him. Flames and acrid smoke billowed outward.

Total elapsed time: Less than seven-tenths of a second.

NEW YORK CITY

IN THAT EXTREME MOMENT WHEN GIL MURRAY DIED, three-thousand miles and three time zones away, executives of The New York Times struggled to understand a letter they had pondered in their offices since early afternoon. The Times's newsroom had spotted it and turned it over to the FBI unopened. Agents conducted x-ray and fluoroscopic scans for explosives, then opened and read it.

It was a blackmail note that had been mailed from Oakland a day after the Oklahoma City bombing. The letter was crudely typed, with phrases crossed out with X's. It claimed to be from the Unabomber, the shadowy figure credited with murdering two people—it would soon be raised to three, now, with Gil Murray's death—and injuring twenty-three others with hand-crafted bombs over a seventeen-year period, generating the longest, most extensive federal manhunt in the nation's history.

It was identical in style to a letter The New York Times had received two years earlier, on June 24, 1993, postmarked San Francisco and bearing the "FC" trademark identifying Unabom, as the Federal Bureau of Investigation tagged him after his university and airline targets. In that letter the Unabomber had spoken to the world for the first time. It claimed that bombs which severely injured two university scientists were the work of an anarchist group. It promised further communiqués and gave a nine-digit code—553-25-4394—to authenticate future writings. The number itself was a dead end lead, identical to the Social Security number issued to a Northern California resident who had been checked out and cleared of all involvement.

The letter that arrived the day Gil Murray died began with the numbers 553-25-4394. It stated:

THIS IS A MESSAGE from the terrorist group FC.
　　We blew up Thomas Mosser last December because he was a Burston-Marsteller executive. Among other misdeeds, Burston-Marsteller helped Exxon clean up its public image after the Exxon Valdez incident. But we attacked Burston-Marsteller less for its specific misdeeds than on general principles. Burston-Marsteller is about the biggest organization in the public relations field.
　　This means that its business is the development of techniques for manipulating people's attitudes. It was for this more than for its actions in specific cases that we sent a bomb to an executive of this company.[3]

The Unabomber's misspelling of Burson-Marsteller was less puzzling than his belief that the firm had "helped Exxon clean up its public image after the Exxon Valdez incident," the disastrous March 24, 1989 oil spill on Bligh Reef in Prince William Sound, Alaska. The repeated spelling error might be a deliberate and grisly bomb joke—"Burst-on." But, in fact, the company had no part in such a campaign. Why did the Unabomber believe it did? The question flickered briefly, then faded as others emerged. For here, after seventeen years of cloaked carnage, an inexplicable change of tactics gave us our first look into the world of the Unabomber:

> In our previous letter to you we called ourselves anarchists. Since "anarchist" is a vague word that has been applied to a variety of attitudes, further explanation is needed. We call ourselves anarchists because we would like, ideally, to break down all society into very small, completely autonomous units. Regrettably, we don't see any clear road to this goal, so we leave it to the indefinite future. Our more immediate goal, which we think may be attainable at some time during the next several decades, is the destruction of the worldwide industrial system. Through our bombings we hope to promote social instability in industrial society, propagate anti-industrial ideas and give encouragement to those who hate the industrial system.
>
> ...For security reasons we won't reveal the number of members of our group, but anyone who will read the anarchist and radical environmentalist journals will see that opposition to the industrial-technological system is widespread and growing.

The Unabomber's sketchy version of anarchy slightly resembled some of today's fashionable anarchist ideologies, for example, that of Murray Bookchin, who felt that domination of the environment had the same roots as the oppression of women and the Third World, and focused on interpersonal growth within a small community while cultivating ecological consciousness. The Unabomber's anarchy sounded sufficiently like technophobe guru John Zerzan's books *Elements of Refusal* and *Future Primitive* for the FBI to investigate. But the short passage about radical environmentalists endorsing "destruction of the worldwide industrial system" and breaking down all society "into very small, completely autonomous units" contained the threads, patterns and cloth of the radical environmentalist philosophy known as "deep ecology."

"Deep ecology," a philosophical perspective initially developed by Norwegian academician Arne Naess, is an ultimate expression of hate for the industrial system. Industrial civilization is completely rejected because it is anthropocentric—human centered—while "deep ecology" is biocentric—nature centered. "Reform environmentalism" in this view is "shallow ecology" in that it seeks to preserve the environment in order to support human life and human use. "Deep ecology," by contrast, is "radical environmentalism" in that it gives human beings no special place in the universe, much less their industrial civilization.[4]

Influential American radical environmentalists Bill Devall and George Sessions elaborated Naess's concept. "The ideal is the 'primitive' society, because it fulfills the needs of individuals and communities and preserves the integrity of the natural world. In such societies, human beings are organized in small, decentralized, nonhierarchical and democratic communities."[5]

Devall and Sessions codified this ideal into a set of action principles in their 1985 book, *Deep Ecology: Living as if Nature Mattered*, including calls for "a substantial decrease of the human population" and deep change in all policies that affect "basic economic, technological, and ideological structures." Those who subscribe to the principles of deep ecology, they asserted, "have an obligation directly or indirectly to try to implement the necessary changes."[6] (The complete statement is on p. 287.)

Dave Foreman, co-founder of the radical environmentalist group Earth First, made clear the depth of those "necessary changes" in his 1991 book, *Confessions of an Eco-Warrior*. His group believed that the earth had to be placed first in all human decisions, "even ahead of human welfare if necessary."[7]

Nature is to be saved for its own sake, even if that means "voluntary human extinction."[8]

"If you'll give the idea a chance, you might agree that the extinction of *Homo sapiens* would mean survival for millions if not billions of other Earth-dwelling species," said an article in Wild Earth, one of Foreman's publications.[9]

Foreman had proposed a start on reducing America to small, separate communities with the "Earth First Wilderness Preserve Plan" in 1983. He wrote, "It is not enough to preserve the roadless, undeveloped country remaining. We must re-create wilderness in large regions: move out the cars and civilized people, dismantle the roads and dams, reclaim the plowed land and clearcuts, reintroduce extirpated species."[10] Foreman revived this "rewilding program" in 1992 as the North American Wilderness Recovery Project, also known as the Wildlands Project, with the goal of developing a "continental wilderness recovery network."[11] In practical terms, that meant depopulating at least 716 million acres (this figure has increased with time)—or about one-third of America's nearly 2.3 billion-acre total area—as a start.

But Dave Foreman is perhaps most notorious as the author of a saboteur manual called *EcoDefense: A Field Guide to Monkeywrenching*—"monkeywrenching" being a synonym for "sabotage" borrowed from Edward Abbey's cult novel *The Monkey Wrench Gang* in which environmental activists plot to blow up Glen Canyon dam and return the Colorado River to its natural state.[12] *EcoDefense* was a sassy, brassy, detailed instruction manual for committing a wide array of crimes to save nature by destroying the worldwide industrial system piecemeal, alone and at night. The book made Foreman a folk hero among radical environmentalists.

EcoTerror:
The Violent Agenda to Save Nature –
The World of the Unabomber
by Ron Arnold

ERRATUM

This first edition of *EcoTerror: The Violent Agenda to Save Nature - The World of the Unabomber*, contains factual inaccuracies with respect to Mitch Friedman, the Executive Director of Northwest Ecosystem Alliance. The editor and publisher apologize to Mr. Friedman, and in the interest of historical accuracy offer the following corrections.

On pages 25, 46, and 77, Mr. Friedman is referred to as an editor and publisher of the journal *Live Wild or Die*. In fact, Mr. Friedman was not an editor or publisher of *Live Wild or Die* and had no role in the writing, editing, or publication of any of its issues or the *Eco Fucker Hit List* that was contained in them.

Also on page 25, Mr. Friedman is referred to as a "convicted felon" for his role in cutting down a billboard in 1986. In fact, the Judgment Order in the case related to the billboard incident reads that Mr. Friedman's conviction is "to be treated as a misdemeanor."

Contact: Ron Arnold
425-455-5038
FAX 425-451-3959
email: ron@cdfe.org

Since its publication in 1987, hundreds of Earth Firsters have followed the manual's instructions, directing their monkeywrenching against timbermen, dam operators, factory owners, road builders, cattlemen, farmers, wool growers, retailers, oil and gas producers, mineral prospectors, miners, off road vehicle users, ski resort owners—virtually everyone who could be considered industrial. A wave of incidents prompted headlines such as the Phoenix Gazette's, "Terrorism fears grow at home—Earth First!, other groups stir fears of injury, deaths."[13]

Earth First had no formal membership and called itself a "movement," not an organization. Earth First had no formal leadership beyond a small unincorporated council that melodramatically called itself "The Circle of Darkness." Earth First's structure was tribal and in ideal, non-hierarchical. However, two social scientists who studied Earth First thought otherwise: Kimberly D. Elsbach and Robert I. Sutton, wrote:

> We must note, however, that Earth First! literature asserts that it is not an organization and thus can have no members. Yet, we consider Earth First! to be an organization and Earth First!ers to be its members because it has many trappings of an organization. Its fund-raising, media, and direct action committees indicate that a differentiated social structure is used to achieve collective goals. Mailing lists of people affiliated with Earth First! are maintained, and most people on such lists describe themselves as Earth First!ers. And local chapters operate under the Earth First! banner."[14]

Earth Firsters singlemindedly pursued their agenda. Their modus operandi in spreading *EcoDefense* was simple: read this book; do what it says. Their anarchic structure and hit-and-run tactics made apprehension of suspects all but impossible.[15]

But not totally impossible. In June, 1989, Dave Foreman was arrested by the FBI in connection with a widely-publicized sabotage case. A federal grand jury indicted Foreman and four other Earth First activists on charges of conspiring to damage power lines and transmission towers at three nuclear sites. All but Foreman were also accused in the sabotage of equipment at the Snowbowl ski facility near Flagstaff, Arizona. Trial was set for early 1991 in Prescott, Arizona.[16]

Two months into the prosecution's case, the federal judge abruptly broke it off. To the surprise of environmentalists everywhere, Foreman and the four others entered into a plea agreement with the U.S. Department of Justice.[17] Foreman signed a guilty plea of felony conspiracy for giving money and his instruction manual to co-conspirators "to illegally sabotage high voltage electrical transmission towers and lines," and for a "planned attack on nuclear facilities in the Western United States." His sentence was deferred for five years, conditioned on his compliance with probation and supervised release terms, when he would be allowed to withdraw the felony plea and enter a misdemeanor plea. In 1996, the court fined him $250 for misdemeanor depredation of government property.[18]

The hatred of the industrial system that drove such acts was openly discussed by Earth Firster Christopher Manes in his 1990 book, *Green Rage: Radical Environmentalism and the Unmaking of Civilization.* It was an angry, discursive history of Earth First with the central theme that industrial civilization is causing "the end of the world as we know it, the meltdown of biological diversity." This apocalyptic vision helped fuel Earth Firster rage against everything industrial.[19]

Activist James "Rik" Scarce wrote such detailed descriptions of the Animal Liberation Front's illegal anti-industry activities in his 1990 book, *Eco-Warriors, Understanding the Radical Environmental Movement,* that he was jailed for 159 days in 1993 for refusing to reveal the identities of his sources to a grand jury.[20]

Animal Liberation Front activist Rodney A. Coronado, a fugitive on the FBI's most-wanted list for several years, pleaded guilty in March, 1995 to aiding and abetting an arson in connection with a February 28, 1992 fire at Michigan State University research facilities. Coronado's attack caused $125,000 in damages and destroyed 32 years of research on the effects of pollution on wild mink in the Great Lakes region and 10 years of research on innovative laboratory procedures. He was sentenced to four years and nine months in prison and ordered to pay $2.5 million in restitution for fires set in five states. In exchange for his plea agreement, U.S. attorneys' offices in the five states agreed not to pursue charges against him.[21] Significantly, he had earlier joined with Earth Firsters to sabotage a hunt in the Mojave Desert, and Earth First Journal published lengthy diatribes he wrote after his arrest.[22]

Animal rights activist Fran Stephanie Trutt had been sentenced to 32 months in prison in 1990 for planting a radio-controlled pipe bomb in the attempted murder of Leon C. Hirsch, chairman of U.S. Surgical Corporation, a Connecticut firm that used anesthetized dogs to teach surgeons to use synthetic surgical staples.[23]

The U.S. Department of Justice had documented 313 sabotage and personal attacks aimed at animal industries from 1977, when the first animal rights-related attacks were recorded, to mid-1993, most of them attributed to the Animal Liberation Front, an FBI top-ten-list terrorist organization that appeared to have substantial cross-affiliation among Earth Firsters.[24]

Captain Paul Watson, leader of the Sea Shepherd Conservation Society, wrote a manual titled *EarthForce! An Earth Warrior's Guide to Strategy* (1993) based on his experiences sinking whaling vessels, blocking seal hunts and harassing fishing boats. Intelligent and witty, it was another detailed instruction manual for committing crimes to save nature by destroying the worldwide industrial system one piece at a time.[25]

An obscure animal rights group published a 1991 book called *A Declaration of War: Killing People to Save Animals and the Environment* under the pseudonym "Screaming Wolf."[26]

The anonymous authors of a page posted on the World Wide Web in late 1994 titled "Gaia Liberation Front" argued for the total eradication of all humans as the only way to save nature.[27]

Radical environmentalists in the Animal Liberation Front, Farm Animal Revenge Militia, Earth First, Earth Liberation Front and dozens of other anarchic cadres had expressed their hate of the industrial system in more than a thousand incidents of violence, including arson, tree spiking, bombings, equipment destruction, livestock shootings and attempted murder.

Did the Unabomber walk among them?

The denouement of the Unabomber's letter drove Times publisher Arthur Sulzberger Jr into a moral dilemma:

```
The people who are pushing all this growth and
progress garbage deserve to be severely punished.
But our goal is less to punish them than to propagate
ideas.  Anyhow we are getting tired of making bombs.
It's no fun having to spend all your evenings and
weekends preparing dangerous mixtures, filing trigger
mechanisms out of scraps of metal or searching the
sierras for a place isolated enough to test a bomb.
So we offer a bargain.
We have a long article, between 29,000 and
37,000 words, that we want to have published.  If you
can get it published according to our requirements we
will permanently desist from terrorist activities.
```

Bottom line: The Unabomber's ransom price was free space in The New York Times, Newsweek or Time. It would take about seven full pages of dense print in a newspaper—and much more space in a newsmagazine.

But there were conditions:

```
Our offer to desist from terrorism is subject
to three qualifications.  First: Our promise to de-
sist will not take effect until all parts of our
article or book have appeared in print.  Second: If
the authorities should succeed in tracking us down
and an attempt is made to arrest any of us, or even
to question us in connection with the bombings, we
reserve the right to use violence.  Third: We distin-
guish between terrorism and sabotage.  By terrorism
```

we mean actions motivated by a desire to influence
the development of a society and intended to cause
injury or death to human beings. By sabotage we mean
similarly motivated actions intended to destroy property without injuring human beings. The promise we
offer is to desist from terrorism. We reserve the
right to engage in sabotage.

This naïve distinction between terrorism and sabotage was significant: it suggested that the Unabomber had been reading his radical
environmentalism mindfully. Earth Firsters, beginning with their mentor
Edward Abbey, emphasized a similar contrast between destroying property and injuring human beings. As Dave Foreman echoed Abbey's theme:

> Monkeywrenching is non-violent resistance to the destruc-
> tion of natural diversity and wilderness. It is not directed toward
> harming human beings or other forms of life. It is aimed at inani-
> mate machines and tools. Care is always taken to minimize any
> possible threat to other people (and to the monkeywrenchers them-
> selves).[28]

Federal law does not agree with Ed Abbey, Dave Foreman or the
Unabomber. Property has always been protected by law on the presump-
tion that damage to a person's belongings, especially one's home or means
of livelihood, is damage to the person. The official definition of terrorism
used by the FBI has remained unchanged on this issue for many years:
Terrorism is officially defined by the bureau as

> the unlawful use of force or violence, committed by a group(s)
> or two or more individuals, against persons or property to in-
> timidate or coerce a government, the civilian population, or
> any segment thereof, in furtherance of political or social objec-
> tives.[29]

The Unabomber's ransom note to The New York Times was part
of a flurry of messages. Nobel laureates Phillip A. Sharp of the Massachusetts Institute of Technology and Richard J. Roberts of New England Biolabs
Inc. in Beverly, Massachusetts received letters deriding technology. The
Unabomber taunted computer scientist David Gelernter of Yale, a victim
he seriously injured in 1993: "People with advanced degrees aren't as smart
as they think they are. If you'd had any brains you would have realized
that there are a lot of people out there who resent bitterly the way techno-
nerds like you are changing the world and you wouldn't have been dumb
enough to open an unexpected package from an unknown source." And he
sent a package to the Timber Association of California.

All this mail was sent at one time, postmarked Oakland, Thursday, April 20, 1995.

Newspapers do not publish under threat of violence. Nor do they willfully doom the innocent to death. Arthur Sulzberger had reporter James Barron quote him directly in a story slated for Wednesday's editions: "While the pages of The Times can't be held hostage by those who threaten violence we're ready to receive the manuscript described in this letter. We'll take a careful look at it and make a journalistic decision about whether to publish it in our pages. But whether we publish it ourselves or not, we'll do all we responsibly can to make it public."[30]

Telephone lines between The New York Times and law enforcement agencies hummed that afternoon as

flames and acrid smoke billowed outward. It had sounded like the pop of a giant M-80 firecracker.

Eleanor Anderson sat gripping her telephone handset, unable to speak the address she had found in her Rolodex.

Jeanette Grimm had just slipped off her pumps beside her desk in the office nearest the lobby, ready to get back to work. She stiffened with the startle reflex as shrapnel shot through her front wall. A wisp of pink fiberglass insulation fluttered to the floor.

Lisa Tuter stood in mid-stride just inside her office door.

Bob Taylor had plopped the mail on his desk and was about to sit down and open it.

Melinda Terry had the telephone to her ear, put on hold by a legislative aide. She felt the air pressure convulse like an airliner decompressing at high altitude.

Caustic stinking smoke surged from the front.

A long silent shock hung in the air.

"Oh shit, it *was* a bomb!" Bob Taylor cried, and rushed toward the reception lobby.

Voices called from office to office, Jeanette, Eleanor, Lisa, Melinda, accounting for each other.

Bob peered into the roiling fumes. He could see only knee-deep rubble where Gil had been standing. He quickly glanced behind the reception desk where he expected to find Eleanor. Not seeing her, he dashed back through the inner office and tried to get into the reception area from the twin corridor.

"Eleanor!" exclaimed Jeanette as the two nearly collided in the dim light. "I thought you were at the front desk."

"I was in my office."

Bob looked back and saw Jeanette talking to Eleanor. He concentrated on finding Gil.

Melinda yelled from her office doorway, "Is everyone okay?"

"No!" screamed Eleanor. "Gil was in there. He opened the package."

The women crammed the corridor. The lights had been blown out in the windowless reception lobby. Jeanette winced as she trod the splintery debris with bare feet.

Bob realized that if Gil was still alive beneath that pile of wreckage, he would need far more than first aid. Sacramento City Fire Department No. 2 was half a block down I Street. He made his way around the ravaged room by a side door into the lobby where the heavy foyer doors lay torn off their hinges. He ran for help.

The women called Gil's name.

There was no answer.

"I'm going to call 911," said Melinda.

Lisa already sat in her office dialing the emergency number.

At the fire department, Bob breathlessly explained the situation.

A firefighter told him, "All our trucks are gone," just as one of them lumbered into view. Bob wasted no time and ran for the truck. In less than a minute firefighters followed him into the damaged entry.

"Get everybody out of here," said the lead firefighter.

Taylor did as he was told, scrambling everyone out the back door and into the dusty alley, barely giving Jeanette time to grab her shoes. Melinda hastily told the 911 dispatcher that help had arrived and hung up. Firefighters closed in behind, making sure everyone was out.

Police and emergency vehicles converged on the office, sirens blaring. Throngs of the frightened and curious from nearby buildings poured out to see what all the noise and fuss was about.

Jeanette cried uncontrollably and a firefighter tried to comfort her. A fire department car slid into a space across the alley and the firefighter guided her to the passenger seat. Fire Department Division Chief Jan Dunbar invited her to sit.

"Can I get your statement, please?"

Jeanette got in the car and tried to speak through her horror as a police investigator approached the remaining four standing in the alley.

"Come on, let's get you folks away from here," he said as his crew began spooling out yellow crime scene tape around the office. He herded them down the alley to 13th Street, but the cameras of every television outlet in town suddenly found them. The officer retreated back up the alley and ushered his charges into the rear parking lot of the Sterling Hotel.

An aid car lurched up the alley. Paramedics leaped out and hustled Lisa and her unborn baby onto a gurney.

"I'm fine," she insisted as an attendant slowly lowered her neck and shoulders.

"You might go into labor."

The media broke through the growing crowd and jammed around her as the paramedics strapped her down for transportation to Sutter Memorial Hospital for observation.

Chatter from reporters drifted to her ears: "Was she hurt? Did the bomb get her? Can you see any injuries? No, but look, is she pregnant? Wow, she's really pretty."

"I feel really stupid," she muttered through clenched teeth as they lifted her into the aid car.

The remaining three stood in a little clutch now, watching the aid car disappear down the sweltering alley, feeling very insignificant, guarded by a cordon of police and firefighters who kept the media and onlookers at bay. Jeanette went on with her statement in the fire chief's car, barely noticing that Lisa was gone.

The trio stood there nearly half an hour, watching the emergency crew scurry around. They watched a crew that did not bring an injured man to an ambulance that did not rush off to the hospital.

They guessed what it meant. They asked. Nobody would tell them anything.

As Jeanette finished reconstructing the incident, a firefighter emerged from the wrecked office and motioned for the chief to roll down the driver's window. He leaned into the opening and the two conferred in low voices.

"What about Gil?" Jeanette pleaded.

"The fellow standing over there?" The man pointed to Bob Taylor.

"No, not him. My boss. He opened he bomb."

The firefighter said nothing. His lips thinned in a stony face and he shook his head.

Jeanette broke down.

Bob heard her sobbing. Eleanor and Melinda stepped close to the passenger door.

Eleanor said slowly and deliberately, "Jeanette, what is going on? Where's Gil?"

She looked up at her co-workers.

"Gil's dead."

2:29 P.M.   *Auburn, California*

RON STOCKMAN of Mother Lode Research Center got a call telling him to turn on the television. The first image of the blast scene jolted him into action. As the bulletin disclosed the meager information available—one person dead, one hospitalized, no names—Stockman went to work drafting a fax. There was no time to be lost. Other organizations might be on the same hit list as the California Forestry Association. At 2:45 exactly, his fax machine broadcast a warning:

> [This afternoon] the California Forestry Association (CFA) was bombed via letter bomb. Initial reports confirm at least one person dead.

CFA is one of the organizations MLRC works closely with on various issues which usually involve opposing certain enviro views. While it is still too early to point to exactly one group with leanings in any direction we would advise all groups involved in public and private land use issues to be on the alert. CFA dealt with mostly timber oriented issues but considering the heated debate over private property rights, ESA [the Endangered Species Act], wetlands, recreational use etc., we urge that all organizations be careful of incoming packages and large letters and to keep an eye on their premises.

A short time later in upstate New York, David Howard of the Alliance for America broadcast a similar warning to several hundred organizations on his fax list.

GRASS VALLEY, CALIFORNIA
DONN ZEA, CFA's VICE PRESIDENT FOR INDUSTRY AFFAIRS, got a call telling him to turn on his television. It was Nadine Bailey, CFA's grassroots coordinator, calling from her home in Redding. She tried to control her sobs. "Donn, this is Nadine. The office has been bombed. I think somebody is dead."

Zea's head jolted as if physically struck, his sharp features drawn into a mask of disbelief.

"What are you talking about, Nadine?"

He loped past the babysitter and his toddler Madeline into his den with the cellular phone to his ear. Nadine was saying something, but he didn't hear. He clicked the TV's remote control.

Fade up: the first image on the screen was the bombed out entry of his place of employment. His hands began to shake.

"Oh my God."

CHESTER, CALIFORNIA
BILL DENNISON HAD TAKEN HALF THE DAY OFF. He was in his back yard when his wife Pat called from the house. It was Barbara Matthews, a friend in Sacramento. She said to turn on the television. The CFA office had been bombed. Bill dashed into the house.

The television image came on.

Bill and Pat just stared.

3:10 P.M.   *Interstate 80 near Auburn*
HALFWAY TO SACRAMENTO, Donn's cellular phone rang. His wife Lisa was driving so he could have his hands free—and he was not too sure of his reflexes yet.

It was Connie Murray, Gils' wife. She had heard about the bombing. How was everybody? Was Bob Taylor okay? How was Jeanette?

Connie asked who was killed. But she didn't ask about Gil. Donn felt her silently screaming it couldn't be Gil.

His stomach knotted. He had spent the past twenty minutes calling his own relatives to tell them he was safe, then association directors to warn them and get any new facts. John A. Campbell, president and CEO of Pacific Lumber, had told him that the fatality was being reported as a male. All the men in the office were accounted for: Bob Taylor, seen in news film; Jim Craine, vice president of resources, at a meeting across town; John Hofmann, vice president of government affairs, in Washington D.C.; Donn, at home. Everyone except Gil Murray. It wasn't confirmed, but there was no other conclusion.

Donn told Connie things were still confused and he would find out more when he got to police headquarters.

When Connie said goodbye he squelched the sickening feeling and watched the landscape go by. He was losing his grip.

But he had more calls to make. Who else might be on the bombing hit list?

The California Forestry Association had for years been highly vocal in defense of timber harvest. Former leader Bill Dennison had met environmentalist confrontations head-on, vigorously entering contentious frays anyplace timber interests were threatened. The organization had also petitioned the U.S. Fish & Wildlife Service in 1993 to remove the Northern spotted owl from the federal endangered species list, charging that the association's scientific computerized models proved that owl numbers had been underestimated and its survival needs overestimated. Dennison's signature was on that petition.

The firms that made up CFA's membership had frequently been targets for radical environmental group sabotage. Logging equipment, mills and offices throughout Western timber country had been vandalized, burnt or blown up. Forest and mill workers had been shot at, received death threats and faced the danger of deadly tree spiking, the terror tactic that could cause chainsaws to kick back and injure loggers in the woods or shatter a high-speed bandsaw blade in the mill to devastating effect.

Donn Zea knew of more than 200 incidents of forest violence attributed to radical environmentalists since 1980. He also knew that there were probably more than twice that many: executives were reluctant to report incidents for fear of encouraging the perpetrators, inviting retaliation and copycats, or lowering shareholder confidence.

Ecoterrorism had become a seriously underreported crime. It had also become a routine cost of doing business. Donn Zea could not remember a time in his six years at CFA without violence and threats from radical environmentalists. They had become complacent.

But now every CFA member was a potential target of whoever bombed the association headquarters.

And the eco-rhetoric had been heating up. Earth Day had been celebrated just this past Friday with acrimonious anti-industry events throughout Northern California. Only days before, Zea had sent out the association's newsletter with a warning about possible violence.

In the continuing debate over the use of natural resources, radical environmentalists maintain their philosophy to confront conflict with physical violence. The changing political climate over our nation's environmental policies apparently is upsetting a few people.

In the December-January 1995 issue of *Earth First Journal*, in an article entitled, "Forest Grump," Mike Roselle commented on the recent elections' results and urged on his fellow members with the following:

"Their Big Ten [environmental group] memberships decline because people no longer believe the braggadocio that saturates their direct mail like the smell of urine in the bathroom of a biker bar. They whine with worry. The mainstream environmental groups are quickly becoming irrelevant.

"Fortunately the grassroots groups with a few exceptions are sticking to their guns. We don't care who is in power in Washington, for whoever stands on the wall of Babylon will be a target for our arrows. When we raze the citadel, it will matter not who holds the keys to the corporate washroom or who has reserved parking at National Airport.... What we want is nothing short of a revolution....

"F___ that crap you read in *Wild Earth* or in *Confessions of an Eco-Warrior* [writings of Dave Foreman, whom Roselle had replaced as head of Earth First after differences split their movement]. Monkeywrenching is more than just sabotage, and your go__am right it's revolutionary! This is jihad, pal. There are no innocent bystanders, because in these desperate hours, bystanders are not innocent.

"Remember tree spiking? ...more spiking is needed to convey the urgency of the situation! Very little action is happening. Too many armchair eco-warriors walking around town in camo. Go out and get them suckers, fill 'em full of steel, and I promise you this: you might get caught; you might do some time; your friends might abandon you. But you will never have to spike the same tree twice."[31]

3:35 P.M.   *Sacramento*

BY THE TIME DONN AND LISA ARRIVED at the Sacramento Police Department, the FBI had tied the Unabomber to the explosion. FBI spokesman Bob Griego in San Francisco told reporters, "We're obviously investigating to see if there is any connection with our Unabom suspect. They're scrambling around down there checking this out."[32]

Zea had to go through security to gain entry to the inner offices. Cars were being restricted near public buildings. The whole city was on alert.

Donn Zea took an officer aside. "I want to give some information. It's important."

"Someone will talk to you as soon as possible."

He and Lisa were taken to the others, who were undergoing trauma counseling in a conference room with Police Chief Arturo Venegas, six chaplains and two investigators. Spouses sat with the CFA employees: Jeanette's husband Lance; Eleanor's husband Rae; Melissa's husband John; Bob's wife Susan and his grown son John, who was working part-time for the association. As soon as they heard about the disaster, each came to the bombed-out office and was redirected to police headquarters downtown.

The employees had gone through individual questioning by homicide detectives for more than an hour on the second floor and brought downstairs for a talk of encouragement by Chief Venegas.

The Chief pointed out that until it was confirmed, he couldn't positively say that Gil Murray was dead. "There was a UPS delivery man nearby at the time of the explosion. He might have walked in and been the blast victim. Mr. Murray could have gone out the back door. We have to be certain."

And upon confirmation, Venegas did not say, the first to know would be the next of kin—those in the room would not be allowed to stray to a public phone and call Connie Murray.

Sacramento Police Chaplain Mindi Russell, a trained Critical Intervention Debriefer, sat with them and dealt with their reactions. She was compassionate and professional. Russell herself had just returned from the International Critical Incident Stress Foundation in Baltimore after spending the previous days counseling survivors of the Oklahoma City blast. She had barely put her suitcase down at 2:40 that afternoon when she responded downtown at the request of the FBI and came to deal with this bombing incident.[33]

4:45 P.M.   *Chester, California*

A NEWSPAPER REPORTER CALLED to ask Bill Dennison what he thought about his name being on the bomb. He dismissed the question the first time. When the second media call came with the same question, he replied curtly, "It wasn't meant for me."

His cryptic answer seemed to carry mystic overtones of destiny. Bill's strong religious belief indeed made him feel that had the bomb been meant for him, he would have received it. But it was also simple common sense to him: The bomber had probably targeted the association and what it symbolized more than Dennison individually—

the killer hadn't bothered to get current information for either the addressee or the association's name, and had probably just looked in some library's old association directories.

But Pat reminded Bill of that morning 365 days ago when they had said their goodbyes to the California Forestry Association staff and left at the pinnacle of his career for Chester to begin a new low-key life of family and community service. The decision to leave had been drawn out and wrenching, but Bill felt a call he could neither explain nor resist.

"Remember what you said?" Pat asked. "You said that I may not understand all the reasons for leaving yet, but when a year was up I would."

5:05 P.M.   *Sacramento*

THE CONFIRMATION CAME IN from the Coroner's office and Russell was ordered to the Murray family. She had only twenty minutes.   Police had arranged a brief truce with the media: they wouldn't contact Connie asking their ghoulish questions until at least 5:25. Russell and a deputy dashed to Roseville and pulled up in front of the nice home on Copper Way.

Gil Murray should have been coming up the steps to get ready for a night at Roseville High School, a proud father watching his oldest son receive one of the highest awards a graduating senior can get for athletic and academic achievement. Instead, Connie and her two sons, Wil and Gil, saw a chaplain and deputy coming up those steps.[34]

Once Russell was inside beginning the long ordeal, the deputy returned to the patrol car and paged the chaplain that Russell had left in charge back in the Police Department conference room: the death notification was done. He could inform the others now.

The chaplains told them that Gil Murray had died instantly in the bomb blast. They told them what to expect during the next hours and days.

You will feel guilty, they were told. You knew it was a bomb and you didn't stop him. You will also experience the guilt survivor syndrome. Why him and not me? I'm glad it wasn't me. Oh, no, how could I feel that way? You will feel anger, anger at Gil. Why did he open it? He knew it was a bomb. These emotions will flood you at first. You will roller-coaster for a while. Then you will begin to heal, to get beyond being victims. We will be there with you every step of the way.

The session broke up after 6:30. After her husband drove home, Jeanette Grimm went with Homicide Detective John Cabrera back to the office to see if she could get her car keys, which were in her purse on her desk.

Donn and Lisa were taken to a waiting room where they remained for half an hour. Finally, a homicide detective took Donn's statement. When asked about possible suspects, he said, "In my personal opinion, this is the work of extreme environmentalists."

He talked for nearly an hour. When he finished, Zea told the detective he had been saving radical environmentalist literature for years. He had several boxes of it in the office. Would someone please examine them?

The police investigator assured him someone would look at them, probably the FBI.

Donn and his wife were ready to leave when Jeanette came in and asked for a ride home to Roseville. "The FBI wouldn't let the detective into the crime scene," she said. "I can't get my car keys until tomorrow afternoon."

The three went out the back way to avoid the media.

Lisa had parked the car on a public street. While the two women watched, Donn approached the vehicle cautiously, got down on his knees, looked under it for any wires, stood up, looked inside to see if anything was leaning against the driver's door. He unlocked the car door and placed the key in the ignition without getting in. He rolled the window down and shut the door. Sticking his arm through the open window he turned the ignition key, thinking, *Okay, God, I may see You in a second.* The engine rumbled to life. Nothing exploded.

Donn Zea did not sleep that night.

TUESDAY, APRIL 25, 1995   *Grass Valley*

AFTER PACING THE HOUSE THROUGH THE DARK HOURS, straining at every sound while his wife and daughter slept, he got dressed in the early light and visited the sheriff and the local police, asking for increased patrols of his home.

Shortly after he returned home, his doorbell rang. The trim, athletic black man at the door showed his badge. Special Agent, FBI. Zea joined the man in the dark suit and they sat outside on the front steps together, talking in the morning quiet. Some of the agent's questions, he marked with a touch of irony, made him feel as if he were a bombing suspect, not a bombing target.

The man had not been gone an hour when FBI Special Agent Cliff Holly of the environmental crimes unit called and asked Zea about the files on radical environmentalists he had been collecting. Would Donn come to the CFA office the next morning? The crime scene tape would be down and they could retrieve the documents.

10:02 P.M.   *Arcata, California*

CANDY BOAK WATCHED THE LATE NEWS on Channel 2 from Oakland. News anchor Elaine Corral came onscreen:

> "There were two major developments today in the case involving the so-called Unabomber, who authorities believe was responsible for yesterday's fatal package bomb in Sacramento.

"Federal agents say for the first time in 17 years, the Unabomber is talking, and may be unraveling.

"Also, the FBI has reason to believe he may be in the Bay Area.

"Rita Williams is in San Francisco with a live report. Rita."
RITA WILLIAMS: "It's kind of an eerie thought. The man known as the Unabomber may even be listening to this report right here in the Bay Area."[35]

Candy Boak had special reason to watch attentively. She was the wife of a third-generation logger. For years she had watched as radical environmentalists vandalized the equipment of her friends in the redwood region. Her family's own logging equipment had been sabotaged. It wasn't just the little family-owned contract logging outfits like her husband's. She had seen big local companies like Pacific Lumber and Louisiana-Pacific beset with sabotage and disruptions by protesters. In response, she had formed a local citizen group called Mother's Watch to track radical environmentalist actions, collect their literature and expose their violence.[36]

She found dozens of troubling rants in radical newsletters and publicized them. Even the radicals' jokes pointed to danger. The Earth First Journal of September, 1989, had mused in a section called "Mirth First," ostensibly with tongue in cheek:

> While Eco defenders are quick to point out that life is sacred and is not a target of Eco-Defense, many doubt that multinational takeover artists who liquidate old growth forests to pay off junk bonds qualify as Life-forms [*a reference to Charles E. Hurwitz, CEO of CFA-member firm Maxxam, and his characterization after taking over Pacific Lumber Company*]. Such Robotoids, they aver, should be classed with dams, dozers and drillers. A "Hit List" is available upon discreet inquiry.[37]

Such gallows humor was lost on Maxxam and Pacific Lumber employees who saw Northern California Earth Firster Darryl Cherney on CBS News Sixty Minutes in March 1990 saying: "If I knew I had a fatal disease, I would definitely do something like strap dynamite on myself and take out Grand Canyon Dam. Or maybe the Maxxam Building in Los Angeles after it's closed up for the night."[38]

The radical response to Boak's activism had been furious. Earth Firsters counterattacked initially with invective, dubbing the women of Mother's Watch "polyester bitches." Groups of radicals soon began sitting in front of the Boak home and staring in for hours at a time. Then came the death threats—"You polyester bitch, we're gonna kill you." Her lawn was salted with the words, "Die Bitch." Someone put a bullet through her car windows—in through the driver's window, out through the passenger's window. A relative was nearly driven off the road. When her five-year-old grandson picked up the phone one day and heard a voice say, "We're going to kill your mommy, little boy," that was enough. She shut up.

Candy Boak felt Gil Murray's death keenly, and forced herself to watch KTVU-TV's news report on his probable murderer.

> RITA WILLIAMS: It's only the second time in 17 years and 16 bombings the Unabomber has communicated other than through his bombs. At least 2 of the 3 letters from the Unabomber himself were postmarked here in Oakland last Thursday and arrived yesterday, the same day as the fatal package bomb. Investigators would not reveal who the Unabomber sent the letters to, but sources say two of them went to former victims here in the bay area and the other to The New York Times.
>
> JIM FREEMAN / FBI SPECIAL AGENT IN CHARGE: "Certainly there is information in the letters that'll perhaps assist a member of the public to ID a suspect because there is information relative to the Unabomber explaining his motives."
>
> RITA WILLIAMS: "In fact, The New York Times in tomorrow's edition excerpts part of the letter it received yesterday. Here are some major points, all in quotes:
>
> "This is a message from the terrorist group FC.
>
> "The people we are out to get are the scientists and engineers, especially in critical fields like computers and genetics. We call ourselves anarchists because we would like, ideally, to break down all society into very small, completely autonomous units…. Our more immediate goal…is the destruction of the worldwide industrial system."

Candy Boak felt a shock of recognition. She grabbed the phone as fast as she could and dialed her friend Mary Bullwinkel, Pacific Lumber Company's director of public relations, at her home in Fortuna. Mary was already in bed.

"Mary," Candy said, "listen to this." She held the telephone handset up to her TV and flipped to Channel 5, KPIX-TV in San Francisco. They were just getting to the same passage of the Unabomber's letter.

"What does this sound like, Mary?"

Mary Bullwinkel listened to the words of the Unabomber. Her skin crawled.

"That sounds like Live Wild Or Die."

"Bingo."

Candy Boak made another call to a Pacific Lumber employee, Gary Gundlach, inventory clerk in the accounting department.

"Gary, this is Candy. Do you have a copy of Live Wild Or Die?"

"It's around here in my stuff someplace. Why?"

"Find it. The FBI is going to need it."

8:15 A.M. WEDNESDAY, APRIL 26   *Grass Valley*

"DONN, THIS IS CANDY BOAK. I just wanted to tell you how terribly sorry I am about Gil."

"Thank you. We all miss him something awful."

"How's Connie doing? And the boys?"

"Horrible. I hate to think how they're doing."

"That's just heartbreaking. How are you bearing up?"

"The truth? Not good. I find myself thinking of something and reaching for the phone to tell Gil. And the little things sneak up on you. You know what a major chocaholic Gil was. Now every time I see a chocolate bar I know he'll never have one again."

"You know you were on a hit list, don't you?"

"What?"

"Pacific Lumber has a copy of it."

"Candy, the FBI has asked me to go to the office and get some radical literature. I'd like to have that for them."

Candy called Pacific Lumber Company and spoke to John Campbell's executive assistant, Kathy Wigginton, asking her to get the Live Wild Or Die tabloid from Gundlach and fax a copy of it to Donn as soon as possible.

Wigginton got Gundlach on the line. He had found his copy the night before, but it was at home—should he go get it? He lived only a few blocks away in the company town of Scotia. Then he recalled that Rich Bettis, property manager for Pacific Lumber's timberlands subsidiary, had his own collection of radical literature right here in the main office building. That might be faster.

Kathy talked to Bettis and they spent the next hour raiding his boxes of material. Kathy called Zea and said, "I have 38 pages of propaganda. Where can I fax it to you?"

Donn thought fast. The CFA office was out: the FBI might not allow him into the crime scene yet.

"How about faxing it to the California Forest Products Commission? I'll call Amy and see if she'll get it to me."

"There's going to be more. Rich found some publications of the I.W.W."

"The Wobblies? Do they still exist?"

"They not only still exist, they seem to have formed some kind of alliance with Earth First."

"Fax that to me next."

"And a page about making bombs."

"Fax that too."

1:45 P.M.   *Sacramento*

HE FOLLOWED THE TWO FBI AGENTS into the wrecked office with a soul in turmoil. In a sudden reflex, Donn's left hand lifted to blinker his view as he walked past the spot where his boss had died. He hurried into his office with Holly and another agent, intent upon retrieving the boxes. He found the items and helped carry them.

On the way out, he no longer felt compelled to avert his eyes. He felt instead a strong need to confront dread. He stared into the lobby. Most of the debris had been cleared away, but the biological cleanup crew had not yet come in. A dark plastic sheet had been hung up on the bowed storeroom wall, concealing whatever remained on it. A white medical blanket lay over the place on the floor where his friend had opened the wooden box forty-eight hours earlier. It was not large enough to cover the dark stain. Sacramento County Coroner Bill Brown had removed evidence in paint cans.

Jeanette Grimm drove up the back alley in time to see Donn and the two FBI agents carrying a cardboard storage box and a large bundle of papers out the front door. Jeanette had brought Connie Murray to gather up her husband's personal effects from his office. Gil's Ford Explorer still stood alone next to the main entry.

Jeanette saw California Forest Products Commission administrator Amy Edwards walk up to Donn in the parking lot and give him a stack of fax transmissions. The FBI agents took the materials to their downtown office where Zea explained things.

The main item of interest from the faxed materials was the Earth First-offshoot journal called *Live Wild Or Die,* a tabloid published in Bellingham, Washington by Michael J. Jakubal (who uses the pseudonym "Doug Fir"), Mitchell Alan Friedman and several others who felt Earth First Journal was not sufficiently radical.[39]

Both Jakubal and Friedman were convicted felons: Mike Jakubal for a 1986 billboard vandalism,[40] plus a misdemeanor criminal trespass arrest in a tree-sitting stunt in the Willamette National Forest's Pyramid Creek timber sale area, 40 miles east of Sweet Home, Oregon on May 21, 1985;[41] Mitch Friedman for the same 1986 billboard felony, plus a misdemeanor criminal trespass arrest for occupying a U. S. Forest Service building in Okanogan, Washington on July 4, 1988 with twenty-three other protesters—when arrested there, Friedman gave his name as "Ben Hull," a Forest Service special agent who had dogged the heels of Earth Firsters.

As described in an Earth First Journal advertisement, *Live Wild Or Die* promised to serve as an outlet for the most radical anarchists in the movement. Friedman said the idea was to cover "revelry and the nihilistic," and the "release of the wild human spirit."[42] Jakubal, characterized by Harrowsmith magazine as "one of the hotter firebrands of the Earth First movement,"[43] wrote of "the necessity of acting out of our own true desires, our own wild subjectivity, our internal wilderness. ... I have no hope, only demands. There is no future, only Now. So why be modest in the face of impending doom?"[44] The second issue, although undated, had come out in early 1990. Printed on the next-to-last unnumbered page of the 26-page publication was the "Eco-Fucker Hit List." The Washington Post would later call it "an unruly compendium of sophomoric humor, obscenity, sloganeering and poorly drawn cartoons."[45]

The scrawled shock-jock title introduced the names of 100 organizations that had supported a 1989 wise use conference in Reno, Nevada, characterized as an "anti-wilderness" conference.[46] An amateurish political cartoon at the top of the page depicted the Exxon Valdez disaster as a goblet spilling crude oil on panicky Alaska sea creatures. It bore the wry caption, "They Annointed Our Waters With Oil; Our Cup Runneth Over."

Prominently featured in mid-page, just under the illustration, was a special list of eleven "national steering committee" wise use organizations, showing not only each group's name, but also a contact person and an address.

At the top of the short list stood the Timber Association of California. The contact was the late Roberta Andersen, the association's communications officer, who was as yet unaware she was dying of cancer. Bill Dennison was her boss. The address, 1311 I St., Sacramento, California 95814, stood out because of its location on the page.

The anarchist slogan "Disarm Authority — Arm Your Desires" slanted up from the end of the Sacramento address, calling further attention to it.

Biting parodies of Mobil and Exxon media ads occupied the rest of the page.

A drawing of an ax-wielding worker smashing the logo of Shell Oil was accompanied by a symbolic wooden shoe—a sabot—with a quotation by I.W.W. leader Bill Haywood: "Sabotage means to push back, pull out or break off the fangs of capitalism."

A hand-lettered note closed the page: "Send us your nominations for the Eco-Fucker Hit List!"

The juxtaposition of real names and addresses with tongue-in-cheek satire sent mixed signals—reminiscent of the celebrated saying of folk singer and Earth First friend Utah Phillips: "The earth is not dying, it is being killed. And those who are killing it have names and addresses!" It was not clear how the intended audience would interpret the "Eco-Fucker Hit List."

Did the Unabomber read it?

Did it guide him to his last murder victim?

APRIL 28, 1995   *Mount Vernon Memorial Park, Fair Oaks, California*
AN OVERFLOW CROWD of 400 people from throughout California left more than 100 mourners listening to the memorial service for Gil Murray outside in the rain.

In the chapel lobby, a display of photographs showed Murray on vacation with his family, skiing at Lake Tahoe, snorkeling, backpacking, standing in front of the White House in Washington, D.C. and clowning around with his sons on Halloween.

"My father was the greatest man I ever met," said Wil Murray, Gil's 18-year-old son.

"He loved my mom, my brother and me more than life itself. He was always there for us. We always came first.

"I only hope I can be half the man he was."

Clasping the edge of a lectern on a stage filled with flowers that surrounded a large color photograph of Gilbert Murray, the young man spoke steadfastly through a breaking voice.

"We loved being with him so much. He loved to go to our football games, our baseball games and our basketball games. He loved to ski with us and to play catch with us and just to be with us.

"Dad, I know you're listening. You always did. You were always there. You gave us the strength to make it through this, but right now we just don't want to do it without you.

"Everybody misses you."[47]

Michelle brought a giant-size chocolate bar and left it for her boss.

Chapter One Footnotes

[1] The events of Monday, April 24, 1995, have been reconstructed through extensive telephone interviews with the involved California Forestry Association staff including Eleanor Anderson, Michelle Goldsberry, Jeanette Grimm, Sophia Runne, Bob Taylor, Melinda Terry, Lisa Tuter, and Donn Zea. Bill Dennison read drafts. Dialog has been recreated with their critical advice as close to verbatim as possible.

[2] The denomination of the Eugene O'Neill stamp in the Prominent American series, coil version (1305C) issued January 12, 1973, was one dollar, for a total postage of $9.96. The Unabomber used the O'Neill stamp on five of the ten bombs he sent through the mails.

[3] "Text of Letter from 'Terrorist Group,' Which Says It Committed Bombings," *New York Times*, Wednesday, April 26, 1995, p. A16.

[4] Arne Naess, "The Shallow and The Deep, Long-Range Ecology Movements: A Summary," *Inquiry* 16 (Oslo, 1973), pp. 95-100.

[5] Martha F. Lee, *Earth First! Environmental Apocalypse*, Syracuse University Press, Syracuse, 1995, p. 38.

[6] Bill Devall and George Sessions, *Deep Ecology: Living As If Nature Mattered*, Gibbs Smith, Publisher, Peregrine Smith Books, Salt Lake City, 1985, pp. 70-73.

[7] Dave Foreman, *Confessions of an Eco-Warrior*, Harmony, New York, 1991, p. 26.

[8] The leading voice for this method is the Voluntary Human Extinction Movement, P.O. Box 86646, Portland OR 97286-0646.

[9] Quoted in *Fortune* magazine, Daniel Seligman, "Down With People," September 23, 1991, p. 215.

[10] Dave Foreman, Howie Wolke, and Bart Koehler, "The Earth First! Wilderness Preserve System," *Earth First!* vol. 3, no. 5, Litha / June 21, 1983, p. 9.

[11] Dave Foreman, John Davis, et al., "The Wildlands Project Mission Statement," *Wild Earth*, Special Issue, 1992, p. 3.

[12] Dave Foreman and Bill Haywood (pseudonym), editors, *EcoDefense: A Field Guide to Monkeywrenching*, A Ned Ludd Book, Tucson, 1987. The identity of Foreman's anonymous "co-editor" Bill Haywood has never been released; some believe it was Mike Roselle, some believe Foreman worked alone. Using Big Bill Haywood's name does, however, reveal Foreman's sympathies. The real William Dudley Haywood was an early leader of the "Wobblies," the International Workers of the World. He was arrested with 164 other IWW members in 1917 and convicted the following year on charges amounting to treason and sabotage. While out on bail during appeal, he jumped bail and fled to Russia, where he died in 1928. *Bill Haywood's Book: The Autobiography of William D. Haywood* was published in 1929. See also, Edward Abbey, *The Monkeywrench Gang*, J. B. Lippincott, New York, 1972.

[13] "Terrorism fears grow at home - Earth First!, other groups stir fears of injury, deaths," *Phoenix Gazette*, Saturday, March 17, 1990, by Ed Timms, Dallas Morning News, p. A7.

[14] "Acquiring organizational legitimacy through illegitimate actions: a marriage of institutional and impression management theories," *Academy of Management Journal*, October 1992 vol. 35 no. 4, p. 699, by Kimberly D. Elsbach and Robert I. Sutton.

[15] A scholarly analysis of the structure and behavior of Earth First can be found in Martha F. Lee, *Earth First! Environmental Apocalypse.*

[16] "U.S. Widens Earth First! Indictments," *Arizona Republic*, Saturday, December 15, 1990, by Lisa Morrell, p. B9. See also, "'Monkey Wrench' Activist Linchpin In Sabotage Trial," *Arizona Republic*, Sunday, June 9, 1991, by Jonathan Sidener and Lisa Morrell, p. A1.

[17] "Earth First! Trial Halted. Activists Accept Plea Deals In Alleged Sabotage Plot," *Arizona Republic*, Wednesday, August 14, 1991, by David Cannella and Jonathan Sidener, p. A1.

[18] Plea Agreement One, felony conspiracy, *United States of America v. David William Foreman*, CR-89-192-PHX-RCB, Filed September 6, 1991, p.12. Plea Agreement Two, misdemeanor depredation of government property, *United States of America v. David William Foreman*, CR-89-192-PHX-RCB, Filed September 6, 1991, p. 2.

[19] Christopher Manes, *Green Rage: Radical Environmentalism and the Unmaking of Civilization*, Little, Brown & Company, Boston, 1990.

[20] "Researcher Freed After 159-Day Term," *Seattle Times*, Wednesday, October 20, 1993, by Associated Press, p. B3.

[21] "Animal Rights Activist Enters Guilty Pleas In Arson, Theft," *Portland Oregonian*, Tuesday, March 7, 1995. Compiled by staff and wire reports, p. B5.

[22] "Open Letter from Rod Coronado: Spread Your Love Through Action," *Earth First Journal*, March, 1995, as posted on the World Wide Web at http://www.envirolink.org/arrs/coronado.html.

[23] "Briefly," *USA Today*, Tuesday, July 17, 1990, p. 2A.

[24] U.S. Department of Justice, *Report to Congress on the Extent and Effects of Domestic and International Terrorism on Animal Enterprises*, U.S. Government Printing Office, Washington, D.C., August, 1993, p. 8.

[25] Captain Paul Watson, *Earth Force! An Earth Warrior's Guide to Strategy*, Chaco Press, Los Angeles, 1993, Foreword by Dave Foreman.

[26] Screaming Wolf [pseudonym attributed to Sidney and Tanya Singer], *A Declaration of War: Killing People to Save Animals and the Environment*, Patrick Henry Press, Grass Valley, California. Patrick Henry Press was established by Sidney and Tanya Singer.

[27] "Gaia Liberation Front: A Modest Proposal," http://www.envirolink.org/orgs/coe/resources/glf/glfsop.html. Posted winter solstice 1994 by GLF, P.O.Box 127, Station P, Toronto, Ontario M5S 2S7, Canada.

[28] Dave Foreman and Bill Haywood (pseudonym), editors, *EcoDefense: A Field Guide to Monkeywrenching*, A Ned Ludd Book, Tucson, 1987, p. 14. Edward Abbey, in response to a magazine article I wrote ("EcoTerrorism," by Ron Arnold, *Reason*, February 1983, vol. 14, no. 10, p. 31), sent me a scrap of a note insisting that ecodefense was "not terrorism" because it only injured property and not people, a distinction he appeared to hold sincerely, however incorrectly.

[29] FBI Terrorist Research and Analytical Center, *Terrorism in the United States: 1994*, Washington, D.C., U.S. Department of Justice, 1995, p. 24.

[30] "Letters From Serial Bomber Sent Before Blast," *New York Times*, Wednesday, April 26, 1995, by James Barron, p. 1.

[31] "Radical Enviros on the Move Again," by Donn Zea, *California Forests Today*, April, 1995, p. 5. The quotes are excerpted from "Forest Grump,"

Earth First!, vol. 16, no 2, Yule, December 1994 / January 1995, by Mike Roselle, p. 23. Expletives were deleted by Zea.

[32] "Explosion Kills 1 in Lobbyists' Office, No Indication That Package Bomb is Linked to Oklahoma, Official Says," *Chicago Tribune*, Tuesday, April 25, 1995, by Associated Press, p. 3.

[33] Telephone interview with Sacramento Police Chaplain Mindi Russell, July 31, 1995.

[34] "Stunned Community Remembers 'Great Guy,'" *San Francisco Examiner*, Tuesday, April 25, 1995, by Tom Abate, Venise Wagner and Steven A. Capps. Scott Winokur and Jane Kay contributed to this report, p A1.

[35] News transcripts, KTVU San Francisco / Oakland, Two Jack London Square, Oakland, CA 94623, programming for April 25, 1995, by Bob Hirschfeld, archive: 1887, p. 1.

[36] Telephone interviews with Candace Boak, July 19, 23 and 25, 1996.

[37] Gula [pseudonym], "Eco-Kamikazes Wanted," *Earth First!*, vol. 9, no. 8, Mabon / September 22, 1989, p. 21.

[38] Darryl Cherney, quoted in *Sixty Minutes Transcripts*, vol. 22, no. 24, March 4, 1990, p. 3.

[39] Six issues have been published in various locations: No. 1, 1989; No. 2, 1990; No. 3, 1991; No. 4, 1994, No. 5, 1995; No. 6, 1996 (48 pages). An advertisement in *Earth First Journal* Vol. 16, No. 2, Yule, December 22, 1995, p. 37, called for editorial contributions for No. 6 to be sent to P.O. Box 2732, Asheville, North Carolina 28802.

[40] "Mother nature's army; guerrilla warfare comes to the American forest," *Esquire*, Feb. 1987, vol. 107, by Joe Kane, p. 98.

[41] "Logging Protester Arrested - 'Doug Fir' Descends From Lofty Perch," *The Washington Post*, May 22, 1985, by United Press International, p. A13.

[42] Martha F. Lee, *Earth First! Environmental Apocalypse*, p. 123, fn 185, 58.

[43] "Mr. Monkeywrench," *Harrowsmith*, Vol. III, No. 17, September-October, 1988, by Kenneth Brower, p. 44.

[44] As quoted in "War of woods: Logging terrorism," *Bellingham* [Washington] *Herald*, Sunday, September 17, 1989, by Leo Mullen, p. A1.

[45] "Madman or Eco-Maniac? Conspiracy Theorists See Environmental 'Hit List' as Unabomber Fodder," *The Washington Post*, April 17, 1996, by Richard Leiby, p. C1.

[46] The Second Annual Wilderness Conference, Friday, April 21 1989 and Saturday, April 22, 1989, John Ascuaga's Nugget, Reno, Nevada.

[47] "400 Bid Farewell To Bomb Victim. Lobbyist Mourned, Praised At Service," *Sacramento Bee*, Saturday, April 29, 1995, by Patrick Hoge, p. B1.

Chapter Two
MANIFESTO

M<small>ONDAY</small>, M<small>AY</small> 1, 1995   *Bellevue, Washington*

I<small>T LOOKED LIKE A TRAVEL BROCHURE AT FIRST GLANCE.</small> But the words above the vivid illustration said, "Bombs by Mail." Opened up, the layout read L<small>ETTER AND</small> P<small>ACKAGE</small> B<small>OMB</small> I<small>NDICATORS</small>: unexpected package, excessive postage, wrong name with title, fictitious return address, hand canceled stamps, oily stains on wrapping, strange odor, restrictive markings such as "*personal*," lopsided weight, protruding wires.

"I'll leave enough of these for your mail room staff," said our instructor, Postal Inspector James D. Bordenet, a middle-aged man of medium stature, narrow face and matter-of-fact, almost cheery manner. He had given this workshop to more than a hundred organizations during his 25 years as a postal inspector.

He stood next to the television stand and handed two sets of government documents around to the half-dozen people in our modest conference room: "Mail Center Training," and "Security Plan for Suspected Letter and Parcel Bombs."

The name of my organization—the Center for the Defense of Free Enterprise—was among those on the *Live Wild Or Die* "Eco-Fucker Hit List." For many years we had been prominent in criticizing environmental radicals and the destruction of private property. Our first Wise Use Strategy Conference in 1988 had been emceed by our favorite personal dynamo, Robbie Andersen, the California Forestry Association staffer

31

whose name topped the *Live Wild Or Die* Hit List. We had allied with Bill Dennison and his association on a number of projects over time. When Gil Murray was killed, the FBI urged me to run our staff through the Postal Inspector's half-day mail bomb detection course.

"The Postal Inspector's Office is the law-enforcement arm of the U. S. Postal Service," said Jim Bordenet. "The Postal Service delivers approximately 177 billion pieces of mail each year. Over the past ten years, an average of 16 mail bombs per year have moved through this system. Approximately one-third of these devices have detonated.

"You can see that mail bombs are extremely rare. But they do happen. All types of organizations and individuals get mail bombs. This includes businesses, political figures, government officials, police agencies, witnesses in criminal trials and private citizens. People involved in romantic triangles remain the biggest targets of mail bombers."

The half-hour lecture session gave us a basic grasp of the problem, followed by a video titled "Mail Bombs" hosted by the late actor Greg Morris, who played technical wizard Barney Collier on the television series *Mission Impossible*, which ran from 1966 to 1973. He guided viewers through key bomb recognition points, screening methods for mail clerks, special equipment and security actions. Very impressive.

A tag-on update, though, contained the most intense scene of all: In a demonstration room, a pipe bomb had been placed on a desk with a mannequin sitting in a chair behind it, just as things would be in a business office. A technician used a remote control device to detonate the bomb. The desk and the mannequin dissolved in a volcano of splinters. It made the point.

Bordenet then spent more than an hour answering our questions: How do we set up a screening program? Who do we call in the event of suspicious packages? What do we do with them while waiting for the bomb squad?

At the close of the session, my wife Janet asked the question that was on all our minds: Will you ever catch the Unabomber?

Bordenet smiled wearily. "Most bombers tell us how to catch them—clues, letters, evidence. This bomber who killed your friend doesn't want to be caught. Not yet, anyway. He's sending letters now, but it's not good enough. We won't catch him until he tells us how."

THURSDAY, JUNE 29, 1995   *New York City*
"THE INDUSTRIAL REVOLUTION and its consequences have been a disaster for the human race."

With those words the Unabomber began the reinvention of a wheel that had already run over us many times, relentlessly flattening our high hopes for perfecting humanity and controlling nature through technological progress. He came across as an earnest thinker who had decided that modern life is tough and felt compelled to explain it to us as if we didn't already know. He actually seemed to think he was the first one ever to work out such ideas.

The 56-page, 34,390-word typewritten essay titled *Industrial Society and Its Future* marched through 232 neatly numbered paragraphs and 36 footnotes, with a shortened alternative for Note 16 in case his use of a long quote from a book might constitute a copyright violation. Quite fastidious legalism for a serial killer.

The manifesto posed many puzzles, the first being why its clearly intelligent and literate author seemed unconscious of the intellectual history of anti-technology and anarchism, yet at the same time expressed ideas so similar to radical environmentalism that thousands saw the connection.

The Unabomber quoted from exactly three books: *Violence in America: Historical and Comparative Perspectives*, edited by Hugh Davis Graham and Ted Robert Gurr, *Chinese Political Thought in the Twentieth Century* by Chester C. Tan, and *The Ancient Engineers*, a non-fiction work by prolific science fiction writer L. Sprague de Camp, whose Conan stories gave Arnold Schwarzenegger his big movie break.

The Unabomber also mentioned but did not quote from *The True Believer* by Eric Hoffer. He quoted articles from Scientific American and Omni magazines and the New York Times. He shoveled through a compost heap of history—the early Christian era, the Middle Ages, the American Revolution and the Nineteenth Century—without mentioning which books had fertilized his imagination. He knew a lot about science and technology. He knew his way around social psychology. It looked like he'd been prowling big libraries. He'd obviously been reading technical journals and academic texts.

But something was missing.

It was hard to believe that there was no mention of the great precursors of anti-technology such as Max Weber, whose 1904 study, *The Protestant Ethic and the Spirit of Capitalism*, railed against economic acquisition having become the ultimate purpose of life.

There was no mention of Sigfried Giedion, whose 1948 book *Mechanization Takes Command* popularized anti-technology.

There was no mention of Jacques Ellul, whose magisterial 1964 work, *The Technological Society*, gave intellectual prestige to anti-technology—and who, a year later, wrote deeply of another pet Unabomber hate, *Propaganda: The Formation of Men's Attitudes*.

There was no mention of Lewis Mumford, whose three-volume 1973 masterpiece *The Myth of the Machine* installed anti-technology as part of the American core academic curriculum.

Nor was there any mention of other major laborers in the anti-technology vineyard: Paul Goodman, the academic whose *Compulsory Mis-Education* (1962) sought to tear down the academy because it had become a gigantic feudal corporation that choked out the creativity and community necessary to deal with modern life; economist E. F. Schumacher, whose *Small Is Beautiful* (1973) preached the downsizing of technological society; or Rachel Carson, whose *Silent Spring* (1962) was the fountainhead of environmental doomsday prophecy.

Nor was there any acknowledgment of contemporary neo-Luddites—Kirkpatrick Sale, Jerry Mander, Chellis Glendinning, Jeremy Rifkin, Bill McKibben, Wendell Berry, Langdon Winner, Stephanie Mills and John Zerzan among them—who shared a great many of his views about the pernicious effect of the Industrial Revolution.

Was he merely hiding his sources to elude detection?

Despite the Unabomber's obviously omnivorous reading, he appeared ignorant of those who should have been his icons.

Perhaps he was influenced by classic literature. He had certainly read Francis Bacon and Isaac Newton and other inventors of science from the Enlightenment. He might have also read reactions to the Enlightenment: William Blake, whose vast epic poetry cursed the "dark Satanic mills" of early nineteenth century industrial England, where "Bacon and Newton, sheath'd in dismal steel," had founded modern science, whose "cruel works / Of many Wheels" enslaved the soul "with cogs tyrannic."

Or Mary Shelley's archetypal 1818 anti-technology novel, *Frankenstein, or the Modern Prometheus*, which, more than a century later, provided wildly popular "cruel works" for the "many Wheels" of Hollywood's merrily "dark Satanic" movie mills.

Or Goethe's *Faust*, in which a techno-nerd studied so much and partied so little he had to sell his soul to the Devil to regain his lost youth and get a girl—and then everybody ends up dead.

It was anybody's guess.

Yet parts of the manifesto seemed disturbingly familiar, as if they had been gleaned from a weed patch right in our own back yard. Earth First Journal and Fifth Estate, a feisty anti-technology paper published out of Detroit for the past thirty years, flashed in many paragraphs. And whole sections seemed just like... What? Baffling.

Two things seemed clear about whatever the Unabomber had been reading: the ability to construct a graceful English sentence had not rubbed off; and he had cobbled his anti-technology together from a hodge-podge of less than seminal sources.

His manifesto was as homemade as his bombs.

The Unabomber sent copies of the manifesto to the New York Times, the Washington Post, Penthouse magazine (which had previously offered to publish it)—and a Berkeley psychology professor. Scientific American got a letter but no manuscript. Time and Newsweek, the two newsmagazines nominated as outlets in his April ransom letter, received neither letter nor manuscript.

The Times and the Post were given a 90 day deadline by which to publish the screed *in toto*. If they agreed, there would be no more bombs.

If they refused and Penthouse agreed, the Unabomber reserved the right to "plant one (and only one) bomb intended to kill, after our manuscript has been published," because Penthouse was less "respectable" than the two newspapers. If everybody refused, the bomb was in the mail.[1]

There was no threat to the professor, Tom R. Tyler. All the Unabomber wanted from him was recognition for the soundness of his arguments, which the professor promptly provided in a front page San Francisco Chronicle story.[2]

Scientific American got a mild reproof for an article it had published on particle accelerators—the letter was so dull that the publishers at first refused to believe it was really from the Unabomber.

All this mail was sent at one time, postmarked June 24, 1995 from San Francisco.

A letter accompanied each of the four manuscripts. The letters tell us more about the Unabomber than his manifesto. The one addressed to the New York Times began:

> This is a letter from FC, 553-25-4594[3]
> If the enclosed manuscript is published reasonably soon and receives wide public exposure, we will permanently desist from terrorism in accord with the agreement that we proposed in our last letter to you....
> Contrary to what the FBI has suggested, our bombing at the California Forestry Association was in no way inspired by the Oklahoma City bombing.

The Unabomber had just revealed in his letter to Penthouse magazine that the enigmatic "FC" monogram stood for Freedom Club, the purported anarchist group behind the bombs. It had mystified law enforcement officials since June 1980. They first noticed it inscribed on a piece of a bomb hidden inside a novel mailed to Percy Wood, then president of United Airlines, in suburban Chicago.

The inscription was found on seven of the next eight bombs. Because of his computer-scientist targets, they thought it might stand for "Fuck Computers."[4]

He also got the name of the forestry association right. And he seemed quite familiar with it.

> We have no regret about the fact that our bomb blew up the "wrong" man, Gilbert Murray, instead of William N. Dennison, to whom it was addressed. Though Murray did not have Dennison's inflammatory style he was pursuing the same goals, and he was probably pursuing them more effectively because of the very fact that he was not inflammatory.

The Unabomber's contempt for Murray's death contrasted sharply with his motherly concern over a possible public misunderstanding of radical environmentalists:

A letter from an anarchist to editors of the NY Times made us realize that we owe an apology to the radical environmentalist and non-violent anarchist movement. Statements we made in our letters to the NY Times would tend to associate us with anarchism and radical environmentalism, and therefore might make the public think of anarchists and radical environmentalists as terrorists. So we want to make it clear that there is a NONVIOLENT anarchist movement that probably includes most people in America today who would describe themselves as anarchists. It's a safe bet that practically all of them strongly disapprove of our bombings. Many radical environmentalists do engage in sabotage, but the overwhelming majority of them are opposed to violence against human beings. We know of no case in which a radical environmentalist has intentionally injured a human being. (There was one injury due to a tree spiking incident, but the spiking was probably intended only to damage equipment, not injure people.)

Why was the Unabomber so intent on giving cover to radical environmentalists? And was he really unaware of their violence against human beings?

The injured person he wrote of was George Alexander, a 23-year-old third-generation mill worker, who was just starting his shift at the Louisiana-Pacific lumber mill in Cloverdale, California on Friday, May 8, 1987. He worked directly behind the headrig sawyer's control booth, directing big slabs of freshly sawn wood down to the edgers and trimmers that would turn them into lumber, and dropping broken chunks into the chipper for use in pulp and papermaking. The log that would alter his life—cut from a company-owned forest in the Cameron Road area of the coastal community of Elk—rolled down the infeed chain deck toward the high-speed bandsaw. When the whining jagged-tooth saw struck an 11-inch bridge timber nail that had been driven into the log, the huge blade exploded. A ten-foot steel section ripped through Alexander's safety helmet and face shield, tore his left cheek, cut his jawbone in half, knocked out upper and lower teeth and nearly severed his jugular vein. He was transported immediately with paramedics to Healdsburg Hospital and at 3:00 p.m. that afternoon transferred by helicopter to San Francisco's University of California Medical Center. Alexander, who had worked for L-P for less than a year and was married only a month earlier, almost died. The spike had been driven in the same manner described in Dave Foreman's *EcoDefense*. County Supervisor Norm deVall had notified L-P that residents in the Cameron Road area opposed the company's logging of its own land.[5]

The Mendocino County Sheriff's Department identified a suspect named William Joseph Ervin, who owned property in the Cameron Road area and was known to have spiked several trees on his own property to deter timber thieves. Sheriff's investigators were never able to prove to the satisfaction of the District Attorney that Mr. Ervin spiked the vandalized tree. The District Attorney's office declined to prosecute.[6]

Earth First denied that the suspect was a member. The denial could not be refuted because Earth First styled itself a "movement," not an organization, and had no members. Dave Foreman said the incident would not stop him from publishing his sabotage manual, adding, "I think it's unfortunate that somebody was hurt, but, you know, I quite honestly am more concerned about old-growth forests, spotted owls and wolverines and salmon—and nobody is forcing people to cut those trees."[7] Concerning the intent behind the spike, the Unabomber could be right.

But he could not be right about the intent behind Earth Firster Lee Dessaux's savage attack on March 17, 1990, against Dan R. Jacobs of Kalispell, Montana and Hal Slemmer of Billings, stabbing each many times with a steel ski pole as part of a Fund for Animals protest against a buffalo hunt near West Yellowstone. Dessaux became so violent that the animal rights video camera operator documenting the scene stopped taping and screamed at Dessaux to stop. Dessaux was found guilty on two counts of misdemeanor assault on February 8, 1991 in a jury trial in Justice Court in Bozeman. Judge Scott Wyckman sentenced Dessaux to 90 days in Gallatin County Jail.[8]

And the Unabomber could not be right about the intent behind the May 25, 1989 Molotov cocktail hurled into the California Cattlemen's Association office in Sacramento, where Executive Vice President John Ross was working late. An intruder smashed the glass front door and entered the reception area to light the gasoline bomb when Ross stopped him. The trespasser tossed the incendiary device as he escaped and it failed to explode. Four months earlier, on January 29, 1989, the cattlemen's office windows were etched with acid, the locks were filled with liquid metal, and slogans painted on the building's front—"Earth First! Agribusiness Kills."[9] At 2:00 a.m. that morning someone set fire to the livestock auction building in Dixon, California. The slogans "Animals Are Not Slaves" and "Earth First" were painted on the walls.[10] The $250,000 sheep ring was a total loss, but the beef ring survived undamaged. An anonymous caller told the Associated Press that Earth First set the blaze. Dixon Fire Department Chief Rick Dorris said witnesses saw a suspicious

vehicle near the scene known to belong to an Earth Firster. Earth First co-
founder Mike Roselle told the Sacramento Bee, "We don't rule out tactics
simply because they involve the destruction of property. I applaud this
type of action, if it proves to be effective in the long run. The only prereq-
uisite to being an Earth-Firster is owning a shirt. So if the guy says he's an
earth-firster, by God, he probably is."[11] Shamelessly, Earth First Journal
in 1996 claimed that this arson is "well known now as an insurance scam."[12]
That's news to Marie Cammerota, the property owner, and to James F.
Schene, the Dixon Auction operator, and to Rick Dorris, Dixon Fire Chief,
and to the Solano County Sheriff's Department: there was no fire insurance to scam, and Schene repaired the damage with his own money.[13]

   And the Unabomber could not be right about the intent behind the
events in Great Britain: in the past year alone, Animal Liberation Front
activists had been responsible for more than 100 attacks against people.
ALF members mailed incendiary devices to William Waldegrave, the agriculture minister, and Tom King, the former defense secretary; and burned
down department stores.[14]

   In Bristol, England on Sunday, June 10, 1990, Dr. Patrick Headley,
of Bristol University's medical sciences department, drove less than a mile
from his home before a bomb attached under his car by the Animal Liberation
Front detonated. Dr. Headley escaped with a cut to the nose, but shrapnel
from the explosion blasted onto the sidewalk and ripped through the push-
chair of a 13-month-old toddler, John Cupper, seriously injuring his spine.[15]

   A few days earlier in Porton Down, Wiltshire, Margaret Baskerville,
a veterinary officer at the Chemical Defense Establishment, escaped serious injury in another animal rights car-bomb attack. The London Times
said of the incidents, "Although there have been previous attacks on scientists, including letter bombs being sent to their homes and firebomb attacks on their vehicles, the campaign of the past six days appears to involve plastic explosive, only previously used when the bar and the restaurant of the senate building at Bristol University were blown up in February last year. That attack was claimed by the Animal Liberation Front and
the hitherto-unknown Animal Abused Society."[16]

   Nor could the Unabomber be right about the intent behind a death
trap constructed by Earth Firsters inside a tunnel on the route of a 1987 desert
motorcycle race. Playboy magazine interviewed Mike Roselle of Earth First
in its April, 1993 issue. Interviewer Dean Kuipers wrote, "When I asked Mike
Roselle to tell me about his favorite action, or ecodefense, he didn't hesitate."
   Kuipers quoted Roselle:

> A band of desert saboteurs from Earth First resolved in 1989
> to put an end to the desert motorcycle race called the Barstow to
> Vegas, which ran through the East Mojave scenic area, a prospective
> national park and habitat of the desert tortoise, kangaroo rat and other
> creatures.

The night before the race, we took a trailerload of railroad ties and four-by-eights down to the track," remembers Roselle, a former oil-field roughneck and one of the five men who cooked up the idea for Earth First on a camping trip to Mexico's Sonora Desert in 1980. "See they had to go under Interstate Fifteen. There was this tunnel about six feet wide, eight feet high and one hundred fifty feet long that was made for water to go through. We built this cube to the size of the culvert, and at night we set it up in the middle of the tunnel."

Kuipers obtained opposing comment, writing: "I want you to picture this," snaps Rick Siemans, senior editor of Dirt Bike magazine and head of the Sahara Club, a race sponsor. "Here are top expert riders going a hundred and ten miles per hour down a sand wash at eleven o'clock, sun directly overhead, coalblack shadows, dust on their goggles, and they're going to dart through this shadow, assumedly, and go to the other side. If our people hadn't spotted that, they would have killed a half-dozen riders." [17]

The roadblock was wedged in so tight it took a winch to pull it out. It was designed to cause a fatal accident, according to race and federal officials. "This was an attempt to get someone hurt. You're coming from the bright sun to a dark tunnel," said Steve Fleming, acting chief ranger for the federal Bureau of Land Management, as he inspected the tangled wreckage of the dismembered roadblock.[18]

A letter to the editor of the Los Angeles Times said of the incident, "Fortunately, no one was injured or killed. However, this does not alter the fact that Earth First! knowingly and purposely erected a dangerous obstacle. Someone should be held responsible for this dangerous and illegal publicity stunt. I invoke the legal system to closely evaluate the treacherous form of protest employed by Earth First! and to repudiate these radicals so that protests like this might never occur again."[19]

The Unabomber could not be right about another incident: In the 1990 poll-tax riot in London's Trafalgar Square, animal rights activist Simon Russell was arrested and sentenced to 18 months' imprisonment for attacking a policeman with a stick. He was sentenced to a further two and a half years after incendiary devices intended for hunt vans were found at his home in Tunbridge Wells. The London Times reported that the Animal Liberation Front, founded in the United Kingdom, was helping to spread terrorism: "its tactic of using cheap and easy-to-plant incendiary devices is increasingly copied by the IRA."[20] The Animal Rights Militia, a 1984 offshoot, took credit for using several such incendiary devices at a shopping mall on Isle of Wight shortly after midnight, August 24, 1994, causing £2 million in damage and endangering countless lives. More than 100 firefighters and all the 16 implements on the island were needed to cope with the inferno, from which 15 people were evacuated without injury. Help was summoned from the Hampshire mainland and four fire engines were ferried across to ensure essential cover.[21]

And the Unabomber could not be right about the intent behind the January 1992 poisoning of 87 Canadian Cold Buster bars—a cold-fighting chocolate bar manufactured in Canada—by injecting them with liquid oven cleaner. The Animal Rights Militia said in a letter that they had poisoned the bars because of University of Alberta scientist Larry Wang's 16 years of animal experiments that led to the invention of the bar. Wang had allegedly "frozen, starved and injected with various drugs, including barbiturates, countless rats." A contaminated bar was discovered after the candy had been on the market about a month. Production was halted and bars were pulled from store shelves in five provinces across Canada, reportedly causing $250,000 in losses to Okanagan Dried Fruits Ltd., a firm near Penticton, British Columbia.[22]

Then there was the question of sabotage, road blockades and site occupations by radical environmentalists to coerce loggers, miners, ranchers and other workers: is such physical duress non-violent?

Is torching machinery non-violent?

Is barricading or digging trenches in roads to prevent workers from earning their living non-violent?

Is criminal trespass to fasten one's body to equipment non-violent?

A long catalog of force against human beings by radical environmentalists had escaped the Unabomber's notice.

His letter to The New York Times held other puzzles. In a paragraph deriding the FBI for failing to keep its facts straight, the Unabomber wrote:

```
It was reported that the bomb that killed
Gilbert Murray was a pipe bomb.  It was not a pipe
bomb but was set off by a home made detonating cap.
(The FBI's so-called experts should have been able to
determine this quickly and easily, especially since
we indicated in an unpublished part of our last
letter to the NY Times that the majority of our
bombs are no longer pipe bombs.)  It was also re-
ported that the address label on this same bomb
gave the name of the California Forestry Association
incorrectly.  This is false.  The name was given
correctly.
```

Jeanette Grimm and Eleanor Anderson both scrutinized the address label and discussed it. It did not say California Forestry Association. It did say Timber Association of California. Just exactly as it appeared on the "Eco-Fucker Hit List" in *Live Wild Or Die*.

The final section of the Unabomber's cover letter dealt with morality and motives:

What about the morality of revolutionary violence? ... Do the revolutionaries' goals outweigh the harm they cause to others? Do the people they hurt "deserve" it?

Such questions can be answered only on a subjective basis, and we don't think it is necessary for us to do any public soul-searching in this letter. But we will say that we are not insensitive to the pain caused by our bombings.

A bomb package that we mailed to computer scientist Patrick Fischer injured his secretary when she opened it. We certainly regret that. And when we were young and comparatively reckless we were much more careless in selecting targets than we are now. For instance, in one case we attempted unsuccessfully to blow up an airliner. The idea was to kill a lot of business people who we assumed would constitute the majority of the passengers. But of course some of the passengers likely would have been innocent people - maybe kids, or some working stiff going to see his sick grandmother. We're glad now that that attempt failed.

But even though we would undo some of the things we did in earlier days, or do them differently, we are convinced that our enterprise is basically right. The <u>industrial-technological system</u> has got to be eliminated, and to us almost any means that may be necessary for that purpose are justified, even if they involve risk to innocent people. As for the people who wilfully and knowingly promote economic growth and technical progress, in our eyes they are criminals, and if they get blown up they deserve it.

Of course, people don't kill others and risk their own lives just from a detached conviction that a certain change should be made in society. They have to be motivated by some strong emotional force. What is the motivating force in our case? The answer is simple: Anger. You'll ask why we are so angry. You would do better to ask why there is so much anger and frustration in modern society generally. We think that our manuscript gives the answer to that question, or at least an important part of the answer.

2:30 P.M. TUESDAY, MAY 9, 1995   *San Francisco*
JIM FREEMAN, HEAD OF THE FBI's UNABOMBER TASK FORCE, with Special Agent Cliff Holly, held a telephone conference with seven people who in

one way or another were active in the wise use movement: Donn Zea of California Forestry Association; Nadine Bailey, CFA's grassroots coordinator; Chuck Cushman of the American Land Rights Association; Chris West of the Northwest Forest Resource Council; Candy Boak of Mother's Watch; and Bill Pickell of Washington Contract Loggers Association. I was the other participant.

The agents asked if we understood that we were considered potential targets. We did. They asked if our groups had completed the Postal Inspector's mail bomb training. Most said yes. They wanted to know if we had ever received death threats or hate mail. Everyone laughed: it comes with the territory when you challenge environmentalism. Had we encountered anyone suspicious at conferences or meetings? That was harder to answer: there are always environmentalists at our public meetings, but were any of them suspicious in the sense of being a serial killer? No, certainly not. We joked that a recent Outside magazine photo of David Helvarg, Sierra Club author of *War Against the Greens*, a book that reviled most of us, looked just like the FBI sketch of the Unabomber, but the agents were not in the mood for humor. They asked if we would give them any physical evidence we might still have from the hate mail. Most of us had already sent them such materials. We had a few questions about the investigation ourselves, and wanted more details.

The FBI asks more questions than it answers.

Cary Hegreberg got a call from an FBI agent in Salt Lake City: Hegreberg, who runs the Montana Wood Products Association in Helena, was told he could be on the Unabomber's lethal roster of representatives of the industrial and technological age. He got the same questions as the rest of us. Then he got the safety lecture. Be suspicious, he was told. If you come home from work and there's a package left by the door, don't touch it. Don't let the children get the mail. Have the Post Office screen your mail. There was no reason to believe environmentalists were sending these bombs, but since Gil Murray had occupied a similar position in a counterpart timber association, it was only prudent to be cautious.

1:35 P.M. TUESDAY, JULY 11, 1995   *Englewood, Colorado*
TOM MCDONNELL PICKED UP THE PHONE. It was Will Verboven of the Alberta Sheep and Wool Commission in Calgary.

"Tom, a package bomb went off at Alta Genetics this morning. We're getting American-style terrorism in Canada now."

"Was anybody hurt?" asked McDonnell. As the associate director of natural resources for the American Sheep Industry Association, he had become well acquainted with the targeted livestock breeding firm—Alta Genetics' major market was the United States.

"Don't know any details, Tom. Just wanted to alert you."

"Thanks. I'll call Gary Smith."

McDonnell dialed Alta Genetics in Calgary. Gary Smith, director of international marketing, confirmed that a suspicious package had arrived and a company officer opened it behind a baffle with a straightened coat hanger. He had not expected the large explosion. It bruised his hand and scattered two-inch nails everywhere. A few workers got hit with debris.

"It's a miracle it didn't kill someone," Smith said.

THURSDAY, JULY 13, 1995   *Toronto, Ontario*
MACKENZIE INSTITUTE EXECUTIVE DIRECTOR JOHN THOMPSON received a package bomb similar to the one that blew up at Alta Genetics. Thompson did not attempt to open it and it did not go off.

The Institute is a non-profit think tank that researches and publishes on organized crime, radical ideologies, propaganda and terrorism. The May, 1995 issue of the *Mackenzie Intelligence Advisory* had been devoted to a critique of "Extremism and Deep Environmentalism," examining sabotage incidents, Earth First in Canada, and assessing the potential for violence in the animal rights and deep ecology movements.

FRIDAY, JULY 14, 1995   *Englewood, Colorado*

> To: ASI RESOURCE MANAGEMENT COUNCIL
> From: Tom McDonnell
> Re: Mail Bombing of Alta Genetics
>
> Gary Smith of Alta Genetics in Calgary, Canada informed me that his facility received a mail bomb on Tuesday, July 11th. The mail bomb was powerful, causing extensive damage to the facility. Employees were extremely fortunate, suffering only minor injuries.
>
> The Royal Canadian Mounted Police is investigating the incident. Apparently the U. S. Federal Bureau of Investigation has ruled out the Unabomber. I have forwarded our files on environmental and animal rights terrorism to Canadian authorities to assist in their investigation. As you are aware, Earth First has launched an aggressive terrorist campaign in British Columbia, and their organization has not been ruled out as a suspect in this bombing.
>
> Alta Genetics has a leading edge in cattle embryo technology in North America. Their laboratory is equipped for in-vitro fertilization and parthenogenesis of ova, embryo sexing and gene injection, and nuclear transplantation with embryos. They have also expanded their work into sheep genetics and presently work with the American Sheep Industry Association on genetic improvement.[23]

McDonnell sent the R.C.M.P. several Internet postings by radical environmentalists showing that Earth Firsters were crossing the border from the United States into Canada. A constable told McDonnell by telephone, "Some of the names on those postings are on our suspect list."[24]

MONDAY, JULY 17, 1995   *Vancouver, British Columbia*
BRITISH COLUMBIA REPORT, a newsmagazine covering Western Canada, noted that in the past three months alone, animal rights radicals had perpetrated ten acts of terrorism, vandalism or theft in British Columbia:

- On April 19, a bear hide worth $4,000 was stolen from the truck of a North Island outfitter.
- On April 28, arsonists destroyed a wildlife museum and a taxidermy shop in Cranbrook.
- Also on April 28, two butcher shops were vandalized, one in Jaffray and another in Fernie. Obscenities were painted on the storefronts, along with the initials ALF.
- On the evening of May 2, a group of anti-hunting activists in Port McNeil took responsibility for stealing a hotelier's outdoor freezer, loaded with four bear hides.
- On June 5, a lodge and other buildings owned by Monashee Outfitting were ransacked. A variety of goods was either stolen or destroyed.
- On June 15, a Spallumcheen taxidermy studio was firebombed, causing about $1,000 in damage before owner Hayes Niemeyer was able to douse the flames.
- Also on June 15, a taxidermist's business in Burnaby was hit with stink bombs.
- On June 19, a cabin formerly owned by Monashee Outfitting was firebombed by a group calling itself the Earth Liberation Army.[25]

TUESDAY, JULY 18, 1995   *Vancouver, British Columbia*
THE ROYAL CANADIAN MOUNTED POLICE issued a news release on the Alta Genetics and Mackenzie Institute explosive devices, urging individuals and organizations "to exercise extreme caution when receiving unexpected / suspicious packages or letters." The public was advised that, "The R.C.M.P., in cooperation with local, national and international agencies, take these incidents seriously and are actively investigating."[26]

TUESDAY, JULY 26, 1995   *Vancouver, British Columbia*
FOUR TYPEWRITTEN LETTERS from a group claiming responsibility for the Alta Genetics explosion were received, one by the Toronto Globe & Mail, one by the Montreal Gazette, one by the Vancouver Sun and the other by the Vancouver Province.[27] The group called itself the Militant Direct Ac-

tion Task Force. The Province's copy was headed by a hand-drawn logo of an assault rifle with the initials MDATF superimposed, the central "A" inscribed in a five-pointed star:

> On Tuesday, July 11/95 at aprox. 10:30 a.m. our parcel-bomb addressed to [name withheld] of Alta Genetics went off, causing extensive damage to an office. Our third action this time against the fascist bio and gene-technology industry.
> (Verification: The parcel, addressed to "[name withheld] PERSONAL" and contained a 1 ¼" diameter pipe containing a high explosive, 2" nails, 9v battery, initiator, and switch)
> This company is responsible for many crimes, namely, though, its abuse of technology in its attempts to create higher levels of cattle and their wish to have greater control over more species. We chose them because of their work in live cattle and reproductive techniques through their use of bull and cow semen [sic], and because agriculture is the fastest growing of all genetic-technological areas. This company alone exports to over 50 countries and last year made $23 million through their genetic-tampering. This is blood money, earned through the exploitation of those things man already controls and yet wants to tighten the reins and control even more. This type of technology constitutes a manipulation of nature simply so as to suit our ever-growing greedy appetites for bigger and better versions of unnecessary items....
> Genetic tampering must be stopped before it is too late. Nature is not there for us to manipulate; it is there for us to live with, together. Thus, we took our first step in stopping this exploitative industry through Armed Revolutionary Action, and we will continue to do so in our fight against the people who wish to have even more control over the rest of society. DIRECT ACTION is the only way...[ellipsis in the original][28]

The MDATF did not explain what the first two "actions against the fascist bio and gene-technology industry" might have been. Three months earlier a letter from the "Anti-Fascist Militia" had claimed responsibility for sending mousetraps primed with razor blades to ten Canadians, including notorious Holocaust denier Ernst Zundel and white supremacist Charles Scott. RCMP Sergeant Peter Montague could not confirm that the Anti-Fascist Militia and the Militant Direct Action Task Force were the same, but he did say "an irresistible inference" could be drawn from common themes throughout the two groups' communiqués. The MDATF had also sent bombs to Zundel and Scott in June.[29]

WEDNESDAY, JULY 27, 1995   *Toronto, Ontario*
THE BOMBERS SENT ANOTHER LETTER to the Globe and Mail in Toronto, this

one castigating Mackenzie Institute executive director Thompson as a person "who hides amongst the upper crust of corporate fascists, capitalists and imperialists." The MDATF charged that the Institute supported neo-Nazis and multinational corporations and "is pro-American, supporting the likes of the CIA and NATO." The Institute was accused of being against bilingualism, being pro-apartheid, homophobic, antifeminist, antichoice on abortion, and of favoring money for military spending over social programs. The bombers said the Institute was spreading lies, supporting right-wing hatred toward certain groups and attacking natives and immigrants.[30]

WEDNESDAY, AUGUST 2, 1995   *Sacramento, California*
THE NEW YORK TIMES AND WASHINGTON POST both published 3,000-word excerpts from the Unabomber's manifesto and the FBI announced that it had sent 75 copies of the whole manuscript to a wide group of scholars, specialists in fields ranging from criminal justice to forensic psychology to the history of science, hoping to find clues.

And Cynthia Hubert's phone rang in the news room of the Sacramento Bee.

It was Barry Clausen, a controversial former private investigator from Montana who had infiltrated Earth First in 1989-90 on behalf of loggers, ranchers and miners who had suffered ecoterrorist attacks.

"Would you like to see some evidence linking the Unabomber's last two victims and Earth First?"

She would. Clausen faxed her a copy of the "Eco-Fucker Hit List." The list had been brought to his attention by conversations with Donn Zea. He thought he remembered it from somewhere in his old undercover files, then ransacked his storage boxes and came up with a copy of *Live Wild Or Die* that had been given to him personally back in 1990, fresh off the press, by one of its editors, Earth Firster Mitch Friedman. After perusing it carefully, he called Zea and said, "Did you notice that Exxon is on that list, too?"

Zea looked and found Exxon Company USA third on the list below California Timber Association. Exxon's Denver, Colorado contact was Fernando Blackgoat, a Native American exploration geologist.

"The Unabomber says he killed Thomas Mosser because he did some work for Exxon," Clausen noted.

The Sacramento Bee wrote its story and published it a day after readers saw the Unambomber's ideas for the first time. They read, "America's most hunted man thinks the industrial revolution has taken the joy out of life, fears dictators will someday get nuclear weapons and says he kills so people will pay attention to his ideas."[31]

Then they read, "The FBI is investigating whether the Unabomber used a 'hit list' published in an underground environmental newspaper five years ago to select some of his recent targets, including the California Forestry Association in Sacramento."[32]

The Times and the Post published nine judiciously edited portions of the Unabomber's long tirade that gave a good overview. Leonard Downie Jr., the Post's executive editor, said his paper wanted readers "to see some of the material the FBI is making available to the academic community." Both papers also said that they had not yet made a decision on whether to publish the entire manifesto. Neither expected that this incomplete edition would stop the killing.

> The Industrial Revolution and its consequences have been a disaster for the human race. They have...destabilized society, have made life unfulfilling, have subjected human beings to indignities, have led to widespread psychological suffering...and have inflicted severe damage on the natural world.
>
> The industrial-technological system may survive...only at the cost of permanently reducing human beings and many other living organisms to engineered products and mere cogs in the social machine.... [I]f it is to break down it had best break down sooner rather than later.
>
> We therefore advocate a revolution against the industrial system.... This is not to be a POLITICAL revolution. Its object will be to overthrow not governments but the economic and technological basis of the present society....[33]

The excerpts from the manifesto made some interesting points about contemporary society. The Unabomber mounted a sophisticated attack on leftist intellectuals, arguing better than Rush Limbaugh that the "politically correct" movement is motivated by self-hatred and low self-esteem. Conversely, the Unabomber rated conservatives as fools for supporting technological progress and then whining because progress erodes traditional values. A plague on both your houses. Alienated and astute.

Between bashing the left and right, the Unabomber explained what is wrong with our society in four easy lessons:

<u>On 'Oversocialization'</u>
The moral code of our society is so demanding that no one can think, feel and act in a completely moral way.... Oversocialization can lead to low self-esteem, a sense of powerlessness, defeatism, guilt, etc. One of the most important means by which our society socializes children is by making them feel ashamed of behavior or speech that is contrary to society's expectations. If this is overdone, or if a particular child is especially susceptible to such feelings, he ends by feeling ashamed of HIMSELF....

<u>On 'the Power Process'</u>
Human beings have a need (probably based in biology) for some-

thing that we will call the "power process." This is closely related to the need for power (which is widely recognized) but is not quite the same thing. The power process has four elements. The three most clear-cut of these we call goal, effort and attainment of goal. (Everyone needs to have goals whose attainment requires effort, and needs to succeed in attaining at least some of his goals.) The fourth element is more difficult to define and may not be necessary for everyone. We call it autonomy and will discuss it later....

On 'Surrogate Activity'

We use the term "surrogate activity" to designate an activity that is directed toward an artificial goal that people set up for themselves merely in order to have some goal to work toward, or let us say, merely for the sake of the "fulfillment" that they get from pursuing the goal....

In modern industrial society only minimal effort is necessary to satisfy one's physical needs. It is enough to go through a training program to acquire some petty technical skill, then come to work on time and exert very modest effort needed to hold a job. The only requirements are a moderate amount of intelligence, and most of all, simple OBEDIENCE....

On 'Problems of Modern Society'

We attribute the social and psychological problems of modern society to the fact that that society requires people to live under conditions radically different from those under which the human race evolved and to behave in ways that conflict with the patterns of behavior that the human race developed while living under the earlier conditions. It is clear from what we have already written that we consider lack of opportunity to properly experience the power process as the most important of the abnormal conditions to which modern society subjects people....

How much of this was the Unabomber's personal biography and how much social criticism was unclear. Who was this educated killer? Aside from psychological profiles of the Unabomber, the only image we had of this enemy of technology was an incongruous sketch of a man in designer sunglasses and a hood, an image you might reasonably expect to see in a hiking outfitter's catalog.

The Unabomber's psychobabble irritated many readers, but that part about petty technical skills and coming to work on time rang some familiar bells that sounded like "rat race." He was raising serious questions about technological society. And he was killing people to get others to listen.

His ideological bottom line was

On 'Revolution'

The technophiles are taking us all on an utterly reckless ride into the unknown. Many people understand something of what technological progress is doing to us yet take a passive attitude toward it because they think it is inevitable. But we (FC) don't think it is inevitable. We think it can be stopped....

The two main tasks for the present are to promote social stress and instability in industrial society and to develop and propagate an ideology that opposes technology and the industrial system. When the system becomes sufficiently stressed and unstable, a revolution against technology may be possible....

We have no illusions about the feasibility of creating a new, ideal form of society. Our goal is only to destroy the existing form of society....

But an ideology, in order to gain enthusiastic support, must have a positive ideal as well as a negative one; it must be FOR something as well as AGAINST something. The positive ideal that we propose is Nature. That is, WILD nature; those aspects of the functioning of the Earth and its living things that are independent of human management and free of human interference and control. And with wild nature we include human nature, by which we mean those aspects of the functioning of the human individual that are not subject to regulation by organized society but are products of chance, or free will, or God (depending on your religious or philosophical opinions)....

The Sacramento Bee interviewed Donn Zea about the *Live Wild or Die* hit list. He told Cynthia Hubert that the timing of the hit list was significant. "It was published at the height of the battle over Proposition 130," a California anti-timber ballot measure backed by environmentalists, he said. "We went toe to toe with many of those kinds of groups during that period."

The measure, which would have authorized the sale of $720 million in bonds to buy commercial timberlands and turn them into nature museums, was rejected by voters in 1990.

The newspaper quoted him: "'Now the language in that "Live Wild or Die" document seems almost surrealistic to me,' said Zea. 'It seems so close to what the Unabomber has said.'"

The Sacramento Bee checked out Barry Clausen. "I can confirm that we have met with this individual and we are very interested in what he has to say," FBI spokesman George Grotz told reporters. Grotz did not volunteer that the FBI had debriefed Clausen in San Francisco in 1990 about a power line bombing in Santa Cruz by environmental radicals and provided him with a contact agent while he was undercover among Earth Firsters.[34]

The Sacramento Bee obtained opposing comment: "Karen Pickett, an organizer for Earth First in Berkeley, said none of the information published in 'Live Wild or Die' represented a call to violence. 'The rhetoric is extreme, but what it is is a list of corporations that are doing harm to the environment and have been the target of direct action and protest, not murder,' Pickett said.

"'Letter bombs and murder are not a tactic that have ever been used or suggested by the environmental movement,' she said. 'There is no commonality of purpose between this bomber and anyone who has ever espoused environmental causes.'"

The reporters were unaware that Karen Pickett had control of an estimated $20,000 in the Earth First Direct Action Fund, earmarked for activists in the 1993 protests against planned road building and timber sales in north central Idaho—the areas of the Nez Perce National Forest known as Cove and Mallard. That money was funneled through an ad hoc group called the Ancient Forest Bus Brigade, organized by former insurance company executive Robert Amon. Amon disbursed several hundred dollars of the thousands obtained from Pickett to Earth Firster Erik Ryberg for supplies, gasoline and a computer modem.[35] Using the pseudonym "Pajama," Ryberg wrote in a 1993 issue of the Earth First Wild Rockies Review:

Bombthrowing: A Brief Treatise
I have a theory. My theory is that if, every time the Forest Service or some other entity commits an act of destruction of the wild, if every time they plow under another roadless area, or murder a wolf, or mangle and plunder and sack a wild place, if every time they do this I take my anger and I place it in a certain compartment inside my brain, then when it becomes time to throw bombs I will be able to access those pieces of anger that I have stored and be a very good bombthrower, perhaps better than the other bombthrowers.

So, I spend my days patiently contriving means to stop the madness which drives the Forest Service and other renegades, and each day I read the mail, perhaps I file another appeal, and then at the end of the day I open up this special compartment inside my brain and I put the anger of some new atrocity in it, in anticipation of the day when I shall need this anger in order to throw bombs.

But a new fear has overcome me. I perceive my anger calling me from inside its compartment, I hear the door unlatching from inside, and this new terrible question approaches me:

How shall I know when it is time to throw bombs?

If the Forest Service decides to cut occupied owl habitat in Oregon, is it time to throw bombs?

Of if the Fish and Wildlife Service decides to trap and kill wolves, or to shoot them from the sky, is it then time to throw bombs?

What if the Park Service decides to imprison Grizzly bears in a zoo for the benefit of tourists, if the Forest Service ignores the appeal process, or if the largest intact grove of redwoods is only 500 acres in size, if the Endangered Species Act is abolished or sidestepped by people with enough money, if corporations continue to wreak havoc upon the ozone layer, if reason is blindly cast aside in favor of profit, if the last remaining herd of wild Bison is slaughtered for following their migratory instincts, if my generation watches the very last Chinook salmon perish in a home choked with silt, if certain nameable parties proceed in a manner which is clearly imperiling the lives of the multitude of glorious and beautiful critters and plants on our fine planet, our only planet, what then? Is it then time to throw bombs?

Think: when the very last wolves on this continent are trapped and caged for captive breeding (as the remaining Condors were, not so long ago), will it finally be time to throw bombs?

Or will it be too late?[36]

On October 31, 1993 a bomb exploded at the Reno, Nevada office of the Bureau of Land Management. A month earlier, the Earth First Journal, Mabon, September-October, 1993, on page 34, published a section in which the Earth Liberation Front of Germany called for an "International Earth Night" on Halloween as part of an "International Action Week," October 31 through November 6, discussing government policy, urging property damage as an effective tactic for change, and recommending that no credit should be taken for ELF actions in order to thwart law enforcement. On March 30, 1995 a bomb blew windows out of a Forest Service office in Carson City, Nevada. No one claimed responsibility for the blasts.[37]

Radical environmentalists blamed ranchers and the wise use movement. The San Francisco Chronicle reported, "Some experts on fringe organizations say that the far right's willingness to resort to violence is directly related to the growing paranoia of its rhetoric." The experts were from far left organizations. The article did not mention environmentalist rhetoric such as that of Ryberg or the Earth Liberation Front.[38]

The Montana Human Rights Network, a left-wing organization that tracks right-wing organizations, said in a 1994 report, "Unfortunately, many of these groups advocate direct, violent action. There have been recent reports of individuals wearing sidearms in public meetings."

On the evening April 14, 1994, Barry Clausen gave a public talk on radical environmentalists at the high school in Potlatch, Idaho. Ric Valois entered the school auditorium in airborne ranger dress uniform armed with a concealed military-style semi-automatic pistol. After audience members confronted Valois about the suspicious bulge in his uniform jacket, a sheriff's deputy ordered Valois to remove the sidearm from the school premises and lock it in his car. Valois complied.

Valois, who lives in a hand-hewn cabin near Vaughn, Montana, is the founder of the armed eco-militia called the Environmental Rangers, several dozen extremists which he describes as "the special forces of the environmental movement."[39]

Valois was arrested in 1995 during a Cove/Mallard timber protest and transported to court in Boise. An Idaho newspaper quoted him as telling the judge, "some people are going to get hurt, others are going to die. The ballistic vests worn by law enforcement officials will not be sufficient, as they could or will be shot in the head."[40] Another newspaper reported Valois saying that to the arresting officer rather than the judge.[41]

When Phelps Dodge Mining Company and Canyon Resources Corporation planned to locate a mine near the Blackfoot River, Valois, who routinely carries a 9-millimeter sidearm holstered on his hip, told the Los Angeles Times, "That mine is not going in. They're not getting these places without a war. And I mean a real war."

The Rangers customarily wear their weapons to public gatherings. Alone, they often talk about the failure of government, and about how the end of government and society as we know it is near.

Valois' armed rangers threatened the Bureau of Land Management's area resource manager, Richard Hopkins, at a hearing in 1995 on proposed mining claims in Montana's Sweet Grass Hills. "They said if you make a decision to allow exploration or eventual mining, we know where you live, and we'll take care of you in our own way," Hopkins said.[42]

The Montana Human Rights Network was not worried about Valois.

THURSDAY, AUGUST 10, 1995　　　　　　　　　　　　*Kalamazoo, Michigan*
RADICAL ENVIRONMENTALIST RODNEY ADAM CORONADO was sentenced to four years and nine months in prison and fined $2,593,000 by federal Judge Richard Enslen for his part in an Animal Liberation Front raid and massive arson at Michigan State University on February 28, 1992.[43]

Coronado had signed a plea agreement to "aiding and abetting" the MSU raid in exchange for the government dropping prosecutions against him for numerous raids against university and business research facilities in what the ALF called "Operation Bite Back." U.S. Attorney Michael H. Dettmer, in pleading for the harshest possible sentence, gave a crystalline statement of the scatter shock of eco-terror:

> A terrorist combines violence and threats so that those that disagree with him are silenced, either because they have been victimized by violence or because they fear being victimized. Since the defendant's indictment and arrest, the firebombings and massive property damage that were a hallmark of "Operation Bite Back"

have ceased. However, the intimidation and fear that these crimes were designed to inflict continues to this day. Scientists, business owners and farmers around the United States still live in fear that a bomb will be waiting for them the next time they go to their offices, farms or laboratories. The defendant's actions on behalf of the ALF may not have ended scientific research, but they have succeeded in making ordinary citizens of this country afraid to respond to ALF's claims that there exist no legitimate reasons to use animals in scientific research. Nowhere is this continued intimidation more evident than in the events that have transpired since the defendant's guilty plea. In several instances, the defendant has appeared in the media to exhort others to take his place as a "hero to the animal and environmental movement." In contrast, the victims of the defendant's crimes remain so afraid of the defendant and others like him that they would not speak to the Court's own presentence investigator unless he guaranteed their anonymity.

In fashioning the appropriate sentence for this defendant, the Court should consider that he has forever brought fear to the lives of ordinary citizens whose only offense was that the defendant did not agree with them. If others are waiting to accept the defendant's invitation to replace him, the Court's sentence must demonstrate that such actions will not be tolerated.[44]

Federal sentencing guidelines for Coronado's crime would normally be between 33 and 41 months. In Coronado's plea agreement, the parties agreed that a sentence between 41 and 53 months would be warranted. The judge gave him 57 months.[45]

TUESDAY, SEPTEMBER 5, 1995   *Los Angeles, California*
STATE INVESTIGATORS PLOWED THROUGH STACKS OF DOCUMENTS in a bizarre animal rights case. It had begun a year earlier. In September of 1994, while Los Angeles County supervisors were honoring Mercy Crusade Inc., an animal-welfare organization, for donating $20,000 to help the county spay and neuter pets, federal agents investigated why it owned $118,687 in firearms and $15,115 in firearms accessories, and why it maintained a private armed force of twelve humane officers who used quasi-police powers of investigation and arrest to enforce its animal rights agenda on the public. The gun-toting, badge-wearing officers were supervised by no government and had little or no formal law enforcement training, allowed by an 80-year old law originally intended to give dogcatchers the power to appoint deputies for animal control.[46]

The Bureau of Alcohol, Tobacco and Firearms had seized 12 semi-automatic Heckler & Koch "assault pistols" in June from James McCourt, a former Pepperdine University economics professor, who was the chairman of the board of directors of Mercy Crusade, Inc., as well as its chief humane officer. Mercy Crusade had bought or ordered 22 other weapons, including five AR-15s, a Bushmaster, a Heckler & Koch .308 and a Fabrique Nationale De Arms .308, plus an unusually powerful Israeli .50-caliber pistol—more firepower for the Crusaders than the L.A.P.D. SWAT team.

The abuses perpetrated by this extremist animal welfare group outraged the public. As part of their agenda to end all pet ownership, a group of Mercy Crusaders confiscated the seeing eye dog of a blind Westchester, California man, leaving him unemployed and mostly homebound for five months. The negligent actions of other Crusaders resulted in ruined crime scenes and tainted evidence. And one bunch broadcast radio appeals for animal lovers to converge at the scene of the 1993 Malibu/Calabasas wildfire at the time firefighters were desperately trying to keep roads clear to evacuate residents.

The Los Angeles City Animal Regulation Commission suspended Mercy Crusade's participation in all programs, including a high-profile spaying and neutering program. Assemblyman Curtis Tucker Jr. (D-Inglewood), introduced legislation to rein in such groups, saying, "To have a self-run, quasi-governmental, vigilante-type group armed to the teeth running around on behalf of animals is not in the best interests of the state of California." The legislature quickly passed the law. The state attorney general's office then filed a suit against the animal rights group, saying it illegally used more than $400,000 in charitable donations to build an arsenal of assault-style weapons, to make high-risk loans and to pay personal expenses.

WEDNESDAY, SEPTEMBER 13, 1995   *Deming, New Mexico*
THOMAS L. KELLY OF TRES LOMITAS RANCH called Tom McDonnell with a long list of cattle shooting incidents for which rewards had been posted. McDonnell compared the dates with a copy of an Internet transmission originating at 3:56 p.m. on May 6, 1994, from Seattle-based Earth Firster Suzanne Pardee. It was titled, "Hunt Cows, not Cougars."

The text read: "That's right, shoot cows. They don't run. They can't bite. They don't charge. They don't maul. They produce only 2% of the beef from 70% of the public lands. A pound of beef requires 2000 gallons of water, a pound of wheat, only 20. There's WAAAY to [sic] many of them. Happy Hunting."[47]

On Saturday, April 15, 1995, nine days before the Unabomber killed Gil Murray, twenty cows and calves were shot and killed with a high velocity rifle on Tom Kelly's Tres Lomitas Ranch near Deming, New Mexico. Each cow was killed with a single shot at relatively close range. All shell casings were picked up and removed by the shooter or shooters. On a neighbor's ranch, eleven more cattle were killed with a semiautomatic SKS style weapon.[48] The Luna County Sheriff's Department could find no suspect in either incident.

Between September 3 and 4, 1995, a three-year-old Brahma-cross cow belonging to Alan Flournoy was shot on late Sunday evening or early Monday morning in a remote area of Government Flats about 16 miles west of Paskenta, California. The animal died the following day.

The Tehama County Sheriffs Department could find no suspect.

More than thirty similar unsolved cattle shootings had occurred since Pardee's posting.

TUESDAY, SEPTEMBER 19, 1995   *Washington, D.C.*
TONY SNOW SAW THAT THE NEW YORK TIMES AND WASHINGTON POST had published the whole 35,000-word Unabomber Manifesto. Snow is a Washington-based syndicated columnist for the Detroit News. He served in the Bush White House as Director of Speech Writing in 1991 and Deputy Assistant to the President for Media Affairs from March, 1992 to the end of the administration. He hosts *Fox News Sunday* on the Fox Network.

Snow read the statement by Donald E. Graham, The Post's publisher, and Arthur O. Sulzberger Jr., publisher of the New York Times, who said they jointly decided to publish the lengthy manuscript "for public safety reasons" after meeting with Attorney General Janet Reno and FBI Director Louis J. Freeh. The papers split the cost of an eight-page pullout, which appeared only in The Post because it had the mechanical ability to distribute such a section in all copies of its daily paper.[49]

It was a hot seller. The Post ran out of copies immediately. The Post said, "Within hours of publication, Time Warner put the entire screed on Pathfinder, its free World Wide Web site on the Internet. Other Web sites followed suit. By day's end, thousands of readers of everything from Joshua Aasgaard's Universe of Knowledge to Stardot Consulting to Wired magazine's Web site HotWired were downloading the text, taking it every bit as seriously as a bomb."[50]

As soon as the manifesto hit the streets and the Internet, readers noticed that the document skipped from paragraph 115 to 117. Computer chat rooms buzzed with speculation: Had paragraph 116 had been left out at the order of federal investigators? Or Post editors? Or had the usually meticulous mail-bomber slipped up?

It turned out to be a simple typesetting mistake by Post copy editors. In addition to paragraph 116, the Post reprint left out one sentence and part of another in paragraph 117. The Post's Managing Editor Robert Kaiser said both were errors.[51] The Post printed the corrections in a later edition.[52]

The Internet versions of the manifesto filled in the Post's omission, but they somehow left out the last two sentences of Paragraph 182: "We have no illusions about the feasibility of creating a new, ideal form of society. Our goal is only to destroy the existing form of society." You had to edit your own to get a really complete version.

Tony Snow saw that the Unabomber's manifesto had quite a bit to say about making wild nature into a new social ideal—much more than the 3,000 word excerpts had indicated:

> 184. Nature makes a perfect counter-ideal to technology for several reasons. Nature (that which is outside the power of the system) is the opposite of technology (which seeks to expand indefinitely the power of the system). Most people will agree that nature is beautiful; certainly it has tremendous popular appeal. The radical environmentalists ALREADY hold an ideology that exalts nature and opposes technology. [30] It is not necessary for the sake of nature to set up some chimerical utopia or any new kind of social order. Nature takes care of itself: It was a spontaneous creation that existed long before any human society, and for countless centuries many different kinds of human societies coexisted with nature without doing it an excessive amount of damage. Only with the Industrial Revolution did the effect of human society on nature become really devastating. To relieve the pressure on nature it is not necessary to create a special kind of social system, it is only necessary to get rid of industrial society. Granted, this will not solve all problems. Industrial society has already done tremendous damage to nature and it will take a very long time for the scars to heal. Besides, even pre-industrial societies can do significant damage to nature. Nevertheless, getting rid of industrial society will accomplish a great deal. It will relieve the worst of the pressure on nature so that the scars can begin to heal. It will remove the capacity of organized society to keep increasing its control over nature (including human nature). Whatever kind of society may exist after the demise of the industrial system, it is certain that most people will live close to nature, because in the absence of advanced technology there is not other way that people CAN live. To feed themselves they must be peasants or herdsmen or fishermen or hunter, etc., And, generally speaking, local autonomy should tend to increase, because lack of advanced technology and rapid communications will limit the capacity of governments or other large organizations to control local communities.
>
> 185. As for the negative consequences of eliminating industrial society — well, you can't eat your cake and have it too. To gain one thing you have to sacrifice another.

That "eat your cake and have it too" reversal was an odd quirk of language that might be of use to a detective. But the footnote for Paragraph 184 revealed that the Unabomber was well read in another aspect of radical environmentalism.

30. (Paragraph 184) A further advantage of nature as a counter-ideal to technology is that, in many people, nature inspires the kind of reverence that is associated with religion, so that nature could perhaps be idealized on a religious basis. It is true that in many societies religion has served as a support and justification for the established order, but it is also true that religion has often provided a basis for rebellion. Thus it may be useful to introduce a religious element into the rebellion against technology, the more so because Western society today has no strong religious foundation.

Religion, nowadays either is used as cheap and transparent support for narrow, short-sighted selfishness (some conservatives use it this way), or even is cynically exploited to make easy money (by many evangelists), or has degenerated into crude irrationalism (fundamentalist Protestant sects, "cults"), or is simply stagnant (Catholicism, main-line Protestantism). The nearest thing to a strong, widespread, dynamic religion that the West has seen in recent times has been the quasi-religion of leftism, but leftism today is fragmented and has no clear, unified inspiring goal.

Thus there is a religious vaccuum in our society that could perhaps be filled by a religion focused on nature in opposition to technology. But it would be a mistake to try to concoct artificially a religion to fill this role. Such an invented religion would probably be a failure. Take the "Gaia" religion for example. Do its adherents REALLY believe in it or are they just play-acting? If they are just play-acting their religion will be a flop in the end.

It is probably best not to try to introduce religion into the conflict of nature vs. technology unless you REALLY believe in that religion yourself and find that it arouses a deep, strong, genuine response in many other people.

Tony Snow has an eye for irony. He noticed that parts of the manifesto seemed disturbingly familiar, and that whole sections seemed just like... What?

Of course.

Al Gore's book *Earth in the Balance.*

He laughed out loud. He had helped policy analysts in the Bush White House write up a critique of Gore's book and knew the text quite well. The Unabomber's long essay contained strikingly similar polemics.

Snow said, "I went through and read the whole turgid, godawful Unabomber manifesto and marked out the phrases, then looked at the index of *Earth in the Balance*, and boom! What was amazing is that some of it was almost verbatim. I did this the day after the Unabomber's manifesto came out. I thought to myself, 'I've got a deadline coming up and I've got to crank out a column.' So I decided I was going to have some fun with this."[53]

The column was a classic. After pointing out the sham of using modern presses and scientific literature to state an anti-technology theme, Snow wrote:

> The Unabomber's manifesto is fascinating for its flashes of brilliance and overall loopiness. But the most striking thing is how much it sounds like Al Gore's book, *Earth in the Balance*.
>
> Gore, like the Unabomber, distrusts unbridled technology. He frets over the fate of the planet and thinks people must embrace revolutionary curbs. While Gore prefers to concentrate power in the hands of a wise gigantic government, the Unabomber prefers anarchy.

He then considered some parallels from their literary works:

> Unabomber: "Technological progress marches in only one direction; it can never be reversed. Once a technical innovation has been introduced, people usually become dependent on it, unless it is replaced by some still more advanced innovation." [¶ 129]
>
> Gore: "Like the sorcerer's apprentice, who learned how to command inanimate objects to serve his whims, we too have set in motion forces more powerful than we anticipated and that are harder to stop than to start." [p. 205]
>
> Unabomber: "No one knows what will happen as a result of ozone depletion, the greenhouse effect and other environmental problems that cannot yet be foreseen. And, as nuclear proliferation has shown, new technology cannot be kept out of the hands of dictators and irresponsible Third World nations." [¶ 169]
>
> Gore: "In the speech in which I declared my candidacy, I focused on global warming, ozone depletion and the ailing global environment and declared that these issues—along with nuclear arms control—would be the principal focus of my campaign." [p. 8]
>
> Unabomber: "...artificial needs have been created.... Advertising and marketing techniques have been developed that make many people feel they need things that their grandparents never desired or ever dreamed of....[¶ 63] It seems for many people, maybe the majority, that these artificial forms...are insufficient. A theme that appears repeatedly in the writings of social critics of the second half of the 20th century is the sense of purposelessness that afflicts many people in modern society." [¶ 64]
>
> Gore: "Whenever any technology is used to mediate our experience of the world, we gain power, but we also lose something in the process. The increased productivity of assembly lines in factories, for example, requires many employees to repeat the identical task over and over until they lose any feeling of connec-

tion to the creative process, and with it, their sense of purpose." [p. 203]

The Unabomber: "'Oh!' say the technophiles, 'Science is going to fix all that! We will conquer famine, eliminate psychological suffering, make everybody healthy and happy!'" [¶ 170]

Gore: "Some people argue that a new ultimate technology, whether nuclear energy or genetic engineering will solve the problem.... We have also fallen victim to a kind of technological hubris, which tempts us to believe that our new powers may be unlimited. We dare to imagine that we will find technological solutions for every technologically induced problem.... Technological hubris tempts us to lose sight of our place in the natural order and believe that we can achieve whatever we want." [p. 206]

Unabomber: "Industrialized civilization's greatest engines of distraction still seduce us with a promise of fulfillment. Our new power to work our will upon the world brings with it a sudden rush of exhilaration. But that exhilaration is fleeting. It is not true fulfillment."

Gore: "Very widespread in modern society is the search for 'fulfillment....' But we think that for the majority of people an activity whose main goal is fulfillment [technology] does not bring completely satisfactory fulfillment."

Tony Snow closed by noting, "that's the difference between a terrorist and a vice president." Then he fired his parting shot:

P.S. I reversed the last two quotes. Gore actually wrote the passage about exhilaration. (Did you notice?)[54]

A new Usenet news group appeared on the Internet called alt.fan.unabomber. High technology had a fan club about the murderer who despises technology. Postings discussed the Unabomber from myriad angles: the crazed Luddite, his "old thinking for an old medium," the details of his bomb making and speculations he might have studied a popular explosives formulary from the early 1970s called *The Poor Man's James Bond*, praise for his love of nature, curses for his failure to understand Deep Ecology's biocentrism properly, hurt feelings that he didn't take the Gaia religion seriously, criticisms of the overhyped *Anarchist's Cookbook* and why it was a fraud the Unabomber would never have used, why the FBI couldn't find him and what not.

To environmental extremists in the Pacific Northwest, said the Washington Post, "the bomber's message rings true enough that some may see him more as seer than as sick killer. 'His critiques of society's failures are right on!' one reader on the Internet wrote."[55]

The World Wide Web continued to sprout Unabomber sites like

mushrooms after the rain, the FBI's own million-dollar reward notice being one of the dullest. Joseph Keeler, a 15-year-old Sunnyvale, California high school student, built his own Unabomber web site with the comment, "Why is this here? Well, I didn't like the FBI's Unabomber site, because it didn't have enough information." A "Freedom Club" Web site was created by something called The Church of Euthanasia, featuring links to the manifesto, Unabomber letters and news articles.

He was all over cyberspace.

The Luddite King had become post-industrial society's Monster from the Id, lurking in our computerized subconscious.

As autumn turned to winter, as commentary piled upon debate piled upon quarrel, the Unabomber's message solidified into a few tight articles of faith:

Humans were causing a biological meltdown of the Earth and its community of life. The time was drawing close when we would destroy nature completely and ourselves with it.

With that much at stake, any rage was proper, any act justified.

Criminologist Brent L. Smith observed:

> In the 1980s in America, increasing concern over environmental issues led some extremists to turn to terrorism. Guided by a philosophy that is still evolving, environmental extremists contend that human efforts to sustain and improve the quality of human life have led to the suffering and extinction of other species. More important, these groups exhibit a world view similar in its fatalism to that of their left- and right-wing cousins, contending that, if left unchecked, humans will eventually bring the world to a cataclysmic end.[56]

Apocalyptic beliefs. Fatalism. Get mad enough and you're likely to kill.

When Dave Foreman quit Earth First in 1989, he understood. His valedictory was a warning:

"How do you keep from hating the people you confront? How can you be an effective activist but not be consumed by hatred?" he asked. "I don't know the answer, but I'm working on it. I'll be damned if I'll let myself fall into that trap."[57]

Foreman himself had set the stage for apocalyptic fatalism when he wrote Earth First's "Statement of Principles" in late 1980:

—Wilderness has a right to exist for its own sake

—All life forms, from virus to the great whales, have an inherent and equal right to existence

—Humankind is no greater than any other form of life and has no legitimate claim to dominate Earth

—Humankind, through overpopulation, anthropocentrism, industrialization, excessive energy consumption/resource extraction, state capitalism, father-figure hierarchies, imperialism, pollution, and natural area destruction, threatens the basic life processes of EARTH

—All human decisions should consider Earth first, humankind second

—The only true test of morality is whether an action, individual, social or political, benefits Earth

—Humankind will be happier, healthier, more secure, and more comfortable in a society that recognizes humankind's true biological nature and which is in dynamic harmony with the total biosphere

—Political compromise has no place in the defense of Earth

—Earth is Goddess and the proper object of human worship[58]

Foreman, like the Unabomber after him, had second thoughts about making a religion of nature. He immediately removed the clause about Earth worship, claiming it had resulted from his temporary fascination with the writings of Starhawk (pseudonym of Miriam Simos), a Bay Area Wiccan priestess and author of the 1979 book, *The Spiral Dance: a rebirth of the ancient religion of the great goddess.*

Foreman's revised statement closed with the no-compromise clause.

And there it has stood for more than fifteen years.

Foreman's principles remain the manifesto of radical environmentalism.

Foreman's questions remain unanswered.

Chapter Two Footnotes

[1] "Murderer's Manifesto: Threatening more attacks, Unabomber issues a screed against technology," *Time*, July 10, 1995 Volume 146, No. 2, by John Elson.

[2] "Professor Replies To Unabomber - UC Berkeley Psychologist Sympathetic To Some Ideas, Critical Of Violence," *San Francisco Chronicle*, Tuesday, July 4, 1995, by Michael Taylor, p. A1.

[3] Photocopy of Unabomber typewritten text provided by *The New York Times*, August 7, 1996.

[4] "Unabomber claims he told motives in '85 note," *San Francisco Examiner*, Sunday, July 2, 1995, by Seth Rosenfeld, p. 1A.

[5] Memo dated May 26, 1987 from Robbie Andersen of the Timber Association of California to members, "Getting serious about environmental terrorism."

[6] Letter to the author from the Office of the Sheriff-Coroner, County of Mendocino, Ukiah, California, signed by Undersheriff Larry Gander, dated August 16, 1996.

[7] "Environment Radicals Target Of Probe Into Lumber Mill Accident," *Los Angeles Times*, Friday, May 15, 1987, by Larry B. Stammer, p. 3.

[8] "Shelter Rejects Bison Activist - Humane Society Won't Take His Help," *Bozeman Daily Chronicle*, Saturday, March 9, 1991, by The Associated Press, p. 1A.

[9] "Animal Activists Blamed in Wide Ranch Sabotage," *Los Angeles Times*, Sunday, November 19, 1989, by Charles Hillinger; Mark A. Stein, p. 1.

[10] "Earth First! Burns Dixon Livestock Auction," *National Wool Grower*, March 1989, by Debora Thomas Hood, p. 30.

[11] "Caller Claims Dixon Fire Set; Livestock Auction Heavily Damaged," *Sacramento Bee*, Monday, January 30, 1989, by Clark Brooks, p. B1.

[12] "Barry Clausen: Flim Flam Man or Private Dick?" *Earth First Journal*, Beltane (May-June), 1996, by James Barnes as posted on the Earth First World Wide Web Site, http://envirolink.org/orgs/ef/beltane96/beltane96a.html.

[13] Telephone interviews with Schene, Dorris and the Solano County Sheriff's Department, September 20, 1996.

[14] "Breaking the Alf; Animal Liberation Front," *London Times*, Sunday, May 7, 1995, by David Leppard, Features Section.

[15] "Police Chiefs Apologize For Bomb Blunder; Animal Rights Campaign," *London Times*, Tuesday, June 12, 1990, by Lin Jenkins and Arthur Leathley, Section: Home news.

[16] "Medical Researchers Offer Reward In Fight Against Terrorism; Animal Rights Bombing Campaign, *London Times*, Tuesday, June 12, 1990, by David Sapsted, Section: Home news.

[17] "Eco warriors," (Interview with Mike Roselle), *Playboy*, vol. 40, no. 4, April 1993, by Dean Kuipers, p. 74.

[18] "Saboteurs Block Path Of Desert Cycle Racers," *Los Angeles Times*, Sunday, November 29, 1987, by George Stein, p. 3.

[19] "Protecting The Environment," *Los Angeles Times*, Tuesday, December 8, 1987, letter from William Freeman, p 6.

[20] "They shoot scientists, don't they? Animal Liberation Front," *London Times*, Saturday, November 7, 1992, by Thomas Quirke, Features section.

[21] "Animal militants hunted after Pounds 2m island inferno; Isle of Wight," *London Times*, Thursday, August 25, 1994, by Michael Horsnell, Section: Home news.

22 "Candy Tainted, Says Animal-Rights Group," *Seattle Times*, Saturday, January 4, 1992, by Times News Services, p. A5.

23 Memo copy provided by Tom McDonnell.

24 Telephone interview with Tom McDonnell August 15, 1996. Internet postings were found in cdp:gea.news, December 30, 1994, written 1:47 PM, evidently originating from Bellingham, Washington.

25 "The dirty war on hunting," *British Columbia Report*, July 17, 1995, by Robin Brunet, p. 12.

26 News Release, Royal Canadian Mounted Police, issued by Criminal Intelligence Branch, "E" Division, signed by R.G. MacPhee, Insp., dated 1995 07 18, one page.

27 "Letter Defends Blast - 'Militant' group writes to Province," *Vancouver Province*, Wednesday, July 26, 1995, by Paul Chapman, p. A4.

28 "Communique #2," undated unsigned letter from Militant Direct Action Task Force to the Vancouver Province, two pages typewritten. Spelling and grammatical errors have not been corrected.

29 "Group tells media why it mailed bombs - Letters outline Militant Direct Action Task Force's anger over genetic tampering, 'corporate fascists.'" *Toronto Globe & Mail*, July 27, 1995, by Robert Matas, p. A4.

30 *Ibid.*

31 "Two newspapers publish Unabomber's thoughts," Reuter, New York, August 2, 1995 at 1:53 p.m. EDT, by Arthur Spiegelman.

32 "'Hit List' Had Unabomber Targets - Terrorist's Possible Link To Underground Newspaper Probed By FBI," *Sacramento Bee*, August 3, 1995, by Cynthia Hubert and Patrick Hoge, p. A1.

33 "Excerpts from Unabom's Manifesto," *Washington Post*, Tuesday, August 2, 1995, p. A16-A17.

34 Freedom of Information Act response dated May 6, 1996, FOIPA No. 394025/190-HQ-1129969.

35 Deposition of Erik Ryberg, *Highland Enterprises, Inc., v. Earth First! et al.*, Case No. CV-28511. District Court of the Second Judicial District, State of Idaho, County of Idaho, April 5 and 6, 1994, p. 240. Deposition of John Kreilick, *Highland Enterprises, Inc., v. Earth First! et al.*, Case No. CV-28511. District Court of the Second Judicial District, State of Idaho, County of Idaho, February 15, 1994, p. 80-81. Deposition of Robert Amon, *Highland Enterprises, Inc., v. Earth First! et al.*, Case No. CV-28511. District Court of the Second Judicial District, State of Idaho, County of Idaho, February 8 and 9, 1994, p. 236, 256, 259.

36 "Bombthrowing: A Brief Treatise," *Earth First! Wild Rockies Review*, Vol. 6, No. 1, 1993, by Pajama [pseudonym of Erik Bowers Ryberg], p. 9.

37 "National News in Brief," *San Jose Mercury News*, Saturday, April 1, 1995, by Mercury News Wire Services, p. 9A

38 "Heated Rhetoric Set Stage For Oklahoma Blast," *San Francisco Chronicle*, Monday, May 1, 1995, by Bill Wallace, p. A13.

39 "Eco-warriors: Packing heat, and a vegetarian lunch," *Washington Times*, National Weekly Edition, Vol. 3, No. 20, May 19, 1996, by Valerie Richardson, p. 1.

40 "Activist warns of future violence," *Idaho County Free Press*, Grangeville, Idaho, July 5, 1995, no author, p. 1.

[41] "Activists say Forest Service is trying to discredit them," *Lewiston Morning Tribune*, Lewiston, Idaho, July 12, 1995, by Kathy Hedberg, p. 1A.

[42] "Desperate Defenders of Nature; Environmental Activists Are Hardening Their Tactics—Including Some Who Don Khaki And Carry Arms—In What They See As A Last-Ditch Effort To Protect The Northwest from A Mining And Logging Boom," *Los Angeles Times*, Tuesday, January 9, 1996, by Kim Murphy, p. 1.

[43] "Across The USA," *USA Today*, Wednesday, August 16, 1995, p. 9A.

[44] "Government's Presentencing Memorandum," in the case *United States of America v. Rodney Adam Coronado*, No. 1:93-CR-116, United States District Court for the Western District of Michigan, Southern Division, by United States Attorney Michael H. Dettmer, pp. 19, 20.

[45] "Animal Rights Case Suspect Is Ordered To Pay Restitution," *Portland Oregonian*, Thursday, August 17, 1995, from staff, wire and correspondent reports, p. D5.

[46] "U.S. Agents Probe Animal Group's Weapons Arsenal Investigation: Van Nuys organization has a $100,000 stockpile, including assault-style guns, authorities say," *Los Angeles Times*, Sunday, January 15, 1995, by Josh Meyer, p. 1.

[47] Posted by spardee@igc.apc.org in igc:talk.pol.guns, cdp: ef.general.

[48] Interview with Thomas L. Kelly at Reno, Nevada, May 11, 1966.

[49] "Industrial Society and its Future," *Washington Post*, Tuesday, September 19, 1995, by the Unabomber, separate pullout, 8 pages.

[50] "The Terrorist Tract That's Hot Reading - Unabomber's Published Manifesto Gets the Attention He Sought," *The Washington Post*, September 23, 1995, by Marc Fisher, p. C1.

[51] "Unabomber's Essay Was Trimmed a Bit; Newspaper erred, left out a paragraph," *San Francisco Chronicle*, Friday, September 22, 1995, Kevin Fagan, Michael Taylor, p. A2.

[52] Correction, *Washington Post*, Friday, September 22, 1995, p. 1.

[53] Telephone interview with Tony Snow, Thursday, August 22, 1996.

[54] "Unabomber Gores Technology," *Detroit News*, September 21, 1995, by Tony Snow, p. A13.

[55] "The Terrorist Tract That's Hot Reading - Unabomber's Published Manifesto Gets the Attention He Sought," *The Washington Post*, September 23, 1995, by Marc Fisher, p. C1.

[56] Brent L. Smith, *Terrorism in America: Pipe Bombs and Pipe Dreams*, State University of New York Press, Albany, 1994, p. 125.

[57] "Wrenches Give Way To Words – Eco-Sabotage Guru Turns To Persuasion," *Arizona Republic*, Saturday, June 1, 1991, by Hal Mattern, p. D1.

[58] Memorandum regarding Earth First Statement of Principles and Membership Brochure, Sept. 1, 1980, by Dave Foreman, p. 1. Cited in Martha F. Lee, *Earth First! Environmental Apocalypse*, p. 39.

Chapter Three
THE BOMBER

2:05 P.M. APRIL 3, 1996   *Bellevue, Washington*
"THEY'VE JUST CAPTURED A UNABOMBER SUSPECT." It was Kathleen Marquardt, executive director of the animal enterprise defender Putting People First, calling from her headquarters in Helena, Montana.

Kathleen had been a friend for years. She stepped in as permanent emcee of our annual Wise Use Leadership Conference in 1991 after Robbie Andersen died of breast cancer. Kathleen's book *AnimalScam: The Beastly Abuse of Human Rights* had exposed the animal rights movement and its hidden agenda to end all human use of animals, including the ownership of pets.

I could hear excited voices and the television sound in the background.

"It's about time!" I said. "Anyone we know?"

A pause. I could hear someone telling her the name.

"Brian Ross," she repeated.

I wrote the name on my notepad. I stared at it blankly. Then it clicked.

"Brian Ross?" I said. "The ABC News reporter is the Unabomber?"

We broke up into howling laughter.

"No, just a little confusion in my relay system. His name is..." She spoke to someone nearby. "Ron, we didn't see the spelling, but it sounds like Theodore Kozinsky."

"Kozinsky? Never heard of him. Ring any bells?"

"Nope. But the reports are saying he was a math professor at U.C. Berkeley during the 'sixties."

"Interesting. Where'd they catch him?"

"Right here in Montana. A little town named Lincoln, up in the mountains from us fifty or sixty miles."

"Montana? That's a surprise."

"Yep, he's being brought to the jail two blocks away from us. I'm just on my way out to go take pictures of him. I thought I'd tell you first."

"Well, Kathleen, looks like Helena is about to become the media capital of the universe."

"Our fifteen minutes of fame."

AN HOUR EARLIER   *Missoula, Montana*

"THAT WAS THE NEW YORK TIMES," said the journalism professor. Patty Reksten spoke to her University of Montana graduate students Derek Pruitt and Steve Adams.[1]

"They need somebody to go get pictures outside of Lincoln. Want the job?"

Adams filled up his Chevy Blazer and took off with Pruitt. They took along a pack of Pepsi.

It was more than an hour's drive up State Highway 200 to Lincoln, a plain little town of 530 souls on the picturesque Blackfoot River, the trout stream made famous by Norman Maclean's novella and the subsequent film, "A River Runs Through It." The quiet hamlet's few streets seemed an afterthought of the highway, blending easily into the forest countryside near the Scapegoat Wilderness Area on the mountain spine of North America, far from the scurry and strife of industrial civilization.

Adams stopped and asked directions to the Unabomber.

Five miles beyond Lincoln, up unpaved Stemple Road at the junction with Humbug Contour Road, they found the crowd near a row of mailboxes. The one next to Joe Brown's had the name Ted Kaczynski neatly lettered on it. An FBI car blocked access to Humbug Contour Road, which was really just a pair of dirt ruts that wound up Baldy Mountain into Stemple Pass, hard by the Helena National Forest with its larches, tamaracks and ponderosa pines.

Pruitt and Adams joined the milling troupe in Stemple Road below the suspect's heavily wooded property. There were a lot of media and a few locals. Rumors floated that the FBI had already taken the suspect away.

Bruce Ely, an undergraduate photographer friend, showed up in the car of another journalism student. He grabbed a few shots of huddled agents standing around in dark blue jackets with large yellow FBI letters on the back and shoulders. Then he stood around with his two friends and waited.

It was about 5:00 o'clock when Greg Rec, another photojournalism student, arrived after getting the assignment from the Denver Post. They barely had time to say hello when an unmarked white Ford Bronco with ordinary Montana plates rolled down Humbug Contour Road from the suspect's cabin. The FBI's barricade vehicle slowly pulled back to let it out. The silhouettes of five occupants were visible as it passed. The center back seat passenger seemed to have shaggy hair. The sun was low and glared off the tinted windows so nobody could get a decent picture. Somebody said, "That's the guy!" The Bronco headed toward Lincoln. The media people did not move.

"Hey, Derek," said Steve Adams. "Come on, let's go!"

Derek Pruitt jumped into Steve's car while Greg Rec took Bruce Ely in his. The photographers roared down the graded road after the white Bronco. The speeds got higher and more dangerous on the rocky gravel road. By the time they passed through Lincoln at more or less legal speeds, the Bronco was far ahead, eastbound on Montana Highway 200. Greg's car began to overheat and fell farther and farther behind.

A few miles out of Lincoln, the white Bronco took a sharp right onto the quarter-mile-long Seven-Up Ranch Road and slipped into the parking lot in front of the guest ranch—the FBI's local command post. The vehicle stopped only long enough for an agent to get out and wave off his companions. The chase car drivers watched the Bronco turn around and followed it back to Highway 200, where it continued eastward. But was the suspect really in it?

"This is too weird, Derek," Steve Adams said. "There's no flashing lights, no sirens, no sheriff's convoy. That Bronco is going to pull off at a Burger King in Great Falls or Helena and it'll be nothing but FBI guys and they'll laugh their heads off at us for chasing them."

Then the white Bronco took the cutoff to Fletcher Pass, State Road 279. So—they were headed for Helena, not Great Falls. The university students stayed glued to the white Bronco all the way into Helena. At dusk it pulled into a Jackson Street parking space across from the long ruddy brick Arcade Building next to Kathleen Marquardt's Putting People First office.

The four students parked on the street and piled out, car doors flying open. They stomped across a few lingering snow drifts and ran toward the white Bronco. Three FBI agents opened the Bronco's right rear passenger door and took out a disheveled man in handcuffs. There was no doubt now. It was the Unabomber suspect.

Bruce Ely pointed his camera and got the first shot of the suspect, eyes half shut, looking weary and weak.

The three FBI agents escorted him into the Arcade Building and pulled the door shut behind them, then stood in the lobby waiting for the elevator. The students took their photos through the glass doors—then watched their quarry vanish.

A man with a little boy came out the door and the four realized it was a public building. Feeling stupid, they walked in, read the directory, saw an FBI office listed on the third floor and took the elevator. They quickly located the office, with its combination lock on the door. No windows, no voices behind the door.

"Lost them," Ely said.

Steve Adams said, "The Pepsi is getting to me. I've got to find the restroom."

He paced down the hall, feeling the letdown after the chase, and found the men's room. He pushed on the door. It was locked. He heard voices inside.

His friends walked up behind him.

He told them, "If I have to pee after all that time in the car, the Unabomber has to take a leak, too."

Steve, Bruce and Greg lined up along the hallway, cameras ready, while Derek backed away toward the FBI office door, setting up his shot.

They heard the restroom door opening. One FBI agent emerged, but immediately stepped back when he saw the cameras. The agents pushed their prize out front and walked him briskly to their office. Derek Pruitt got the classic capture photo of a gaunt, dirty-faced man with snarled hair, dressed in ripped jeans, a stained black T-shirt pulled over a filthy long-sleeved shirt, head held back in contempt, mouth implacable and silent, unreachable blue eyes coolly examining his camera.

While the FBI questioned the suspect in the Arcade Building, the students souped their film in the lab of the Helena Independent Record and transmitted Derek's image to the New York Times, where it made the final edition.[2]

All this time veteran CBS News and ABC News professionals and other media stood around the jail two blocks away, listening to a frustrated Sheriff Chuck O'Reilly tell them repeatedly the suspect would be there any minute. Kathleen Marquardt waited among them, holding her camera at the ready. They would not see Kaczynski for nearly four hours.

As they waited, *World News Tonight with Peter Jennings* aired its first report on the Unabomber suspect's capture:

PETER JENNINGS: Tonight a suspect is in custody. ABC's Brian Ross is in our San Francisco bureau.

BRIAN ROSS: The FBI raid began just after noon at Kaczynski's cabin in a remote mountainous area called Stemple Pass, about five miles outside the town of Lincoln, Montana...[3]

Viewers learned that Theodore Kaczynski had not been arrested, but "was physically removed in handcuffs from the area." They learned he was a 53-year-old native of south suburban Chicago, a Harvard graduate, Class of 1962, and went on to receive a Ph.D. in mathematics at the University of Michigan. It was Dr. Kaczynski. There has never been a serial killer in the United States with such academic credentials.

Most remarkable, Ross reported:

> Agents apparently learned that Kaczynski had been a suspect early on but had not been fully investigated.

Kaczynski's family contacted them and the FBI finally believed they had their man. Kaczynski's brother had regularly sent money that Theodore used to travel to northern California on dates that seemed to match many of the bombings there.

> Agents are now searching for hard evidence of explosives that would link Kaczynski definitively to the series of deadly bombs and give them enough reason to put him under arrest and bring to an end an almost 18-year-old search for the Unabomber.

Brian Ross, ABC News, San Francisco.

10:58 A.M. THURSDAY, APRIL 4   *Helena, Montana*

AFFIDAVIT

```
STATE OF MONTANA          )
COUNTY OF LEWIS AND CLARK )
```

Your affiant, Special Agent (SA) Donald J. Sachtleben, Federal Bureau of Investigation, states as follows:

1. I, SA Donald J. Sachtleben, have been a Special Agent for 12 years. I graduated from the FBI Hazardous Devices School and the FBI Post Blast School. I have investigated bombing cases for 10 years and taught classes on the investigation of improvised explosive devices (IED). I have participated in the on scene investigation of bombing cases.

2. On April 3, 1996 your affiant and other agents of the FBI, ATF and United States Postal Service began the execution of a search warrant on the residence of Theodore John Kaczynski, located in Lewis and Clark County. The premises is a one-room cabin, approximately 10 feet by 12 feet with a loft and without electricity or running water. I am informed by other agents that records of Lewis and Clark County indicate that this property was purchased by Theodore John Kaczynski and another person in 1971. I am also informed by other agents that interview of neighbors revealed that Kaczynski has lived at this residence by himself since that date.

3. When agents knocked on the door, Theodore John Kaczynski answered and was removed so that the search could begin. The initial entry was to ascertain the presence of any explosive devices....[4]

It had cost over fifty million dollars to find that door. The FBI had gone through 200 suspects, 20,000 calls to 1-800-701-BOMB, thousands of interviews, visits with clairvoyants—one of whom said the bomber lived in Boston and drove a Volkswagen—and investigations of dozens of environmental radicals. Their "initial entry" ended the longest and most frustrating manhunt in FBI history. The presence of explosive devices was ascertained, along with containers of bomb chemicals, batteries and electrical wire, logs of explosives experiments, metal pipe, ten 3-ring binders filled with meticulous drawings of explosive devices, and numerous tools necessary for making bombs.

Helena sprouted gardens of mobile television trucks with satellite dishes bent heavenward like metal sunflowers.

4:20 P.M.　　　　　　　　　　　　　　　　　　　　*Englewood, Colorado*
"IT MUST BE CRAZY IN HELENA, KATHLEEN," said Tom McDonnell.

"Crazy is too mild, Tom. This Unabomber thing is drowning us in media. Oh, say, before I forget it, we just came back from the court house with a copy of the FBI complaint against Kaczynski. I'll fax it to you."

"I'm interested in seeing it. But you know, Kathleen, the Unabomber is just one small segment of this overall issue of ecoterrorism. Think of all of the violence, all of the sabotage, all of the grief our industries have had to deal with, dating clear back into the mid-1980s. If you've got so much press there, shouldn't we try to make them aware, give them a more general background, on all the other ecoterrorism that's been going on, and that's still occurring?"

"Are you talking about a press conference?"

"Yes. Could you hold one there?"

"Sure, if I had all the materials."

"Let me do this: I'll take all my files on environmental terrorism and copy them. I'll overnight them to you. And I'll call Henry Lamb and ask if he's got his quotes finished."

"What quotes are those?"

"He's writing up side-by-side quotes from Earth First and the Unabomber."

9:45 A.M. FRIDAY, APRIL 5, 1996　　　　　　　　　*Seattle, Washington*
THE DOOR WAS OPEN, so Barry Clausen entered room 414 of the Alexis Hotel. He stared bewildered. The furniture had been shoved into corners and stacked on end. A thicket of lights, reflectors, camera gear and the ABC News production crew crammed the space.[5]

"I'm here for an interview."

"They're down the hall," a sound man said. "You'll see it. Their door's open, too."

Clausen found the room, actually another part of the same large suite.

"You must be Sarah Koch," he said to the ABC News producer.

"And you must be Barry Clausen," she said. "Thanks for breaking off your trip to Canada for us."

"Well, you guys flew up here from San Francisco."

"Barry, this is David Rummel, our senior investigative producer, and this is Brian Ross. There's coffee and pastries here if you'd like some. Now, you have some things for us to look at?"

Clausen gave them a stack of materials with several lists, including the *Live Wild Or Die* "Eco Fucker Hit List," the same one he had given the media—including ABC News—the previous August. Clausen told them he'd been trying since then to get the FBI to realize that the Unabomber had to have some kind of connection to Earth First or other radical environmental groups. They talked about the last two Unabomber victims and the possibility that they had been targeted from this list.

Rummel studied Clausen. He spoke bluntly: "What makes you think the Unabomber could have used this list?"

Clausen looked at him in near-surprise. "Look how visually prominent this list is on the page. Your eye is drawn to it. What are the odds that two of the top three companies on that list were hit by accident?"

"But the name on the timber association is Roberta Andersen, not the guy the package went to."

"Association directories have listed Bill Dennison as president for years. You can find them in any library. And small libraries still have old directories with the old name, Timber Association of California. Same with Burson-Marsteller. Thomas Mosser hadn't worked there for a year, but his name was still in old business directories. And his home address was listed in the phone book."

They took Clausen to the cameras and wired him for sound.

Ross asked him to repeat their conversation.

"There's eleven names, or eleven company names, on the hit list. Two of the top three are the last two victims of the Unabomber. The number one name on the list is the Timber Association of California, with the current address."

It was a rare nice day in Seattle, so the crew took Clausen for an outside shot walking side by side with Ross along the Elliott Bay waterfront. Ross asked Clausen why he thought the Unabomber was somehow linked to radical environmentalists.

"Based on his beliefs and what he's put in print with his manifestos and his letters, his way of thinking—his ideologies—conforms to that of Earth First."

Off-camera, he told the news crew he thought the Unabomber had been at a big radical environmentalist conference in Missoula, Montana in November of 1994. "Thomas Mosser was killed 30 days after that meeting," he told them. "I think that's a weird coincidence." He had compiled a partial list of attendees that suggested Kaczynski was at that meeting.

Ross was unimpressed. He asked if he knew for sure.

"No, I don't know. I wasn't there. I didn't see him. But it's still a weird coincidence."

It was nearly noon when they all went by taxi to the newsroom of the Seattle ABC network affiliate, KOMO-TV. Rummel asked Clausen to view some file footage of Earth First in a screening room. Barry watched a patchwork of scenes showing conflicts between loggers and Earth Firsters.

When the tape was over, he walked out to the newsroom where Koch, Rummel and Ross were busy on the phone or at their computers. He stopped to look over Sarah Koch's shoulder, watching the story she wrote as it unfolded on her computer monitor.

"Barry," she said, "I don't mean to be rude, but it's really distracting to have you there."

"Sure, no problem," Clausen said, and backed off. He stood awkwardly near the newsroom entry for some time, probably ten minutes.

Brian Ross got off the phone and called excitedly to the three others, "Come here, c'mere, c'mere."

Koch, Rummel and Clausen collected around him.

"I just had a long talk with my government contact. Get this: the FBI had Kaczynski's name in their files two years ago as an environmental radical. They believe he was at that Missoula conference."

6:00 P.M.   *New York City*

PETER JENNINGS: (voice-over) The Unabomber suspect Ted Kaczynski—he was on the FBI's list of suspects years ago....

ANNOUNCER: From ABC, this is World News Tonight with Peter Jennings.[6]

Jennings traced new twists in the Unabomber story: The FBI had Kaczynski's name in its files at least a couple of years before his arrest. It was there in 1994 as having a connection to a radical environmental group, and again in 1995 in the files as a possible suspect in the Unabomber case itself.

Brian Ross informed viewers that authorities familiar with the case told ABC News that Kaczynski's name was on a list of suspects in the Chicago FBI office in late 1995 based on an extensive examination of local high schools and students specializing in advanced mathematical theory.

And Kaczynski's name appeared in FBI files in November of 1994, "in connection with an FBI investigation of a radical environmental group called Earth First, which is active in Montana."

File footage showing Earth Firsters scuffling with loggers ran as Brian Ross stated that Earth First had been best known over the years as a violent group, "spiking trees and blowing up logging equipment," noting also that in many respects its anti-corporate philosophy "parallels that of the Unabomber."

They ran Barry Clausen's remarks about the *Live Wild Or Die* hit list, followed by Leslie Anne Hemstreet, 31, of Earth First in Eugene, Oregon: "Earth First can't take responsibility for what the Unabomber has done with some information he might have gotten from our publication. Because if he had read our publication very thoroughly, he would have seen that we only espouse nonviolence."

Hemstreet, a native Texan, was arrested with more than two dozen Earth Firsters at 2:40 p.m. on August 19, 1996, in a violent riot in the lobby of the Lane County Jail, and convicted of criminal trespass at trial on October 29, 1996.[7] The riot was over federal agents breaking up an eleven-month blockade of a Forest Service logging road leading to the Warner Creek timber sale near Oakridge, Oregon.[8]

But what about that conference?

"It was in November of 1994," Ross said, "on the campus of the University of Montana in Missoula that authorities believe Kaczynski was at a meeting attended by top Earth First members."

The gathering was officially called the Native Forest Network Second International Temperate Forest Conference. A first "International Temperate Forest Conference" had been held in Tasmania in the fall of 1992, and the first "Native Forest Network International Temperate Forest Conference" occurred in Burlington, Vermont in the fall of 1993.

The theme of the 1994 Missoula conference was "Focus on the Multinationals." It was attended by more than 500 people from many parts of the world.

The Native Forest Network is a group founded in 1991 by long-time Montana Earth Firster John Frederick Kreilick, aka Jake Kreilick, aka Jake Jagoff, with two Australian activists, Tim Cadman and Beth Gibbings. Kreilick spent a little more than a year in Australia after earning his Master of Science degree in environmental science from the University of Montana in 1990. His Australian sojourn was largely in the island state of Tasmania. During this time he ventured with a group of international Earth Firsters to Malaysia to disrupt the export of logs in the state of Sarawak, for which he spent two months in jail there.[9]

The Native Forest Network has over a thousand individual members and some 80 environmental organization members. Its purpose is to internationalize temperate forest issues using tactics identical to those of Earth Firsters. Its members are loosely coupled with Earth First, meaning they are responsive to one another but can take stands that the other organization leaders do not support and can dictate their own agendas and tactics and require no endorsement from the other organizations. In practice, they are indistinguishable from Earth Firsters.

"Focus on the Multinationals" began Wednesday, November 9, 1994 and ran through the afternoon of Sunday, November 13.[10] Barely a month later, Thomas Mosser was dead. The Unabomber claimed that

Mosser died "because he was a Burston-Marsteller executive," a firm he erroneously believed to have cleaned up Exxon's image.

If Kaczynski was at the conference, did he hear anything erroneously connecting Burson-Marsteller to the Exxon image cleanup that might ignite his anger and launch one of his bombs?

ABC News asked Thomas Fullum, one of the organizers of the Missoula meeting, about Burson-Marsteller. "You know," Fullum said, "there probably was some discussion of it. I don't think there was a real formal discussion."

Could Kaczynski have been there? Fullum said, "He could easily have been there." The Native Forest Network knew only about 200 of the 500 people who were present.[11] Among the many publications available to those who attended the meeting was The Earth First Journal. Brian Ross told his viewers that it was clear now that the FBI was very interested in pursuing a possible connection to Earth First, as a way of explaining what had been a baffling pattern of violence.

> BRIAN ROSS: What's not so clear is how Kaczynski's name could twice show up in FBI files, first as a possible environmental radical and then as a possible Unabomber suspect without the FBI closing in on him sooner. And today, the FBI said it would have no comment on that.
> Brian Ross, ABC News, Seattle.

9:13 A.M. SUNDAY, APRIL 7, 1996   *Eugene, Oregon*
> "ON FRIDAY, APRIL 5, THE ABC NEWS PROGRAM World News Tonight with Peter Jennings aired a report linking the non-violent environmental group Earth First! with Theodore Kaczynski, the alleged Unabomber. The piece was riddled with distortions and inaccuracies, and can only be described as a hit piece on Earth First! and the environmental movement. ABC's sensationalistic coverage has done serious damage to the reputation of the Earth First! movement, based on the word of Barry Clausen, an individual employed by the timber industry."[12]

Thus began a long electronic letter to ABC News posted on the Internet by the Earth First! Journal staff. The Earth Firsters asserted that ABC News had based its allegation of a Unabomber-Earth First! link "on two flimsy pieces of information," Kaczynski's alleged presence at the Missoula conference, and the "Eco-Fucker Hit List."

How Earth First dealt with this alarming linkage to the Unabomber illustrates a strategy they pioneered in gaining legitimacy. Two social scientists, Kimberly D. Elsbach and Robert I. Sutton, studied Earth First's method of dealing with illegal acts such as tree spiking and equipment sabotage and discovered a four-step strategy.[13]

1. *Institutional conformity*: This involves using practices similar to those of legitimate organizations, with spokespersons, press releases and formal responses to criticism. Conformity implies that the organization and its spokespersons are credible, rational, and legitimate. In its efforts to conform, Earth First sometimes misses the mark: for example, it frequently uses the words "nonviolent and peaceful" to describe itself, a statement legitimate organizations have no need to make.

2. *Decoupling*: separating legitimate organizational structures and practices from members' illegitimate actions. Decoupling is achieved through the use of independent affinity groups or anonymous individuals who carry out illegitimate actions, such as tree spiking, sabotage and arson, but are not formally linked to the organization.

3. *Impression management*: Institutional conformity and decoupling increase overall credibility and pave the way for two specific impression management tactics: *defenses of innocence* and *justifications*.

Defenses of innocence are claims that one is not responsible for an event or that the event did not occur. Spokespersons can defend their organization's innocence by asserting that it did not endorse an illegitimate action—citing decoupling—and thus isn't to blame.

Justifications are claims that an event was not bad, wrong, inappropriate, or unwelcome because of the positive outcomes it led to or the extreme circumstances it was performed under.

4. *Shifting attention*: Defenses of innocence and justifications shift attention away from negative aspects of an event and toward positive aspects of the event and the organization considered responsible, setting the stage for *enhancements* and *entitlings*.

Enhancements are attempts to improve the perceived merit of an event. Spokespersons typically emphasize the progress made toward socially desirable goals as a result of illegitimate actions.

Entitlings are attempts to gain credit for a desirable event. Spokespersons may assert that the organization deserves credit for the action because of the socially desirable goals that it achieved.

Former activists point out something Ellsbach and Sutton missed in the legitimation process: threats that silence dissidents who object to escalating violence perpetrated by fellow activists. Numerous disenchanted Earth Firsters and animal rights activists have been told to keep quiet about crimes or face retaliation ranging from lawsuits to having their legs broken to being murdered.[14]

Ellsbach and Sutton's analysis rings true in what happened next.

Earth First did not deny that Kaczynski was present at Missoula. Instead, they issued a non-denial denial, decoupling themselves from the Missoula conference, asserting that it was not connected to Earth First, but entirely the work of the Native Forest Network. The letter stated, "Activists associated with Earth First! also attended the conference, but had no role in the proceedings."

Earth First's statement is a decoupling maneuver. It is not the whole truth.

There was no mention that the man who convened the Missoula meeting and co-founded the Native Forest Network, Jake Kreilick, was instrumental in the 1987 revival of Earth First in Montana and was a core organizer of Wild Rockies Earth First.[14]

There was no mention that Jake Kreilick was involved in Earth First direct actions in 1988 at the Okanogan National Forest supervisor's office and at the Kalmiopsis area in southern Oregon.[15] Or that he was on the 1990 Earth First Round River Rendezvous committee.[16] Or that he helped organize a 1991 Earth First road show in England.[17] Or that he wrote articles for and assisted in publishing some 1993 issues of the Earth First Journal.[18]

Nor was there any mention that the notice publicly announcing the 1994 Missoula conference was posted by Jake Kreilick in the Earth First Internet bulletin board.

Many if not most of the other radical environmentalists on Kreilick's Missoula program have similar personal biographies, but were decoupled from Earth First. For example, Darryl Cherney, a prominent California Earth Firster, was to give a session on behalf of the "Redwood Action Team," which Professor Martha Lee calls the Earth First! Redwood Action Team—it even received a $500 donation from Patagonia Clothing as an Earth First entity.[19] Michael Marx was slated to appear for the Rainforest Action Network, a group founded by Earth Firsters and still led by Earth Firsters. Philip Randall Knight (no relation to Phil Knight, the Portland, Oregon-based Nike sneaker executive), who was identified with the Native Forest Network, is a seasoned Earth Firster, arrested in the 1988 Okanogan National Forest supervisor's office occupation—and also a co-founder of the radical environmental group, the Predator Project, in Bozeman, Montana. Michael N. Christensen, a veteran Earth Firster using the alias Asanté Riverwind, was down for a workshop in forest activist skills.

Earth Firsters, as part of their tribalist philosophy, create decoupling groups, not only to conform to their non-organizational beliefs, but also gain legitimation from new constituencies. Non-existent factions are not unknown—Mike Jakubal promoted one called "Stumps Suck" in the late '80s. A small clutch of Arizona Earth Firsters took on "EMETIC" as their nom-de-sabotage before being arrested and convicted on federal property destruction and aiding-and-abetting charges (EMETIC was a play on words as well as an acronym for Evan Mecham Eco-Terrorist International Conspiracy, Evan Mecham being a former Arizona car salesman and governor who was impeached and left office with the reputation of a buffoon).

In addition, old-line Earth Firsters have been decoupled by environmental king-makers for more sedate roles. Mitch Friedman, currently of the foundation-funded Northwest Ecosystem Alliance, once entered a guilty plea to first-degree criminal mischief, a felony, and was among those arrested in the Okanogan occupation as an Earth Firster. Another alumnus of that arrest is Peter Jay Galvin, also arrested in July 1988 on criminal trespass charges for trying to chain himself in the headquarters of the Mount Hood National Forest, currently of the Southwest Center for Biological Diversity in Silver City, New Mexico—who also appeared in the ABC News clip of the scuffle with loggers. There are many other Earth Firsters in similarly changed circumstances.

The organization names are different. They may or may not represent a genuinely distinctive constituency. They decouple illegitimate actions from Earth First and help it build legitimacy. The network of actual Earth Firsters grows. The radical agenda grows.

The letter to ABC News likewise did not deny that the Unabomber used the "Eco-Fucker Hit List" to target his last two victims. Instead, their non-denial denial asserted that Earth First Journal had no part in publishing *Live Wild Or Die*. It was not the whole truth.

Two major creators of *Live Wild Or Die*, Mitch Friedman and Mike Jakubal, were important Earth Firsters when they put out the "Eco-Fucker Hit List." A 1989 Washington State newspaper said

> A recent, first-time, national Earth First newspaper, Live Wild or Die!, was edited by Jakubal and assembled in Bellingham. The publication, which advocates destructive responses to industrialization, was supported in part by national Earth First money provided by Earth First co-founder Mike Roselle.[20]

Years later, Roselle told the Washington Post that

> he funded the first issue of "Live Wild or Die," which contained the "Eco-(expletive) Hit List." He provided between $250 and $500, but didn't like the "childish" results and cut off funding. About 1,000 copies of that issue were printed, according to Roselle. He lost track of its editors, but believes one is now "running a hippie sawmill" somewhere in the Pacific Northwest.[21]

Roselle certainly has not "lost track" of its editors: Mitch Friedman, in addition to being a high-profile leader of the Northwest Ecosystem Alliance in Bellingham, Washington, is a board member of Dave Foreman's "Wildlands Project," both of which are well-known to Roselle. Mike Jakubal has a shack in Redway, near Garberville, California, where he is a vocal anti-timber activist occasionally getting money from Earth First's Direct Action Fund for specific Nomadic Action Group anti-timber projects. Roselle knows exactly where they are and what they're doing.

Among the more inane passages of the Earth First Journal's letter to ABC News is one beginning, "ABC's portrayal of Earth First! as violent is totally contradicted by the history of Earth First! activism."

That history is replete with arrests and convictions for felony conspiracy to sabotage, for actual felony sabotage, for felony criminal mischief, felony vandalism, various misdemeanors and for vast numbers of criminal trespass arrests. Top of the list is Earth First's co-founder, Dave Foreman, who entered a guilty plea on a felony conspiracy charge in 1991, which was withdrawn after five years, leaving Foreman sentenced to a $250 fine for misdemeanor depredation of government property.

Earth First's brand of "non-violence" is based on physical coercion of workers to prevent them from working by erecting barriers, trenching roads, chaining bodies to equipment and other physically coercive tactics.

Ordinary people do not consider such acts non-violent. Even Earth Firster Erik Ryberg, author of the notorious article on "Bombthrowing," (see p. 50) stated under oath that he disagreed with actions that physically blocked roads to prevent timber harvest because "I didn't think they were non-violent actions. Because the purpose of blocking the road is to coerce someone into not working, and the purpose of non-violence is to make a person decide for himself to stop the action; in this case, the action of building the road."[22]

The remainder of the letter to ABC News was spent denouncing Barry Clausen, "ABC's source of information." The Earth Firsters wrote, "Clausen is not a credible source," then cited a litany of reasons why: he was a paid informant of the timber industry who infiltrated Earth First, he was rejected by every law enforcement agency he has tried to work with, and an FBI agent no longer with the bureau said that some other agencies told him Clausen was not reliable.

An FBI memo dated 23 August 1990 relating to the destruction of power lines in Santa Cruz, California, on April 22, 1990, by environmental radicals stated:

> Clausen was debriefed in San Francisco August 21, 1990, and provided the following information:
> For the past seven months he has been a private investigator on the payroll of various unnamed timber industry groups. He was tasked to infiltrate the radical environmental movement and provide law enforcement with information about criminal acts past and future. He has established contacts and relationships with activists in Montana, Washington, Oregon and California to include [names deleted]. Clausen is starting a direct action hot line in Seattle funded by Mike Roselle. He states that the leaders network via computer links. He has attended the Montana Earth First (EF) Rendezvous where he spent time with [names deleted].
> Clausen has agreed to be debriefed by Seattle FBI agents and open his files to them. This shall be coordinated through San Francisco Division...

> At Seattle, Washington, provide San Francisco with hello number for Clausen to use in contacting SSA [Supervisory Special Agent] [name deleted].[23]

An FBI memo dated 30 August 1990 from the Director of the FBI to FBI San Francisco, Portland and Seattle, warned them in their dealings with Clausen not to exceed their authority under the "266 Classification" which deals only with single specific terrorist acts:

> Receiving offices are reminded that the 266 Classification can only be used to investigate a specific criminal activity which is terrorist related. The 266 Classification cannot be used to gather intelligence of a domestic terrorist organization. Thus, the captioned investigation, classified 266-SF-91574, can only investigate the people involved in the destruction of the electrical transmission towers in Santa Cruz County....
>
> Therefore, receiving offices should not report any information received from Clausen under the captioned 266 matter unless the information is related to that investigation. Receiving offices can maintain contact with Clausen and if he provides information concerning past, present or future criminal activity by an environmental group, a DS/T [Domestic Security / Terrorism] preliminary inquiry can be initiated under the 100 Classification.[24]

Did the FBI reject Clausen? The documents show that the FBI arranged to get his information about other radical environmentalists properly while avoiding illegal investigations of domestic organizations such as Earth First. When Clausen gave the FBI the "Eco-Fucker Hit List" in August 1995, FBI spokesman George Grotz of the San Francisco Unabomber Task Force told reporters, "I can confirm that we have met with this individual and we are very interested in what he has to say."[25]

It was pointless to argue Clausen's credibility in the first place: ABC's April 5 report wasn't "based on the word of Barry Clausen." Like all reputable news organizations, ABC News obtains "double source" verification for everything it reports. Clausen told ABC News things that could not be verified, and ABC News did not report them. Brian Ross first obtained positive confirmation from an independent source of known reliability for everything Clausen said that ended up on-screen—and most of the story came from other sources to begin with.[26]

The Internet letter closed: "Earth First! Journal is asking for people to call Rhonda Schwartz, Senior Producer of World News Tonight, to complain about ABC's irresponsible, sensationalistic reporting. You may reach her at [home telephone number]."

11:00 A.M.   *New York City*
HALF OF EASTER SUNDAY'S "ABC NEWS THIS WEEK WITH DAVID BRINKLEY"

dealt with the Unabomber. After updates on the latest discoveries in Theodore Kaczynski's cabin and efforts to trace his travels, Sam Donaldson, sitting in for David Brinkley in Washington, called upon Brian Ross, standing by in New York, to explain the link between the suspect and radical environmentalists:

> BRIAN ROSS: Well, Sam, there's one interesting and intriguing link, an apparent meeting held in Montana, November 1994, of a group loosely connected with a radical environmental group called Earth First, where Kaczynski seems to have been in attendance.[27]

The focus, Ross said, was on multinationals, and among the companies and people discussed was the New York public relations company Burson-Marsteller. One month later a former top executive of Burson-Marsteller became a Unabomber victim. What's more, Ross said, authorities were intrigued by possible connections between Kaczynski and radical environmental groups. The top organization on a hit list put out by a radical environmental journal was the last target of the Unabomber.

> BRIAN ROSS: He seems to have become some kind of environmental radical, and much as Timothy McVeigh may have been inspired and inflamed by the militia, it's possible, perhaps, that Kaczynski was inspired and guided by the radical environmental groups."

EARLY MORNING, MONDAY, APRIL 8, 1996   *New York City*
THE NEW YORK TIMES BULLDOG EDITION HIT THE STREETS reporting that law enforcement officials were trying to establish a link between the victims of the Unabomber and the heaps of material that FBI agents took from the cabin of the suspect. Finding evidence that either of the two latest bombs— the one that killed Thomas J. Mosser and the one that killed Gilbert R. Murray—was built by Kaczynski would allow federal prosecutors to pursue the death penalty. Reporter Neil MacFarquhar wrote:

> It is plausible that he drew the idea from what he was reading... While exactly what the bomber read has not been pinned down, his writing seems to parallel strident environmentalist periodicals like the Earth First Journal.[28]

10:40 A.M. EDT   *Hollow Rock, Tennessee*
"I HAVE MOST OF THE QUOTES FINISHED, TOM," said Henry Lamb, executive director of the Environmental Conservation Organization, a landowner group defending property rights and land development. "I hope they're what you want."
"Can you fax me a copy?" asked McDonnell.

"It's on the way. I gave it the title, 'Kindred Spirits: An Analysis of Values and Visions Shared by the Unabomber, Earth First, Dave Foreman, and the Wildlands Project.' See if you can find any mistakes."

Unabomber: "The industrial revolution and its consequences have been a disaster for the human race." (Manifesto, ¶ 1)

Dave Foreman: "In looking at human history, we can see that we have lost more in our 'rise' to civilization than we have gained." (*Confessions of an Eco Warrior*, p. 28)

Unabomber: "They [the industrial revolution and its consequences] have inflicted severe damage on the natural world. The continued development of technology will worsen the situation...and inflict greater damage on the natural world.... We therefore advocate a revolution against the industrial system." (Manifesto, ¶ 1, 4)

Dave Foreman: "Industrial workers, by and large, share the blame for the destruction of the natural world." (*Confessions*, p. 31) "It's time to get angry, to cry, to let rage flow at what the human cancer is doing to Earth, to be uncompromising." (p. 20) "We are warriors. Earth First! is a warrior society. We have a job to do." (p. 33) "The ecologist Raymond Dasmann says that World War III has already begun, and that it is the war of industrial humans against the Earth. He is correct. All of us are warriors on one side or another in this war; there are no sidelines, there no civilians." (pp. *viii-ix*)

Unabomber: "Industrial-technological society cannot be reformed." (Manifesto, Heading at ¶ 111) The only way out is to dispense with the industrial technological system altogether. This implies revolution....(Manifesto, ¶ 140)

Dave Foreman: "There is no hope for reform of the industrial empire. Modern society is a driverless hot rod without brakes, going ninety miles an hour down a dead-end street with a brick wall at the end. Bioregionalism is what is on the other side of that wall. (*Confessions*, p. 45) How, indeed, can you fight the dominant dogmas of Western civilization? ... A monkeywrench thrown into the gears of the machine may not stop it. But it might delay it, make it cost more. And it feels good to put it there. (*Confessions*, p. 23)

Earth First! Journal: "We don't care who is in power in Washington, for whoever stands on the walls of Babylon will be a target for our arrows. When we raze the citadel, it will matter not who holds the keys to the corporate washroom.... What we want is nothing short of a revolution. Monkeywrenching is more than just sabotage, and your goddamn right it's revolutionary! This is jihad, pal." (Mike Roselle, "Forest Grump," *Earth First Journal*, Dec. 94/Jan. 95)

Unabomber: "We would like, ideally, to break down all society into very small, completely autonomous units." (Letter to The New York Times, April, 1995)

Unabomber: "The positive ideal that we propose is Nature. That is, WILD nature; those aspects of the functioning of the Earth and its living things that are independent of human management and free of human interference and control." (Manifesto, ¶ 183)

Dave Foreman: "Bioregionalism, then, is fundamentally concerned with ... becoming part of a community already present — the natural community of beasts and birds and fish and plants and rivers and mountains and plains and sea. It means becoming part of the food chain, the water cycle, the *environment* of a particular natural region, instead of imposing a human-centered, technological order on the area." (*Confessions*, p. 44)

The Wildlands Project: "I suggest that at least half of the land area of the 48 conterminous states should be encompassed in core reserves and inner corridor zones.... Eventually, a wilderness network would dominate a region and thus would itself constitute the matrix, with human habitation being the islands." (Reed F. Noss, "The Wildlands Project: Land Conservation Strategy," *Wild Earth*, Special Issue, 1992, p. 15)

Unabomber: "[A]fter the demise of the industrial system, it is certain that most people will live close to nature, because in the absence of advanced technology there is not other way that people CAN live. To feed themselves they must be peasants or herdsmen or fishermen or hunter, etc." (Manifesto, ¶ 184)

Unabomber: "There is good reason to believe that primitive man suffered from less stress and frustration and was better satisfied with his way of life than modern man is." (Manifesto, ¶ 45)

Dave Foreman: "We can see that life in a hunter-gatherer society was on the whole healthier, happier, and more secure than our lives today as peasants, industrial workers, or business executives." (*Confessions*, p. 28)

Unabomber: "Our immediate goal...is the destruction of the worldwide industrial system." (NYT letter, April 1995)

Wild Earth: "Does all the foregoing mean that *Wild Earth* and The Wildlands Project advocate the end of industrial civilization? Most assuredly. Everything civilized must go...." (John Davis [former editor of *Earth First Journal*], "WE Role in the Wildlands (The Role of *Wild Earth* in the Wildlands Project)," *Wild Earth*, Special Issue, 1992, p. 9)

9:04 A.M. MDT   *Englewood, Colorado*
"HENRY, I'M FAXING YOU SOMETHING FOR YOUR QUOTES," said Tom McDonnell. "Kathleen called this morning and asked if I had anything on Burson-Marsteller. I found an article in the February 1994 issue of Earth First Journal. I think you'll want to use it."

Henry Lamb got the fax transmission and studied the article.

"The International PR Machine: Environmentalism á la Burson-Marsteller," by Carmelo Ruiz-Marrero, began:

> Burson-Marsteller is one of the largest public relations firms on Earth. With offices in 27 countries and a list of customers that includes national governments and transnational corporations, B-M is an extremely powerful institution....
>
> Burson-Marsteller promotes an elite form of "environmentalism" that serves the needs of the corporate world. The main purpose of this shallow environmentalism is to make the public believe that 1) the environmental crisis has been exaggerated by sensationalist and irresponsible activists, and 2) that "responsible" environmentalists work with, and not against the corporate establishment.
>
> B-M's clients have included:
>
> ● Union Carbide of Bhopal tragedy fame. This corporation admits keeping files on activists, and alleges (in a leaked memo in 1991) that grassroots activists are linked to communists.
>
> ● Exxon, which hired B-M to counter the negative publicity from the Valdez oil spill. [29]

There it was, the erroneous link. How eerily it resonated with the Unabomber's words:

> We blew up Thomas Mosser last December because he was a Burston-Marsteller executive. Among other misdeeds, Burston-Marsteller helped Exxon clean up its public image after the Exxon Valdez incident. But we attacked Burston-Marsteller less for its specific misdeeds than on general principles. Burston-Marsteller is about the biggest organization in the public relations field.
>
> This means that its business is the development of techniques for manipulating people's attitudes. It was for this more than for its actions in specific cases that we sent a bomb to an executive of this company.

Where did the Earth First journalist get his misinformation?

Carmelo Ruiz-Marrero, a Puerto Rican free-lance writer living near San Juan, wrote "The International PR Machine" while studying at the Institute for Social Ecology in Plainfield, Vermont, a school co-founded in 1974 by noted American anarchist Murray Bookchin.

When Ruiz-Marrero wrote his piece, he was immersed in the world of Bookchin, whose influence on radical environmentalism is incalculable. Bookchin was born in New York City January 14, 1921, to immigrant parents who had been active in the Russian revolutionary movement. As a teenager in the early 1930s he entered the Communist youth movement, but he grew disillusioned by its authoritarian character and was expelled in September 1939—still in his teens—for "Trotskyist-anarchist deviations."

He became a libertarian socialist and worked with several U. S. labor organizations during the 1940s. His 1952 article "The Problem of Chemicals in Food" was one of the earliest on the subject. In the 1960s he was deeply involved in countercultural movements, and championed the ideas of social ecology, sometimes using the pen name Harry Ludd.

His book, *Our Synthetic Environment*, written under the pseudonym Lewis Herber, was published by Alfred A. Knopf in 1962, preceding Rachel Carson's *Silent Spring* by nearly half a year. His 1971 collection titled *Post-Scarcity Anarchism* comprised such pioneering essays as "Ecology and Revolutionary Thought" (1964) and "Towards a Liberatory Technology" (1965), both of which advanced the radical significance of the ecology issue. "Listen, Marxist!" (1969), his critique of traditional Marxism, profoundly influenced the New Left.

His many books since then, *Remaking Society, The Philosophy of Social Ecology, Reenchanting Humanity*—including one published by the Sierra Club, *The Rise of Urbanization and the Decline of Citizenship*—have explored virtually every niche of radical environmentalism.

His Institute for Social Ecology earned an international reputation for its advanced courses in ecophilosophy, social theory, and alternative technologies.[30] Bookchin's stamp is clearly visible on Ruiz-Marrero's writing.

Ruiz-Marrero wrote the Burson-Marsteller piece as part of his master's thesis for Goddard College, a school founded in Plainfield, Vermont in 1938 on the educational principles of John Dewey and other progressives. "The International PR Machine" was first published in the fall 1993 issue of No Sweat News, newsletter of the Atmosphere Alliance in Olympia, Washington, an affiliate of David Brower's Earth Island Institute. Ruiz-Marrero's source for the erroneous Burson-Marsteller assertion was a 1992 publication, "The Greenpeace Book of Greenwash."

A similar connection between the Valdez spill and Burson-Marsteller was made in a 1993 article in the Washington, D.C.-based far-left magazine, Covert Action Quarterly, which cited a Canadian newspaper, the Vancouver Sun, as its source. The connection was talked about around radical environmental circles, but virtually nowhere else.

NOON   *Helena, Montana*
MARK LAROCHELLE, KATHLEEN MARQUARDT'S PRESS SECRETARY, sent out faxes to the media announcing a Putting People First news conference for Wednesday morning, to be held at the Park Plaza Hotel across Last Chance Gulch from their office. The news conference was to publicize the widespread problem of ecoterror, crimes committed in the name of saving nature. The Unabomber case was the lens through which a far worse problem would come into focus.

The press kit was to include copies of the *Live Wild Or Die* Hit List, the Earth First Burson-Marsteller story, Henry Lamb's compilation of radical environmentalist / Unabomber quotes, and extensive news clips of radical environmentalist criminal acts dating back nearly two decades.

Barry Clausen agreed to drive the 600-plus miles from Puget Sound to Montana and speak at the news conference. I agreed to participate by telephone link from my office, offering further contacts and background material.

2:15 P.M. TUESDAY, APRIL 9, 1996   *Helena, Montana*
KATHLEEN MARQUARDT'S OFFICE HAD BECOME THE UNABOMBER MEDIA CENTER. Reporters fresh off the plane guided each other to her, seeking story leads, asking where the new angles were and who could give them the best background material.

Thus it was no surprise when Rhonda Schwartz, ABC News Senior Producer, called Kathleen and asked for some of her time.

"Have you seen the letter Earth First put up on the Internet?" she asked on the phone.

"Yes, " said Kathleen.

Schwartz had received dozens of calls from angry Earth Firsters. The ABC News main office in New York got them too. Some of the calls were civil, some were nasty, some were downright sick.

Schwartz came into the office for a visit. The slight woman with chin-length dark hair took the art deco wicker chair offered her. Mark Larochelle sat in. Schwartz was angry, not fearful—she took the threats seriously, but she'd never knuckle under to extremists, no matter how dangerous. The only threat she completely shrugged off was that of a lawsuit against ABC News—her news team would stand by their story, which experience told her could easily meet any legal challenge.

"I don't mean to sound crass, Rhonda," said Kathleen, leaning into her desk, "but I'm glad you got those calls. I'm glad one of you news people is finally seeing the kind of environmentalist threats and hate calls we've lived with for years. When I started Putting People First back in

Washington, D.C., we got death threats from animal rights people just about every week, we never got in our car without looking under it, our office was constantly harassed. It was like we lived under siege."

"Kathleen, I would never believe this if I hadn't seen it."

Schwartz was pressed for time on this assignment—she had to return to home base in Georgia to meet a family commitment: taking her daughter camping. She needed two things: information on any new developments in the Kaczynski case and continuing contacts for her crews that would remain to cover the ongoing story.

Kathleen said, "Our network is helping everybody and your people seem pretty good about taking care of themselves. But we do have something new. Have you seen the Earth First Journal with the phony link between Burson-Marsteller and Exxon?"

Schwartz looked at the tabloid Marquardt handed her. She scanned the Ruiz-Marrero piece quickly. She hadn't seen it.

"Can I take this and have my crew film it?"

"Would you take a copy instead? It's my only original."

"I need the original. I'll get it back to you."

7:00 A.M. WEDNESDAY, APRIL 10, 1996   *Helena, Montana*
BARRY CLAUSEN ARRIVED AT THE PUTTING PEOPLE FIRST OFFICE just as Kathleen opened it. He had finished the long drive last night and lodged at a friend's place outside town. Mark Larochelle walked up at the same moment and told them he'd seen an Internet posting by Earth Firster Phil Knight urging radical environmentalists to converge on Helena and ask questions at their news conference—nonviolently, of course.

"They obviously don't want this information to get out," Mark said.

"I've been getting threats on my car phone, too," said Barry. "Do we have any crowd control?"

Kathleen said, "The first thing on my list this morning is a call to the police. Mark, will you take care of it?"

Larochelle quickly got on the phone and spoke to Chief Troy McGee of the Helena Police Department, telling him about the threats and Internet posting. Chief McGee was aware of the news conference and its venue in the Park Plaza Hotel's meeting room.

"That room is in the basement and it can't be secured," said McGee. "If you have any trouble, call 911."

Mark relayed the news to Barry and Kathleen.

"That's it?" Barry sputtered.

Kathleen looked resigned. "Bureaucracy. We should have known. Well, let me show you the layout anyway."

Barry followed Kathleen downstairs and across Last Chance Gulch to the Park Plaza Hotel. They inspected the small meeting room in the basement where the news conference was set.

"This is the only door?" Clausen said. "There's no back exit or other way out?"

"Nope. Not good, is it?"

"With an unruly crowd?" he sighed.

"It could be a death trap."

8:10 A.M.

"RON, WE'VE HAD TO CANCEL THE NEWS CONFERENCE. There have been too many threats and law enforcement can't promise us protection. But we'll give reporters the same press kit and interviews in our office, one-on one. Can you still be available for phone interviews?"

I told her I could.

Kathleen and Barry spent the morning in session after session with reporters from the national media. At noon, attorney Bill Wewer, Kathleen's husband, joined them for lunch at The Windbag Saloon, a converted whorehouse on Last Chance Gulch left over from boomtown days. They commiserated about losing the opportunity for the news conference.

"At least I can still drive up to Lincoln and see Kaczynski's cabin," said Barry glumly. "For whatever that's worth."

When they finished and went forward to the cash register, Bill Wewer looked out the window toward their office. Police, rescue and fire vehicles pulled up and swarms of officers crowded the street.

Wewer shook his head and laughed bitterly. "Can you believe this? After telling us to take a flying leap, the cops sent enough people to start their own riot."

That afternoon Kathleen talked to Cary Hegreberg, executive director of the Montana Wood Products Association. Cary heard that Kaczynski had registered at least 24 times since 1980 in Helena's Park Hotel, three blocks down Last Chance Gulch, and walked to the Helena Public Library across the street from his office. "I felt a chill go down my spine," he told her. He had long kept a sign from a wise use group in his window that informed passersby, "We Support the Timber Industry."

4:00 P.M. THURSDAY, APRIL 11, 1996   *Alexandria, Virginia*

LINDA CHAVEZ, PRESIDENT OF THE CENTER FOR EQUAL OPPORTUNITY in Washington, D.C., wrote her Wednesday USA Today column, titled, "Want Unabomber Motive?" It began:

> Is Unabomber suspect Theodore Kaczynski just a brilliant mathematician turned hermit with a mysterious grudge against technology, or is he at least loosely affiliated with an organized radical group that advocates a war on modern technology?
> The Unabomber may well have taken his inspiration from the writings of Earth First!'s radical fringe.

The column was so detailed in its Unabomber-Earth First linkage it prompted threats of lawsuits. It contained a few sizzling details about Dave Foreman:

> Earth First! leader Foreman also once published a how-to manual that provided "detailed, field-tested" instructions for eco-sabotage, including making explosives, and suggestions on harassing "villains." Foreman pleaded guilty to a felony conspiracy charge for having distributed copies of this manual to a group of Arizona Earth First!ers who sabotaged a power plant in 1989.
>
> Foreman actually gave a speech to an Earth First! meeting in 1983 in which he inveighed: "The blood of timber executives is my natural drink, and the wail of dying forest supervisors is music to my ears."

In fairness, Foreman ostensibly gave the speech in humor, part of a burlesque for the road show crowd. Friends of Gil Murray did not think it was funny. Foreman resented the accusation about explosives.

Linda Kanamine, deputy managing editor of USA Today, called asking if I had radical environmental documents that could corroborate the details of the Chavez column. I provided the newspaper's lawyers with pages from Foreman's manual: p. 189 ("Smoke Bombs"), p. 195 ("Stink Bombs"), p. 197 ("Stink Grenades") and the "Pajama" article on "Bombthrowing" from Earth First Wild Rockies Review. Chavez, a former Reagan administration official long accustomed to controversy, based her column in part on a scholarly study by Professor Martha F. Lee, *Earth First! Environmental Apocalypse.* Chavez had interrupted Professor Lee's sabbatical at Cambridge to make sure she understood correctly what the book was saying. The column's accuracy could not be faulted and the legal saber-rattling came to nothing.

6:00 P.M.   *New York City*

PETER JENNINGS SAID, "In the Unabomber investigation, tracing the roots of Theodore Kaczynski's radical views about the environment..."

Brian Ross reported on the riddle of why Kaczynski may have formed such extreme views on the environment.

Panning across spectacular Montana scenery, the report suggested that logging and mining interests had begun closing in on him. A few miles from Kaczynski's cabin, a timber sale had been clearcut. Seven miles down the Blackfoot River was the proposed site of a huge open-pit gold mine that many saw as an ecological disaster.

Kaczynski was worried about the mine, neighbors told ABC News, and at the Lincoln school, "the reclusive Kaczynski apparently ventured out to attend a public meeting about the environmental impact of the gold mine."

ABC News interviewed Jeff Hagener, administrator of the Trust

Lands Management Division of Montana's Natural Resources and Conservation Department: "He looked very familiar to me from a meeting we had last fall. I came in a little late and I recollected seeing him just outside the meeting at that time."

I called Hagener to confirm the facts. Hagener saw Kaczynski in the foyer of Lincoln School (Lincoln schools are all in one complex) on October 12, 1995, standing behind the crowd and listening through the open doors to the meeting room.[31]

ABC News did not find a witness to confirm Kaczynski's concerns over timber cutting, but Kathleen Marquardt located Larry Brown, who, while employed by the Water Quality Bureau of the Montana Department of Health and Environmental Sciences, received a telephone call from Kaczynski August 14, 1986, asking about a proposed timber harvest on some private mining claims in the Poor Man Creek watershed within the Helena National Forest several miles above Kaczkynski's cabin. Brown arranged for a tour of the proposed cutting area and on August 29 escorted Kaczynski and several other locals to the site. Brown says Kaczynski was quiet the entire trip, but sharply attentive to what was shown him. Kaczynski was certainly taciturn, but perhaps not so reclusive as we thought when it came to the environment, and for longer than we thought.[32]

Brian Ross concluded his report by pointing out that "all of this could be significant because the Unabomber a few years ago began selecting his targets based on environmental concerns."

> BRIAN ROSS: In April last year, the head of the California Forestry Association. And in December of 1994, Thomas Mosser, a former executive of Burson-Marsteller, a public relations company, mistakenly identified in radical environmental journals as having worked for the Exxon Company in the Valdez oil spill—a mistake the Unabomber incorporated in one of his letters.[33]

AFTERNOON DRIVE TIME, FRIDAY, APRIL 12, 1996   *North America*
THIS IS ALL THINGS CONSIDERED. I'm Robert Siegel.

> NOAH ADAMS: And I'm Noah Adams. Late this afternoon, it was learned that law enforcement officials believe they may have found the original 35,000 word Unabomber manifesto, in the cabin of suspect Theodore Kaczynski. The anti-technology themes expressed in the Unabomber's writing have raised questions about the role radical environmental groups may have played in motivating the Unabomber. NPR's Howard Berkes reports that the Unabomber has become a point of contention between some environmentalists and their opponents.[34]

Berkes sketched the deaths of Mosser and Murray, then introduced Barry

Clausen and his focus on Kaczynski's presence at the Missoula conference:

> BARRY CLAUSEN: I think that you have to look at what his whole profile was, for all his reign of terror, and at what point did he change his profile and go after natural resources providers. It's my opinion that it could have been at that conference, yes.

Berkes commented that Clausen had "made a career out of exposing what he calls environmental terrorism. He even came up with a list of conference attendees, and a name similar to Unabomber suspect Theodore Kaczynski. He says he's turned the list over to the FBI, but refuses to show it to reporters. The organizers of the conference say the name does not appear on their sign up sheets."

Clausen's concealed list became a *cause celebre* among radical environmentalists, not only as an indicator of Barry Clausen's lack of credibility, but also as a refutation that Kaczynski was present at the Missoula conference. Alexander Cockburn and Jeffrey St. Clair wrote in The Nation:

> In the wake of Ross's first ABC segment, there were endless stories that a list of Missoula conferees, now supposedly in the hands of the F.B.I., included the name "T. Casinski." This was being put about by Clausen, who refused to show reporters the document in question. Tom Fullum and Jake Kreilick, who organized the conference, tell us they've been over the attendance rosters several times and have found no name even remotely resembling Kaczynski or Casinski.[35]

The list was one of the items Clausen had shown to Ross, Rummel and Koch before their taping at the Alexis. It was part of a forty-page report on every environmentalist who appeared on Jake Kreilick's Missoula meeting agenda. The last few pages of the report contained lists of other attendees, both those who registered and those who did not but were observed there. The "T. Casinski" list itself was a plain piece of paper with a list of names typed on it by Clausen, compiled from other lists he said were faxed to him anonymously. Sarah Koch said, "It had no letterhead, no source, nothing official."[36] ABC News did not use it.

Other media considered it newsworthy. Newsday made the "T. Casinski" list its lead in a major story.[37] Clausen remained convinced of its importance. He also believed that Kaczynski was at the Missoula meeting: Brian Ross's government source confirmed it.

FBI agents were seen checking locations in Missoula before and after Kaczynski's arrest. The FBI had maintained a presence in Missoula during radical environmentalist protests, logging equipment sabotage and tree spiking in the Nez Perce National Forest at the Cove-Mallard timber site from 1993 through 1995. That they would have compiled lists of potential Unabomber suspects attending a meeting of radical environmentalists in Missoula is unremarkable.

I received a call from CNN reporter Christine Sharp on April 10, 1996, asking if I knew why FBI agents were swarming over Missoula. I had heard rumors they were looking for a possible accomplice of Kaczynski's, but suggested she speak to Sherry Devlin, environmental reporter for The Missoulian. Curious, I called Devlin myself. She believed the agents were tracing Kaczynski's movements at the bus station and hotels—and possibly the university, but she had heard nothing about any accomplice.

Howard Berkes continued his All Things Considered report with a man who agreed with Clausen about environmentalist rhetoric: Jim Geisinger of the Northwest Forestry Association in Portland, Oregon.

> JIM GEISINGER: Well, I can't help but believe that is true. I mean, that the rhetoric that you referred to is so inflammatory in many instances and usually 90 percent wrong, that I could see where people who may be prone to be violent to begin with could be pushed over the edge by reading that kind of material.
>
> HOWARD BERKES: Geisinger has more than a passing interest in this—the FBI called him this week. His name appears in some handwritten notes found in Ted Kaczynski's cabin. "Be careful with your mail," he was told. Geisinger's group has been the target of an Internet campaign, urging people to disrupt the group's annual meeting. "Put your lives on the line," the e-mail urged.

A number of timber people got the same message from the FBI. David Ford, president of the Independent Forest Products Association, also based in Portland; Fibreboard Corporation of Walnut Creek, California, which sold 80,000 acres of California timberland to Sierra Pacific Industries; James Eisses, former executive vice president at Louisiana-Pacific Corporation, a company Earth First Journal frequently savaged in print; and the American Forest Resource Alliance, an industry group aimed at gaining grassroots support, which came in for special denunciation by radical environmentalists.[38]

With every media call that came in, I reinforced Geisinger's message: This is not just about Kaczynski and the Unabomber case. This is about apocalyptic beliefs, fatalism and desperate acts done by many underground radical environmentalists. This is about hate for civilization, environmentalism gone awry. Kaczynski is the burning glass that kindles our awareness of a greater threat to society.

Many reporters asked the obvious questions: Did the Unabomber act alone? Was the Freedom Club an actual terrorist group? Were radical environmentalists in on the bombings? Was it a conspiracy with environmental groups? Did environmental rhetoric push him over the edge?

My answer was always the same. Yes, there were others in the Freedom Club, but they didn't know it. Theodore Kaczynski clearly felt a solidarity with radical environmentalists and got substantial motive reinforcement from radical environmentalist literature. Many anarcho-environmentalists felt solidarity with the Unabomber after reading his manifesto. Many more shared his hate of industrial civilization.

Did he act alone? Probably—the truth is we don't know and perhaps never will. The FBI won't say whether they were looking for a possible accomplice in Missoula. Even if some environmentalist aided in the crimes, all Kaczynski's acts remain his own. No version of "the devil made me do it" will wash in a murder trial. The same would be true of anyone aiding and abetting.

My point was always the same. This is not to blame radical environmentalists as a whole for the Unabomber. This is to show that the apocalyptic beliefs shared by the Unabomber and radical environmentalists can be used to justify desperate acts by anyone—the preaching of hate for industrial civilization is an incitement to violence.

This is to declare that radical environmentalists have plenty of crimes of their own to answer for. Their apocalyptic beliefs, fatalism and hate for civilization are far more dangerous than the Unabomber.

Radical environmentalists have a First Amendment right to their extreme rhetoric, no matter how reprehensible, no matter who may use it to kill or maim or coerce or intimidate.

But the public has a similar right to scrutinize every word they say for its influence on criminal behavior.

And a right to prosecute every desperate act that flows from their extreme rhetoric and apocalyptic beliefs.

In the days that followed, commentators chimed in on the Unabomber-radical environmental link. Jeff Jacoby, Boston Globe staff columnist, asked, "Are Environmentalists Responsible for the Unabomber?"

> It would be absurd to blame decent environmentalists for the Unabomber's murders. Just as it would have been absurd to blame decent conservatives for the horror in Oklahoma City...
> Whoops. Did somebody say ... "double standard?"[39]

Cal Thomas, a syndicated columnist and former publicist for Jerry Falwell, recalled President Clinton and Vice President Gore denouncing right-wing talk show hosts such as G. Gordon Liddy for creating a "climate of hate" that produced the violence in Oklahoma City. Now the shoe was on the other foot. But why weren't the big newspapers doing big takeouts on Earth First and other radical environmental groups that had been the darlings of the left wing?

At this point The American Spectator magazine entered stage right with an embarrassing revelation: among the books found in Kaczynski's cabin was Vice President Al Gore's own *Earth In The Balance*. "Many sections were underlined in pencil, and there were copious notes in the margins."

That was too funny. I contacted managing editor Mark Carnegie to verify the report. The American Spectator stood by their story. It came from a reliable source. They had no reason to believe it was false. And why, the American Spectator wondered, wasn't Gore's among the handful of titles listed in press references to the books found in the cabin? The assumption was that it was suppressed to avoid embarrassing Gore and the administration.[40]

Payback time. It was all suddenly so clear: the Vice President of the United States had created a climate of hate for industrial civilization that produced the violent acts of the Unabomber suspect! Conservatives chuckled, chortled and horse-laughed at this nutty bit of comedy vérité.

Tony Snow smiled when he found out that what he had surmised in fun turned out to be true in fact.

Other insouciant items showed that America was coming to terms with its long nightmare: A "Unabomber for President" site appeared on the World Wide Web, with the campaign slogan, "Can I keep my car phone when we go back to nature?" A flier went out for a "Unabomber Benefit Concert" at Icky's Tea House in Eugene, Oregon, where anarchist John Zerzan gave a reading while punk bands played. Nobody much showed up.

Still, there was no solid link showing that Kaczynski used the Earth First Burson-Marsteller story to target Thomas Mosser, or the *Live Wild Or Die* Hit List to target Gil Murray. The documents were not listed among those found in Kaczynski's cabin.

The most that could be made of the link was best summarized in the words of Henry Lamb: kindred spirits.

The Los Angeles Times ran a haunting page one story that illuminated another facet of that kinship, headlined, "Adrift in Solitude."

> It had come to this.
> Sometimes he smelled. His hair was matted....
> He lived in a cabin. It was smaller than a lot of closets....
> He had no running water; he dipped plastic jugs into a stream 75 feet from the cabin. He had no electricity; he read by candlelight. He had no outhouse; he used the outdoors....
> He grew parsnips and potatoes, and he fertilized them with his own waste. He killed deer, coyotes, squirrels, rabbits and porcupines, and he broiled them over a fire in the yard....
> ...Theodore Kaczynski's life had come to this: the classic denouement for a person who kills with bombs. Someone involved in the case of Los Angeles' own Alphabet Bomber, who

like the Unabomber has slain three people, notes that controlling contact with the outside world is extremely important to these murderers, so important that they often remove themselves some-how from everything they cannot control—even if it means taking themselves out of society.[41]

Another eerie kinship with a radical environmental journal, the fourth edition of *Live Wild Or Die*, published in 1994, the year of the Missoula conference:

> If you are an active warrior, you should consider ending your public life and begin your private one.... A private life as an activist means not going to any public political rallies and demos, meetings or similar events. It means dropping out of a public existence as much as possible.
> In this private life many new issues will come up that are sometimes not so easily dealt with. This includes loneliness and isolation, especially if you make a complete break from traditional friends and family. This is why the idea of a tribe is so important. We need each other for our psychic and emotional well-being, to enable us to cope and survive. This break from those friends and family does become necessary when you realize that those who lead public lives often do not fully understand your security needs....
> You likely know someone who is involved in illegal actions. Help to maintain their security. How? By not asking them questions such as "Where are you going?" or "Where have you been?" By not talking about these people to your friends or strangers, casually, over the phone, or otherwise. By refusing to answer questions about them posed to you by anyone, either a close friend or the FBI. By opening your door to them when they need a place to crash, no questions asked. By offering them money when they come through town, because they need it, and their sources are few and far between.... Aside from stashing food and money, you might also want to consider acquiring a gun and lots of ammo. Personally, I have no use for such things in 1994, but who's to say what might happen in 5-10 years. Every year it's getting harder to get a weapon for personal defense, so the smart thing to do would be to get that stuff now while you still can and simply stash it in one of those army ammo boxes buried in a spot in a national forest or somewhere. I would suggest either a 9mm or 38 with as much ammo as you can afford....[42]

Among the items found in Kaczynski's cabin were five guns: a .25-caliber gun (Raven Arms), magazines with bullets; a bolt-action .22-caliber rifle; a Remington model .30-06 rifle; a .22-caliber black-handle revolver and nine rounds of ammunition; and a hand-made gun with spent cartridge.[43]

The closest thing to a big takeout by the big newspapers on radical environmentalists was a Washington Post article that asked, "But how plausible is it that the Unabomber was goaded by particular eco-fringe writings?" and then answered:

> It's entirely possible that the terrorist got ideas about the last two of his 26 victims — the head of a timber industry association and a public relations executive — from environmentalist sources. But a directory of associations could just as easily have served as his guide. As one federal official pointed out to the Dallas Morning News, the FBI has seized "zillions and zillions" of pages of notes and published material from Kaczynski's shack.

It was essentially a 1,663-word justification of the radical environmentalists, the reporter removing Kaczynski from them as far as possible:

> In another letter to the Times, which was also sent to The Post last June along with his infamous manifesto, the Unabomber took pains to correct a misimpression. The public, he wrote, shouldn't link environmental extremists or even anarchists to his deadly acts.
>
> "It's a safe bet that practically all of them disapprove of our bombings," he wrote. "Many radical environmentalists do engage in sabotage, but the overwhelming majority of them are opposed to violence against human beings."
>
> In other words: Don't blame them, blame me.

The Washington Post is not usually so eager to let people off the hook. Or so naïve. Kaczynski's Unabomber writings were full of sly misdirections and calculated manipulations. He has an IQ of 170, remember.

The "we" of FC was a loner's conceit not even the Post believed.

But consider his mention of "searching the sierras for a place isolated enough to test a bomb" in his April 1995 letter to the New York Times. He tested his bombs in Montana's Rockies, not California's Sierra Nevada, but knew the FBI thought he lived in Northern California and fed their illusion.

And his ruse in the same letter, "It's no fun having to spend all your evenings and weekends preparing dangerous mixtures," as if he had a day job.

He used symbolically meaningful aliases to register in hotels when he traveled, one of which was "Conrad," after an author he read many, many times over the years in his cabin: Joseph Conrad, the Polish-born novelist, author of "The Secret Agent," which derides science as a false idol. Anarchists in the novel use the initials "FP" for "Future of the Proletariat," in their leaflets, as Kaczynski used the initials "FC" for "Freedom Club" in his bombs. Coincidentally, Conrad's birth name before emigrating to England was Teodore Jozef Konrad Korzeniowski. Compare Theodore John Kaczynski.

And Kaczynski's assertion that he had correctly addressed the Gil Murray bomb to "California Forestry Association," not the old name of "Timber Association of California," as he insisted in his June 1995 letter to the New York Times. It was a flat lie. Did he fear that the old address would lead investigators to his source, the *Live Wild Or Die* Hit List?

And why did the FBI find pipe bombs in his cabin after his capture, when he made such a point in his June letter that "the majority of our bombs are no longer pipe bombs"? John Douglas, the retired FBI agent who developed the first psychological profile of the Unabomber, believes Kaczynski was going to send more bombs to the same type of targets, but change his signature to make investigators think it was a copycat. He clearly had no intention of honoring his promise to kill no more if his manifesto was published.[44]

And former FBI agent Clint Van Zandt noted that 10 years ago in a letter to his brother David, Ted Kaczynski wrote, "I hate to admit it, but as I believed I mentioned to you once before, I would be incapable of premeditatedly committing a crime." He protested his innocence so much that it leaped off the page at Van Zandt, telling him there's no reason to say that except that you are the Unabomber.[45]

However, on one significant occasion the Unabomber told an unvarnished truth that most investigators doubted. Jealousy that the Oklahoma City bombing upstaged him had nothing to do with his murder of Gil Murray. His deadly package was postmarked from Oakland on Thursday, April 20, 1995, some time before midnight. The bomb that destroyed the Alfred Murrah Federal Building in Oklahoma City exploded at 9:02 a.m. Central Time on Wednesday, April 19, 1995. Assuming that Kaczynski somehow heard about it the instant it went off, 8:02 a.m. Mountain Time— a neighbor would have to run and tell him, because he didn't own even a battery operated radio—he would have less than 40 hours and 58 minutes (allowing for the hour he would gain crossing into the Pacific time zone) in which to get to Oakland and drop that bomb in a mail box. The drive time alone by common carrier bus is about 32 hours. That leaves about 9 hours for him to 1) create everything needed for the trip, including four long letters and a neatly wrapped bomb; 2) hitch a ride to Helena; 3) catch the next Trailways bus to Missoula; 4) transfer to another Trailways bus to Salt Lake City; 5) transfer to a Greyhound bus to Oakland; 6) run to a mail box and drop his five items so they would be collected before the postmark stamp changed to Friday, April 21, 1995. No, Kaczynski was telling the truth. He was already on his way to Oakland when the Oklahoma City bomb went off.

Once he mailed those items in Oakland, though, he traveled back to Sacramento to hang out until his bomb went off. Frank Hensley, the day clerk at the century-old Royal Hotel in Sacramento, places Kaczynski there around mid-April. At a Burger King restaurant next to the Sacramento bus depot, manager Mike Singh said he also saw Kaczynski.

On Sunday morning, the day before Gil Murray was killed, the answering machine of the Association of California Insurance Companies, an industry group at 1121 L Street in downtown Sacramento—a few blocks from the bombing site—received a message from a man speaking in a gravelly, strained voice: "Hi. I'm the Unabomber, and I just called to say 'Hi.'"[46]

Donn Zea has heard several people say they thought they saw a scruffy looking person fitting Kaczynski's description in the crowd after the bomb killed Gil Murray, but nobody at the time knew who they were looking for.

Kaczynski could have caught the next bus for Salt Lake City, made his transfers to Montana and got safely back to his cabin above Lincoln near the corner of Stemple and Humbug less than a week after he left. No one would even know he had been gone.

The Washington Post's apologetics on behalf of radical environmentalists held up one intractable fact about the "Eco-Fucker Hit List" and the Earth First Burson-Marsteller story:

> As for whether this or any other relevant eco-list was in the paper hoard of the suspected Unabomber, only the FBI knows for sure.

Two questions: Was the *Live Wild Or Die* "Eco-Fucker Hit List" in Kaczynski's cabin? Was the Earth First Journal containing the erroneous Burson-Marsteller story in Kaczynski's cabin?

Kathleen Marquardt and I vowed to find out. It was not a matter of criminal evidence—although the two documents would help in establishing how Kaczynski selected his last two targets, his personal journals would supersede these two items in importance. It was a matter of proving decisively the linkage between Earth First writings and the Unabomber.

Establishing that linkage would be important because it would show beyond question the correlation between their philosophies. But correlation is not causation. Earth First writings did not cause the Unabomber to kill. Absent those writings he would have killed anyway, but he might not have killed Thomas Mosser and Gil Murray. If such a linkage actually existed, it would show the pernicious reinforcement Earth First gave to the Unabomber's own pre-existing fanatical fatalism.

I pressed all my sources close to the case to determine whether or not the two documents were in Kaczynski's paper hoard. Nothing. I even asked the journalists who confirmed that Al Gore's book was in Kaczynski's cabin to press their source. No luck. I asked friends with high inside contacts to inquire on my behalf. One answer came back: "Wait until the trial." Kathleen Marquardt pursued her sources close to the case. No one working on the case would even return her calls.

There had to be a way to get the answer to those two simple questions. But we could not think how. Between the two of us we had put out perhaps fifty quiet inquiries. Nothing came back.

Until one day a source appeared. It was a surprising source, but clearly had full knowledge of the contents of Kaczynski's cabin.

"You wanted the answers to two questions."

"Yes."

"I will deny it if you tell anyone where you got this information, but the answer to both of your questions is yes."

"Have you personally seen both documents in the Kaczynski materials?"

"Yes."

"Both documents?"

"Yes."

"There is no doubt?"

"There is no doubt."

Chapter Three Footnotes

[1] Telephone interview with Prof. Reksten, September 4, 1996. Telephone interview with Steve Adams, September 7, 1996. Telephone interview with Bruce Ely, September 4, 1996.

[2] All four students obtained valuable first photographs of Kaczynski's capture. They contracted with agency Gamma-Liaison in New York and pooled their earnings, each receiving one-fourth of the income.

[3] ABC News Transcript #6067, *World News Tonight with Peter Jennings, EST Edition*, April 3, 1996, pp. 2-3.

[4] Affidavit of Special Agent Donald J. Sachtleben, FBI, to U.S. District Judge Charles C. Lovell, United States District Court, Helena Division, District of Montana, Criminal Complaint in the case of *United States of America v. Theodore John Kaczynski*, filed 96 APR 4 AM 10 58, p. 1 of 4.

[5] Interview with Barry Clausen in his office, May 30, 1996.

[6] ABC News Transcript #6069, *World News Tonight with Peter Jennings, EST Edition*, April 5, 1996, p. 1.

[7] Lane County Sheriff's Office, mugshot profile, Hemstreet, Leslie Anne, event #940911, dated 8/19/1996, Court records of recognizance release, initial plea of not guilty, trial plea of no contest, conviction and sentencing, case number 96-21634A.

[8] "Federal agents searching for vandals," *Eugene Register-Guard*, Wednesday, October 30, 1996, p. A1.

[9] Deposition of John Kreilick, *Highland Enterprises, Inc., v. Earth First! et al.*, Case No. CV-28511. District Court of the Second Judicial District, State of Idaho, County of Idaho, February 15, 1994, pp. 21-23

[10] "NFN 2nd International Temperate Forest Conference Schedule," Internet posting by Jake Kreilick in Earth First bulletin board cdp:ef.general, 4:54 PM Oct 17, 1994.

[11] "The Unabomber Case: Logging, Mining Issues in Montana Raise Concerns About Kaczynski Aftermath: Some wonder if collisions between nature, technology got Unabomber suspect's attention. Environmental controversies hit close to his home," *Los Angeles Times*, Saturday April 6, 1996, by Kim Murphy, p. A12.

[12] "An Open Letter to ABC Network News from the Earth First! Journal," written 12:37 AM April 7, 1996 by Earth First staff, Eugene, Oregon, Internet posting at cdp:ef.general, 3 print pages. Also posted on Earth First World Wide Web site.

[13] "Acquiring organizational legitimacy through illegitimate actions: a marriage of institutional and impression management theories," *Academy of Management Journal*, Oct 1992 v35 n4 p699(40), by Kimberly D. Elsbach and Robert I. Sutton.

[14] "When a Utah Animal-Rights Activist Told Authorities Of a Suspect in a Bombing by a More Radical Group Misery Was Her Only Reward; 'Reward' For Tipster Is A Strip-Search," *Salt Lake City Tribune*, Monday, January 13, 1997, by Stephen Hunt, p. A1. Activist Anne Davis received a warning saying, "You talk, you die." Also, interview February 24, 1997, with Joanna Logan, former newsletter editor (1987) for Northwest Animal Rights Network, Seattle. Logan said six members claimed hearing others in the group discuss threats and intimidation toward dissidents and others.

[15] Deposition of John Kreilick, *Highland Enterprises, Inc., v. Earth First! et al.*, Case No. CV-28511. District Court of the Second Judicial District, State of Idaho, County of Idaho, February 15, 1994, p. 43.

[16] *Ibid.*, p. 47-48.

[17] *Ibid.*, p. 56.

[18] *Ibid.*, p. 62.

[19] *Ibid.*, p. 92, 97. Kreilick used the pseudonym Jake Jagoff.

[20] Martha Lee, *Earth First! Environmental Apocalypse*, p. 125. See also Greg King, "Redwood Action Team Report," *Earth First* vol. 9, no. 4 (Eostar - March 21, 1989), p. 19.

[21] "Whatcom County's activists take crusades far afield," *Bellingham Herald*, Monday, September 18, 1989, by Leo Mullen, p. A5.

[22] "Madman or Eco-Maniac? Conspiracy Theorists See Environmental 'Hit List' as Unabomber Fodder," *The Washington Post*, April 17, 1996, by Richard Leiby, p. C1. The story contains an error: in fact, the "Eco-Fucker Hit List" appeared in the second issue of *Live Wild Or Die* in 1990. The first issue was published in 1989.

[23] Deposition of Erik Ryberg, *Highland Enterprises, Inc., v. Earth First! et al.*, Case No. CV-28511. District Court of the Second Judicial District, State of Idaho, County of Idaho, April 5 and 6, 1994, p. 126.

[24] Clausen provided me with the names, which included an individual under investigation for participating in both the 1990 Santa Cruz power line sabotage and for the 1987 $5.1 million arson at the University of California at Davis, Animal Diagnostics Laboratory.

[25] Both memos are in Freedom of Information Act request file FOIPA No. 394025/190-HQ-11299669, dated May 21, 1996.

[26] "'Hit List' Had Unabomber Targets - Terrorist's Possible Link To Underground Newspaper Probed By FBI," *Sacramento Bee*, August 3, 1995, by Cynthia Hubert and Patrick Hoge, p. A1.

[27] Telephone conversation with ABC News producer Sarah Koch, September 6, 1996.

[28] *ABC News This Week with David Brinkley*, Transcript #754, April 7, 1996, p. 1-2.

[29] "One Focus of Inquiry: The Selection of Targets," *The New York Times*, April 8, 1996, by Neil MacFarquhar, p. B8.

[30] "The International PR Machine: Environmentalism á la Burson-Marsteller," *Earth First Journal*, vol. 14, no. 3, Brigid, February-March 1994, by Carmelo Ruiz-Marrero, p. 9.

[31] Biographical essay on Murray Bookchin by Janet Biehl, in Anarchist Archives at http://www.miyazaki-mic.ac.ip/faculty/dward/ANARCHIST_AR-CHIVES/archivehome.html.

[32] Telephone interview, September 13, 1996. Mr. Hagener established the date of the meeting in his scheduling log.

[33] Telephone interviews September 9 and 13, 1996. Mr. Brown established the dates of his contacts with Kaczynski in his appointment books.

[34] *World News Tonight with Peter Jennings*, ABC News Transcript #6073, April 11, 1996, EST Edition, p. 3.

[35] "Kaczynski May Have Followed Environmentalist Rhetoric," *All Things Considered* (NPR) Transcript 2181, April 12, 1996, Segment #14, pp.15-16.

[36] "Earth First!, the Press and the Unabomber," *The Nation*, May 6, 1996, by Alexander Cockburn in his bi-monthly column, Beat The Devil, this column written with Jeffrey St. Clair, p. 9.

[37] Telephone conversation with ABC News producer Sarah Koch, September 6, 1996.

[38] "Hit by a List / Matching up targets and attendees," *Newsday*, Wednesday April 10, 1996, by Stephanie Saul and Knut Royce; special correspondent Jane Meredith Adams in Sacramento contributed to this story, p. A7.

[39] "Suspect's List of 70 Names – Corporations, UC Professors, Scientists on Alert," *San Francisco Examiner*, Friday, April 12, 1996, by Seth Rosenfeld, p. A1.

[40] "Are Environmentalists Responsible for the Unabomber?" *Boston Globe*, Tuesday, April 23, 1996, by Jeff Jacoby, Op-Ed Page, p. 15.

[41] "On the Prowl," *American Spectator*, June 1996, p. 15.

[42] "Adrift in Solitude, Kaczynski Traveled a Lonely Journey," *Los Angeles Times*, Sunday, April 14, 1996, by Times Staff Writers, p. 1.

[43] "What's It Gonna Take?" reprinted with [anonymous] author's permission from "Live Wild Or Die!" #4 (1994), posted on the World Wide Web page, "Animal Liberation Frontline Information Service," http://envirolink.org/ALF/articles/ug/ug2_19.html.

[44] "Tools, weapons among items in Kaczynski's cabin" *USA Today Online*, posted June 5, 1996.

[45] "ABC News This Week With David Brinkley," April 7, 1996, Transcript #754, p. 3.

[46] "World News Tonight With Peter Jennings," April 8, 1996, Transcript #6070, p. 1.

[47] "Unabom Probers Study Clues - Voice Message, Letters May Point To Terrorist," *Sacramento Bee*, Wednesday, April 26, 1995, by Cynthia Hubert and Patrick Hoge, p. A1.

Chapter Four
TERRORISTS

1:16 P.M. TUESDAY, FEBRUARY 13, 1996   *Bellevue, Washington*
FOUR STACKS OF DOCUMENTS sat on my desk:

One, a heap of manila folders stuffed thick with reports of "monkeywrenching" crimes, mostly arsons, equipment sabotage and livestock shootings, with a few bombings and conspiracies sticking out here and there.

Two, a pile of news clippings, police reports and federal documents on animal rights crimes, running heavily to bombings, massive arsons, personal assaults, theft of research animals and data, with cases of sea piracy and ship sinkings.

Three, a mound of papers on coercive protest demonstrations that resulted in criminal arrests and public costs in lost employment, lost production and extra law enforcement.

And four, a mountain of death threats, harassment incidents and unsolved crimes against persons ranging from murder to assault.

Over a thousand incidents in all.

The crimes had one motive in common: they were all done to save nature.

Ecoterror.

The crimes had one target in common: they were all aimed at those who extract or convert nature's resources into products that foster human beings.

Ecoterror victims.

103

The crimes had such an overwhelming variety of perpetrators they defied classification.

Ecoterrorists. Lots of them. Hundreds, perhaps thousands.

Questions.

I called the noted law professor Brent L. Smith, author of *Terrorism In America,* because he had personally analyzed the records of 170 individuals indicted for domestic terrorism or terrorism-related activities—he had first-hand knowledge of most of the terrorist investigations in America during the 1980s, including seven acts of ecoterrorism. He agreed to explain his work to me.

Brent Smith's gift is insight, seeing what it all means by carefully examining the details. He actually traveled to the places where most of the 170 terrorist suspects had been tried. He plowed through nearly all the court records. From this staggering mass of empirical information, he was able to separate out the major types of terrorists, finding that Left-Wing terrorists, Right-Wing terrorists, and Special Interest terrorists (ecoterrorists) each had distinctive characteristics.

Smith is also able to explain what he found with engaging simplicity, as in this table from his book (see opposite).

By tracing the personal histories of indicted left- and right-wing terrorists, Smith found other differences:

Left-wing terrorists were younger, an average age of 35 when indicted (only 18% were over 40); 27% were female; most (71%) were minorities; more than half (54%) had college degrees and included many professionals such as physicians, attorneys, teachers and social workers.

Right-wing terrorists, by contrast, were older, an average of 39 when indicted (36% over 40); 93% were male; virtually all were white (3% were Native American); only 12% were college educated and most were unemployed or impoverished self-employed workers.

It made a difference in the outcome of cases. Every right-wing terrorist in the study was apprehended (one was killed resisting arrest) when Smith's study went to press in 1994; a number of left-wing terrorists remained fugitive, some believed to have received sanctuary in Cuba.

Are ecoterrorists left-wing, as most wise use movement advocates think? Smith's study found little publicly available evidence linking the extreme Left to the single-issue environmental activists who find terrorism a viable weapon. "It may very well be that animal rights and ecoterrorists are politically leftist," he wrote, but at the time he finished his study he couldn't say positively. Is Earth First co-founder Dave Foreman, a "right-wing thug," as fellow co-founder Mike Roselle once called him?[3]

Professor Smith recommends that our understanding of ecoterrorists should not be clouded by the political rhetoric of the Left or Right: "The activities of special interest terrorists in America are analyzed best by accepting them at face value—as attempts to change *one* aspect of the social or political arena through terrorism."[4]

CHARACTERISTICS OF LEFT-WING AND RIGHT-WING
TERRORIST GROUPS IN AMERICA

	TYPE OF GROUP	
Characteristic	Left-Wing	Right-Wing
Ideology	Political focus; primarily Marxism[1]	Religious focus; ties to Christian Identity Movement[2]
Economic Views	Pro-communist/socialist; belief in Marxist maxim "receive according to one's need"	Strongly anti-communist; belief in Protestant work ethic, distributive justice
Base of Operations	Urban areas	Rural areas
Tactical Approach	Cellular structure; use of safehouses	National networking; camps and compounds
Targets	For funding; armored trucks preferred	For funding: armored trucks preferred
	Terrorist targets: seats of capitalism/government buildings	Terrorist targets: federal law enforcement agenices; opposing racial or religious groups

Reprinted from *Terrorism in America: Pipe Bombs and Pipe Dreams* by Brent L. Smith
by permission of the State University of New York Press © 1994

If saving nature is what ecoterrorists say they're doing, then that's how we will eventually understand them, not by tagging them as left-wing or right-wing. In ecoterrorism, we are looking at something distinctive. "The 1980s...witnessed the first indictments and trials of environmental terrorists," wrote Brent Smith. "Environmental extremists added monkey-wrenching to the repertoire of terrorist tactics."[5]

From his office at the University of Alabama in Birmingham, Professor Smith told me the key problems law enforcement has with terrorism in general and ecoterrorism in particular.

"The FBI's official definition of terrorism contains essentially two halves," he said, "the first defining the criminal acts, the second defining the motive."

Officially, terrorism is:

> the unlawful use of force or violence, committed by a group(s) or two or more individuals, against persons or property to intimidate or coerce a government, the civilian population, or any segment thereof, in furtherance of political or social objectives.[6]

I told Professor Smith that the bureaucratic language sounded as if it meant something it wasn't saying.

Not really, said Professor Smith, but it took some explaining. "The use of the term *unlawful* restricts the application to criminal conduct, not political or social motivation. You can't criminalize social or political goals—they are protected by the First Amendment to the Constitution of the United States."

Therefore, he said, suspects arrested and indicted for acts of terror are not formally charged with "terrorism." No federal crime called "terrorism" exists.

What the FBI really does is to first establish a criminal investigation, then use the motivation of the perpetrator to determine the *intensity* of that investigation. Prosecutors charge indicted terrorists with a laundry list of common crimes such as arson, illegal possession of explosives, armed robbery or murder—whatever the evidence of the case shows. The courtroom result is longer sentences for those convicted of common crimes that have social or political motivation.

Okay, then, back at the investigation stage, how does the FBI separate terrorists from common criminals?

Professor Smith referred me to the 1989 "Attorney General's Guidelines," which made minor revisions to those put in place in 1983 by U. S. Attorney General William French Smith, which replaced earlier guidelines.[7] Actual domestic terrorism investigations are carried out under their provisions. Four characteristics are currently used to identify terrorists:

1. *Use of violence*—The suspected group must endorse and use "activities that involve force or violence." Nonviolent dissident political groups do not qualify. However, under the Guidelines, the violence does not have to actually occur, but it is necessary that the suspected group endorse activities that involve force or violence.[8]
2. *Political motivation*—The 1983 Guidelines combined criminal enterprises and terrorism in the same set of directives. Thus, a criminal investigation may be started before a political motive is discovered. If a political motivation is found, investigators proceed under the Guidelines' Domestic Security/Terrorism Subsection rather than their Racketeering Subsection. This

makes a big difference in the investigation. Terrorism investigations may remain open even if a group "has not engaged in recent acts of violence, nor is there any immediate threat of harm—yet the composition, goals and prior history of the group suggests the need for continuing federal interest."[9]

3. *Focus on groups*—Terrorism investigations are "concerned with the investigation of entire enterprises, rather than individual participants." A terrorism investigation may not be initiated unless "circumstances indicate that two or more persons are engaged in an enterprise for the purpose of furthering political or social goals...that involve force or violence and a violation of the criminal laws of the United States." In practice, the FBI concentrates on organizers and leaders—the "brains." Destroying the organization that spawns violence is more effective than merely convicting terrorists that happen to get caught committing a crime. The idea is to "decapitate" the leadership of terrorist organizations for "early interdiction of unlawful violent activity."[10]

4. *Claimed responsibility*—Crimes will generally not be designated as terrorism unless a terrorist group claims responsibility or the FBI can positively identify such a group as responsible. Distinguishing terrorist acts from common crimes can be difficult, especially when investigators know that most mail bombs are sent because of a love triangle and that most acts of industrial vandalism are the work of disgruntled former employees. Acts of sabotage can turn out to be insurance scams by the owner of the sabotaged property. Ecoterrorist claims of responsibility can be faked. Each case requires all the investigative skill law enforcement can bring to bear.

I found this explanation useful for understanding what happens in the real world. The Unabomber, for example, was not investigated as a terrorist because the FBI profile said he was a loner, yet everybody considered him to be a terrorist. The same was true of the bombing that killed federal judge Robert Vance in 1989, because the assassin, Walter Moody, acted independently of the influence of any organization. Bombings of abortion clinics likewise appear to be the work of individual criminals without the conspiratorial support of others, but common sense tells us they are terrorist acts.

Cases of monkeywrenching in which no one claims responsibility—sabotage of logging sites in Maine or torching a bulldozer in Illinois or a cattle shooting in New Mexico—likewise are not officially classified as terrorism, which has upset and embittered many victims.

Professor Smith told me that it sometimes takes the FBI a long time to officially designate a crime as an act of terrorism, particularly ecoterrorism. He pointed me to a passage in his book:

> Extremist members of Earth First, founded by David
> Foreman, and the Animal Liberation Front (ALF) accounted
> for all seven acts of terrorism attributed to environmental groups
> in the United States [during the 1980s]. They committed their
> first officially designated acts of terror in 1986 and 1987. It
> was not until 1988, however, that the FBI officially recognized
> these crimes as terrorist incidents and reclassified them as such.
> So little was known about the Evan Mecham Eco-Terrorist In-
> ternational Conspiracy (EMETIC) that it took two years for the
> 1986 sabotage at the Palo Verde Nuclear Generating Station,
> near Phoenix, and the 1987 vandalism of the Fairfield Ski Bowl,
> near Flagstaff, Arizona, to be recognized as acts of an orga-
> nized group planning additional sabotage to nuclear facilities
> in the Southwest. Similarly, the 1987 arson of a veterinary
> research facility at the University of California at Davis by
> members of the ALF was not officially designated until its pat-
> tern of destruction continued over the next three years.[11]

The problem with the official definition is simple: average people
consider many crimes to be terrorist acts that the FBI does not count in its
terrorism statistics.[12]

To compound the disparity between public perception and FBI
policy, political officials in charge of federal law enforcement in recent
years have ordered field agents not to investigate specific ecoterrorist
groups and suspects.[13] The Attorney General Guidelines give the government that flexibility. Even though the "goals and prior history of
the group suggests the need for continuing federal interest," the agency
may decide to place its emphasis elsewhere.

Internal problems have made the investigation of ecoterrorist
suspects even more difficult. In early May of 1995, U. S. Attorney
General Janet Reno convened a quiet, unpublicized meeting of representatives of the Bureau of Alcohol, Tobacco and Firearms, the Bureau
of Land Management, the Forest Service, the U. S. Marshals Service,
the FBI and other agencies that have been involved in questionable
raids and confrontations in the West.[14]

The reason for the meeting was that untrained personnel in many
of these agencies have been acting more like out-of-control cowboys than
the cool professionals government publicists describe. The public line is
that only right-wing elements are threatening federal employees, but pro-
vocative behavior by environmental radicals is behind numerous incidents.
In addition, specific federal employees have sparked confrontations with
loggers, ranchers, motorized recreationists and others, as is well known
within the agencies. However, to investigate environmental radicals would

dilute the political intent of the public line. Some of the worst trouble-prone federal personnel were being assigned to distant locations, but the problem remained unsolved.

The solutions discussed at the meeting were designed to discourage the confrontational attitude of some of these agencies, particularly the BLM and ATF. Senior Justice Department officials suggested that one of the ways to do this would be to put some of the worst actors through rigorous FBI training in how to avoid confrontations and resolve them peacefully when they occur.

However, the FBI has declined to train troublesome employees of other agencies.

To confuse matters further, left-wing environmentalist Jeffrey DeBonis formed two organizations with official sounding names, "Association of Forest Service Employees for Environmental Ethics" in 1990—which he reportedly left because the directors became alarmed at his intemperate accusations against opponents—and "Public Employees for Environmental Responsibility" in 1993—where he is now executive director—whose main function is to propagate the public line that only right-wing elements are a threat to federal employees. DeBonis has been funded by an array of foundations with long-term left-wing and anti-wise use agendas.[15]

So: Less than half the incidents in those four stacks of documents on my desk fit the official definition of terrorism—about 600 of more than 1,300 incidents—and only a handful were actually designated terrorism. Yet each act evoked real terror in the hearts of its victims.

9:05 A.M. THURSDAY, FEBRUARY 15, 1996   *Englewood, Colorado*
"HAVE YOU SEEN THE ANIMAL LIBERATION FRONT'S WEBSITE?" asked Tom McDonnell.

"I didn't know terrorist groups had websites."

"This one does."

"What's in it?"

"You won't believe it."

He gave me the URL and I went surfing.

Netscape took me to the EnviroLink website, which first came online in 1991, just a mailing list of 20 student activists set up by environmentalist Joshua Knauer while he was a freshman at Carnegie Mellon University. Since that time, EnviroLink has grown into a big non-profit organization and one of the Web's largest environmental information clearinghouses. I found the link to the Animal Liberation Front, a group the FBI had officially designated a terrorist organization.

The page name in the blue line at the top of my screen said, "Animal Liberation Frontline Information Service." I read the disclaimer:

The Animal Liberation Frontline Information Service is NOT itself part of any existing Animal Liberation Front Supporters Group and is not intended to replace their work. The Information Service exists in the interest of; free speech, freedom of information and public interest. The Information Service does not exist to incite people to commit any immoral and illegal acts.

I found a link to a page called "North America — Diary Of Actions '92" and followed it.[16]

It was a list of 26 crimes committed in 1992 by animal rights activists—times, places, names of groups claiming responsibility. It said:

1. 1/1/92-Edmonton, Alberta; Ouellette Packing Plant had slogans spray painted and splashed with paint. A van was spray painted, had its tires slashed, and an incendiary device left on the seat which failed to ignite. -A.L.F. (Animal Liberation Front).
2. 1/1/92-Alberta; Animal Rights Militia claims to have poisoned 87 Canadian Cold Buster Chocolate bars in Edmonton and Calgary because of the University of Alberta vivisector Larry Wang's 16 years of experiments on rats that led to the invention of the bar. Production halted and bars pulled from store shelves across Canada, $250,000 damage. -A.R.M.
3. 1/4-7/92-Calgary, Alberta; Saks Furs had windows smashed, Rupps Meats had windows smashed and spray painted, three Kentucky Fried Chicken shops spray painted, fur shop on 17th Ave. had windows etched, fur shop on 4th St. had windows etched, fish shop had windows etched and spray painted, fur shop had windows smashed, and a butcher on McLeod Trail spray painted. -A.L.F.
4. 1/8/92-Edmonton, Alberta; A delivery truck of Ouellette Packers had its tires slashed. -A.L.F.
5. 1/9/92-Edmonton, Alberta; Billingsgate Fish had three of their rental replacement trucks spray painted and 18 tires slashed. -A.L.F.
6. 1/13/92-Walnut Creek, California; Etching fluid thrown on fur shop windows. -A.L.F.
7. 2/6/92-Calgary, Alberta; Six fur stores including RC International Furriers were spray painted and had their locks glued. -A.L.F.
8. 2/27/92-East Lansing, Michigan; Michigan State University Experimental Fur Farm, files taken, two mink liberated and later released. An incendiary device was left setting fire to the offices, $200,000 damage. -A.L.F.
9. 4/24/92-Vancouver, British Columbia; A University of British Columbia vivisector Dr. Fibiger has his house spray painted with slogans and threats. -A.R.M.
10. 6/92-Memphis, Tennessee; Fur stores had windows broken and spray painted.

11. 6/92-South Carolina; Fur store damaged. -Vegan Front.
12. 6/1/92-Edmonton, Alberta; University of Alberta, Ellerslie Research Station, 29 cats liberated and $100,000 damage done, documents taken. -A.L.F.
13. 7/28/92-Memphis, Tennessee; TMX Fur store had a truck spray painted with slogans, tires slashed and locks glued shut, $934.58 damage.
14. 8/4/92-Memphis, Tennessee; TMX Fur store was spray painted with slogans, windows broken and locks glued shut, $2,994.74 damage.
15. 8/5/92-Memphis, Tennessee; Motes Furs was damaged with paint bombs and spray painted slogans $3,086.65 damage.
16. 8/9/92-Memphis, Tennessee; JP Holloway Furriers had their rolldown shutters spray painted with slogans, $800 damage.
17. 10/92-Minneapolis, Minnesota; Simeks **Meats** and Seafood had its locks glued and spray painted with slogans. -A.L.F.
18. 10/11/92-Minneapolis, Minnesota; Swanson Meats trucks spray painted with slogans and windshields broken. -A.L.F.
19. 10/24/92-Logan, Utah; USDA Predator Research Station, 29 coyotes released and one building set on fire. Slogans spray painted, $500,000 damage. -A.L.F.
20. 10/24/92-Millville, Utah; Frederick Knowlton's office on campus also entered files taken and a fire started, $10,000 damage. -A.L.F.
21. 11/8/92-Minneapolis, Minnesota; Swanson Meats Inc. has five trucks spray painted with slogans including "Meat Is Murder" and set on fire, $100,000 damage. The building also had its locks glued. -A.L.F.
22. 12/15/92-Miami, Florida; Two fur stores spray painted with "Fur Shame" and "P.P.". -P.P. (Paint Panthers).
23. 12/17/92-Denver, Colorado; Lloyds Furs, Irv Ringler Furs and Marks Furs spray painted with slogans "Fur Kills" and "Paint Panthers" and damaged with paint bombs. -P.P.
24. 12/18/92-Aspen, Vail, Breckenridge, Keystone, Denver, Colorado; 30 fur coats damaged after having red paint squirted on them. -P.P.
25. 12/20/92-Washington, DC; Saks, Jandel Furs, Furs of Kiszely, Miller Furs, and Roendorf Evans Furs spray painted with slogans "Fur Shame", "Fur Scum", "Blood $" and "P.P." and damaged with paint bombs. -P.P.
26. 12/25/92-Victoria, BC; McDonalds on Pandora Ave. had its windows smashed and was spray painted. -A.L.F.

The posting was neither hoax nor prank. I recalled some of these incidents from nationwide media coverage. I checked out each one in news files. Aside from a few technical quibbles (the "Roendorf Evans Furs" was Rosendorf-Evans, and most "Washington, DC" fur outlets were actually in suburban Maryland or Virginia) and except for a few damage-cost estimates that appeared to be guesswork, the page was correct.

I looked to see if a 1993 page had been posted. It had. Fifty-eight more diary entries.

1. 1/93-Guelph, Ontario; Red Lobster Restaurant; windows smashed, "Killing the oceans to feed your greed", "Meat is Murder", "Red Lobster - Rapists of the sea" painted on building. -A.L.F.
2. 1/93-Ottawa, Ontario; Furs stores have had their windows smashed. -A.L.F.
3. 1/93-Los Angeles, California; Fur Store has its locks glued and spray painted.
4. 1/1/93-Victoria, British Columbia; Williams Quality Meats had its lock glued.-A.L.F.
5. 1/13/93-Cleveland, Ohio; Cikra Furs splashed with paint and slogans spray painted "Fur Kills", "Scum", and "P.P." on windows. -P.P.
6. 2/93-Washington, DC; Five fur stores damaged by paint. -P.P.
7. 2/7/93-New York City; The Fur Vault, Ritz Thrift Shop, Elizabeth Arden, Bloomingdales, Fendi Bergdorf Goodman, and Harold J Rubin Furs spray painted with slogans "Fur Scum", "Blood Money", "Murderers" and "P.P." and splashed with paint. -P.P.
8. 3/26/93-Oakland, California; Butcher shop spray painted with slogans. -A.L.F.
9. 4/93-Berkeley/Oakland, California; A large number of animal abuse billboards spray painted with slogans such as "Go Vegan!", "Vegan Power", and "Animal Torture".
10. 4/14/93-Bethesda, Maryland; Three McDonalds, Honey Baked Ham, and two Kentucky Fried Chickens were spray painted with slogans "Meat Is Murder", "Don't Be a Fowl Mouth", "Pig Killers", and "Blood $". -Meat Free Mission
11. 4/21/93-Bethesda, Maryland; McDonald's, Honey Baked Ham and three other stores were spray painted. -Meat Free Mission.
12. 4/27/93-Montgomery County, Maryland; Five vivisectors had their homes and cars spray painted with slogans. "Animal Torturer" and "Animal Killer". $5000 Reward offered by Americans for Medical Progress. -Animal Avengers.
13. 5/93-Berkeley/Oakland, California; More animal abuse billboards spray painted with slogans.
14. 5/4/93-Montreal, Quebec; Paradise Furs and another fur store were spray painted with slogans.-P.P.
15. 5/5/93-St. Paul, Minnesota; Chainlink fences cut at the Como Zoo to release three timber wolves, but they didn't leave there [sic] pen. -Organization for the Liberation of the Animals.
16. 5/24/93-USA; Letters threatening "bloodletting" sent to 3 vivisectors and 2 consumer products manufacturers. - Animal Lib-

eration Action Foundation.

17. 5/26/93-San Francisco, California; Three chicken restaurants, two butchers, and two leather shops spray painted with slogans and locks glued shut. -A.L.F.

18. 5/31/93-Baltimore, Maryland; Johns Hopkins University School of Public Health, 10 rats, 5 dogs, and 3 cats liberated. -Students Against In Vivo Experiments and Dissection (SAVED).

19. 6/93-Memphis, Tennessee; Two fur shops damaged. -A.L.F.

20. 6/10/93-Memphis, Tennessee; Hathaways Taxidermy and Mid-Town Meats spray painted with slogans and windows smashed. -A.L.F.

21. 6/10/93-Memphis, Tennessee; The home of Gilbert Kirschner owner of Gilbert Kay Furs had his home spray painted with slogans, locks glued, and splashed with paint. -A.L.F.

22. 6/22/93-Marin County, California; Geneticist injured by mailbomb. -Unabomber.

23. 6/24/93-Grosse Pointe, Illinois; Lee's Fashion and Furs and two other fur shops spray painted with slogans, posters glued to windows and locks glued shut. -Vegan Action League.

24. 7/93-Manchester, Tennessee; A meat market was spray painted with slogans "ALF". An empty stockyard was set on fire and burned to the ground $100,000 damage. -A.L.F.

25. 7/93-Memphis, Tennessee; An unsuccessful arson attack on Jack Lewis Furs, the store was spray painted with slogans. The store had a new $3000 security camera, but proved worthless as the attackers were well disguised. It did record the time the attack was made. -A.L.F.

26. 7/93-L'Oreal and Gillette receive threatening damage if all animal experiments are not stopped.-Animal Liberation Action Foundation.

27. 7/6/93-Maryland; Fake bombs left on Lawrence Cunnick, Sharon L. Juliano and two other vivisectors' doorsteps, each contained a threatening letter, a brick, a rubber rat, and a fuzzy bear slipper. -Unclaimed.

28. 7/9/93-Albany, California; A Kentucky Fried Chicken had all its locks glued, slogans spray painted on three walls, and all its windows etched with acid. -A.L.F.

29. 7/16/93-East Lansing, Michigan; Rod Coronado indicted (in his absence) on five counts; malicious destruction of property, interstate travel to commit arson, extortion, arson, and interstate shipment of stolen research materials, all related to the 2/27/92 Michigan State University Experimental Fur Farm raid.

30. 7/16/93-San Francisco, California; City Meats had its front lock glued and slogans spray painted, a Burger King had its front lock glued shut. -A.L.F.

31. 7/22/93-Corvallis, Oregon; Oregon State University at Corvallis Mink Center shuts down.

32. 7/23/93-Berkeley, California; University of California at Berkeley, slogans spray painted calling an unnamed vivisector a "Killer".

33. 7/30/93-Bloomington, Indiana; Sims Poultry Company has there [*sic*] building and four trucks spray painted with slogans "Meat is Murder", "Meat is Death" and "ALF" and their refrigerator units unplugged. -A.L.F.

34. 7/31/93-San Francisco, California; Two trucks at Roberts Corned Beef had all their locks glued and slogans spray painted. United Meats had a back fence was [*sic*] cut through and three trucks had all locks glued and spray painted with slogans. -A.L.F.

35. 8/93-Madison, Wisconsin; University of Wisconsin, the A.L.F. broke through a door at the building housing all campus vivisection offices and set files on fire. -A.L.F.

36. 8/93-Milwaukee, Wisconsin; A meat delivery van set on fire. -A.L.F.

37. 8/93-Milwaukee, Wisconsin; A chicken restaurant set on fire. -A.L.F.

38. 8/93-Memphis, Tennessee; A Kentucky Fried Chicken and another animal abuse shop attacked. -A.L.F.

39. 9/15/93-Alameda County, California; Butcher shop and meat jobber, several windows broken and slogans sprayed. -A.L.F.

40. 9/20/93-Alameda County, California; Butcher shop, poultry shop had their locks glued and slogans etched on the windows with acid. -A.L.F.

41. 9/26/93-Des Moines, Iowa; Dixon's Meats trucks and building spray painted with slogans "ALF", "Meat is Murder" damage into the thousands (Des Moines Register 9/28/93). Smith's East End Meat Market has three Molotov cocktails thrown at it, minimal damage. -A.L.F.

42. 10/17/93-Walnut Creek, California; Fur store: Spraypainted with slogans, locks glued. -A.L.F.

43. 10/28/93-Alameda County, California; A pumpkin with "Happy Halloween from the A.L.F." written on it was thrown through the window of a chicken restaurant. Locks glued also. -A.L.F.

44. 10/31/93-Alameda County, California; Butcher shop: locks glued, windows etched with acid, slogans painted on the walls.

45. 10/31/93-Alameda County, California; Ham retailer: locks glued, windows etched with acid, large front window smashed.

46. 10/31/93-Alameda County, California; McDonald's: window broken. -A.L.F.

47. 10/31/93-Alameda County, California; Fur store: locks glued, slogans painted on walls, windows etched with acid. -A.L.F.

48. 11/93-Los Angeles, California; Two cars belonging to fur store owners had windows broken. -A.L.F.

49. 11/22/93-Los Angeles, California; Van of a veal distributor: windshield etched with acid, tires slashed. -A.L.F.

50. 11/22/93-Los Angeles, California; Ham retailer: windows broken,

slogans painted, locks glued. -A.L.F.

51. 11/22/93-Los Angeles, California; Two butcher shops had locks glued, slogans painted on walls, windows etched. -A.L.F.

52. 11/22/93-Los Angeles, California; McDonalds had "#1 Killer in the World" spray painted on its walls and locks glued.

53. 11/26/93-Oakland/San Francisco/Walnut Creek, California; Z Furman Fur Service (private home): paint bombed.- Bernard's Fine Furs (private home): front windows: smashed, paint bombs thrown.- Saga Fur and Leather: paint bombed, locks glued, most windows etched.- Herbert's Furs: locks glued.- Michelle's Furs: locks to building glued. - Sheepskin store: paint bombed.- Kane's Furs: locks glued, slogans painted, front windows broken.- J.E. Harl Furs: locks glued, windows etched, slogans painted, expensive light fixture destroyed.-Middent's Furs: locks glued, windows etched, slogans sprayed.- California Fur Industry skyscraper: locks glued, windows etched. -A.L.F.

54. 11/27/93-Marietta, Ohio; Leather Shoe store had slogans spray painted with "Blood Money", "Cow Killers", "Don't wear Animals" and "Murderers". -A.L.F. Mid-Ohio Valley Unit.

55. 11/28/93-Chicago, Illinois; Nine timed incendiary devices were placed in four department stores Neiman-Marcus, Saks on 5th Avenue, Marshall Fields and Carson Pirie Scott all of which have fur departments. Five of the nine devices at three stores, ignited after hours, causing small fires and water damage. - A.L.F.

56. 12/14/93-Whitehorse, Yukon; Department of Renewable Resources (DRR): 54 tires slashed, locks of 17 trucks glued, barbed wire of security fence cut. The DRR is carrying out a wolf kill program. - A.L.F.

57. 12/23/93-San Francisco, California; Superglue, razor blades, and scissors used to damage 25-30 fur garments in 5 department stores and on the streets. -A.L.F.

58. 12/26/93-Brookfield, Wisconsin; Chudiks' West Inc. (fur store): At least $125,000 damage caused by arson and spraypainting "A.L.F." and "We skin you alive". Prior to this action the store and the owner's van have been attacked repeatedly with acid, painted slogans, and glued locks. -A.L.F.

Fifty-seven of these 58 entries for 1993 described criminal acts in at least 12 states and 3 Canadian provinces. Again, they all checked out. Some contained details not published in news stories. I noted that the Animal Liberation Frontline Information Service included a Unabomber attack in its Diary of Actions—presumably because the target was a geneticist who used laboratory animals, not because the Unabomber was an A.L.F. agent. A 1994 Diary of Actions had also been posted, listing 163 entries, of which 162 described animal rights activist crimes. Selected offenses:

1. 1/94-Ohio; Unspecified action against animal abusers.
2. 1/5/94-San Jose, California; Tarlow's Furs: Several windows smeared with etching fluid. A.L.F.
3. 1/7/94-Stockton, California; Mansoor Furs and Chuck E. Cheese restaurant: Locks glued, slogans spray painted. Mansoor Furs is the only furrier left in Stockton. A.L.F.
4. 1/8/94-Seattle, Washington; Eddie Bauer: $5,000 worth of merchandise slashed, including down-filled parkas and leather chairs; the store manager reported to police that Eddie Bauer had received several letters telling it to stop selling items made from animal material. On the street: A woman wearing a parka with a fake fur collar had red paint squirted on the back of her coat. She remembers seeing two men walking past her carrying a sign reading "Fur is dead"....
14. 1/30/94: San Diego, California; San Diego Meat Company, boarded-over window broken into, building set on fire (flammable liquid splashed around, fire started in two rooms). "FARM", "Meat is Murder" painted at scene. Estimated damages $75,000. - F.A.R.M. ....
39. 3/94-San Francisco / Bay Area, California; Hanes Furs, locks glued, sign partially smashed, slogans painted in front of the store. Honey Baked Hams, front door lock glued, windows etched, slogans painted on store front. Butcher shop, windows etched, slogans spray painted. Milk ad on street, smashed. McDonald's, windows smashed. Sheepskin store, two windows broken. Robert's Furs, lock glued, slogans painted. Harris Steak House, locks glued, slogans painted, window etched. Robert's Corned Beef, two trucks -slashed tires, painted slogans, etched or broken windows; one had its locks glued. -A.L.F. ....
50. 4/21/94-San Jose/San Francisco, California; EARTH DAY "Durham's Meat in San Jose and Columbus Sausage and Meat in San Francisco were hit by the A.L.F. in an effort to radicalize Earth Day. The attack was an attempt to radicalize Earth Day actions, opposing the usual 'green' corporate bullshit that the general public is being duped into believing will make a difference." - A.L.F. ....
83. 7/21/94-San Francisco, California; Pacific Cafe had windows etched, including with slogan "Lobster Liberation". According to Ross Warren, owner of the restaurant, approximately four months ago his windows were etched (no slogans or legible words), and a month later a man called asking if the restaurant sold live lobsters; when Warren replied "no" the man said killing lobsters was cruel and hung up. Warren also told the newspaper that the morning after the cafe was hit, a man called and said "We hit your place last night. We are the Crustacean Liberation Front. We're protesting your sale of live lobsters. Stop serving live lobsters." According to Warren, The Pacific Cafe sold live lobsters for three

weeks in March but stopped because customers were not buying. -Crustacean Liberation Front....

110. 9/6/94-Abbotsford, British Columbia; Arson destroyed the McClary family's rear stock-yard barn. The hundreds of animals that would normally be kept in the barn for grading and auction were not in the barn, as the long weekend resulted in a postponement of the auction. The four horses and two cows that were in the barn were freed before the fire consumed the building. This arson is one of a recent string of arsons in the Abbotsford-Matsqui area; none of the other fires were animal-related. -Unclaimed....

123. 10/11/94-West Valley, Utah; Jordan Meat Company had a pipe bomb thrown through the front window, approx. $1,500 damage. -Vegan Revolution....

135. 11/7/94-Memphis, Tennessee; New location of Jean Benham Furs firebombed before it opened. The BATF investigated, questioning nearby shop owners and trying to question at least one local activist, who refused to speak to them. It appears that the investigation has gone nowhere, but the fur store is now open for business. -A.L.F. ....

151. 12/2/94-Indianapolis, Indiana; Lazurus [sic] department store had slogans spray painted and windows broken. -A.L.F.

152. 12/2/94-Henrietta, New York; Conti Packing Co. (a meat processor) had windows to seven trucks and three trailers smashed, building spray painted with slogans "Veggie Power" and "This is just a warning", etc., $5,000+ damage. -A.L.F. ....

157. 12/22/94-Lower Mainland, British Columbia; A letter claiming frozen turkeys had been injected with rat poison at Safeway and Save-On-Foods stores. 40,000+ turkeys returned to stores and all frozen turkeys pulled from the shelves. $2 million+ in damages and lost sales. -A.R.M. ....

160. 12/24/94-Memphis, Tennessee; Tandy Leather had windows smashed. -A.L.F.

161. 12/25/94-Oakland, California; Barney's Gourmet Hamburgers, front lock glued. -A.L.F. Golden Gate Unit.

162. 12/94-1/95-outside Philadelphia, Pennsylvania; Eight billboards altered.

163. late 94; San Francisco, California; Herbert's Furs, "Fur Scum" burnt into glass front door with acid, during their going out of business sale. -A.L.F.

The Diary of Actions web page for 1995 listed 52 criminal incidents.

There was a Diary web page for 1996. Although it was only mid-February, there were three entries, all incidents in Canada, one only a week old.

1. 1/6/96-Victoria, British Columbia; A furrier who works from

his home had slogans "Murderer", "Death", "Fur", "Killer" and "A.L.F." spray painted on his home, garage and truck. Tires on a truck and trailer also slashed. -A.L.F.

2. 1/9-/96-British Columbia; 65 envelopes with rat poison covered razor blades, taped inside the opening edge to guide outfitters across B.C. and Alberta. The letter enclosed said "Dear animal killing scum! Hope we sliced your finger wide open and that you now die from the rat poison we smeared on the razor blade. Murdering scum that kill defenseless animals in the thousands every year across B.C., for fun and profit do not deserve to live. We will continue to wage war on animal abusers across the world. Beware scum, better watch out, you might be next! Justice Department strikes again." -Justice Department.

3. 2/9/96-Guelph, Ontario; Guelph Fashion Furs was damaged and spraypainted. -A.L.F.

I searched backward now, to see how many years worth of web pages had been posted in all. The Diary of Actions pages went back to 1977. The tally:

1977-1982: 15 entries.
1983: 19 entries.
1984: 24 entries.
1985: 18 entries.
1986: 28 entries.
1987: 45 entries.
1988: 39 entries.
1989: 51 entries.
1990: 26 entries.
1991: 22 entries.

Two-hundred-eighty-seven actions from 1977 to the end of 1991. Combined with those extending to 1996, there were more than 585 criminal acts posted in what amounted to a "brag book" on behalf of the Animal Liberation Front—hosted by EnviroLink's Web server.

Who posted these pages? A notice at the bottom of the page read: "Maintained by Anon - for Animal Liberation - How to contact us." I followed the contact link, and discovered that the Animal Liberation Frontline's anonymous e-mail site had been shut down:

Anon.penet.fi has closed

Yes, it's true. The most popular pseudonymous service on the Net has finally closed. To quote from the original message at Penet's Web site:

"Due to both the ever-increasing workload and the current uncertain legal status of the privacy of e-mail here in Finland, I have now closed down the anon.penet.fi anonymous forwarding service until further notice."

The Finland-based anon.penet.fi server was the most famous anonymous service on the Internet. It has slightly less than 700,000 registered users and handled about 10,000 messages every day.

So unfortunately that leaves you with out a way to contact us. We may in the future set up something with another anonymous remailer, but for now we will go without.

For those who use e-mail we suggest you follow these suggestions.

1) Use PGP encryption for all e-mail.

2) Preferably send your e-mail from an anonymously acquired account.

3) Use one of the following W3 anonymous re-mailers to send us e-mail.

4) Use an Cypherpunk or Mixmaster anonymous remailer. For more information read an Anonymous Remailer FAQ.[17]

All but a handful of these crimes met the FBI's official definition of terrorism. Yet only a handful were investigated by the FBI as acts of terrorism. The FBI was unaware of the postings.

Conspicuous by their absence from these diaries on the Web were mentions of the mail bombs sent by the Militant Direct Action Task Force to Alta Genetics and the Mackenzie Institute in July 1995, among many others. Animal rights terrorist attacks were under-reported even on their own brag site.

The U.S. Department of Justice had acknowledged the under-reporting problem in their September 2, 1993 *Report to Congress on the Extent and Effects of Domestic and International Terrorists on Animal Enterprises*. The two policy analysts assigned to perform the study contacted literally hundreds of individuals in 28 different types of animal enterprises ranging from farms to ranches, from zoos to county fairs, from product research laboratories to circus acts, from restaurants to race tracks, from high school labs to pet breeders, from universities to rodeos—anyone that used animals. They discovered that many victims of animal rights attacks would not talk to them. They wrote in a footnote, "In fact, it is generally believed that many animal rights-related incidents—especially those involving relatively minor acts of vandalism such as graffiti—go unreported, and therefore are numerically underestimated in this analysis."[18]

Even so, the report documented 313 animal-rights related incidents between 1977, when the first such incident in the United States was recorded, through June 30, 1993. The fact that the report existed at all pointed up the seriousness of the problem: It was mandated by law, The Animal Enterprise Protection Act of 1992.[19]

The law was enacted to protect a broad range of professional and commercial animal enterprises and individuals in response to an increasing number of attacks by animal rights activists. It made it a federal offense, punishable by fine and/or imprisonment for up to one year, to cause physical disruption to the functioning of an animal enterprise resulting in economic damage exceeding $10,000. Those convicted are also required to pay the cost of replacing property, data, records, equipment, or animals destroyed in the attack, as well as reasonable costs of repeating any experimentation that may have been interrupted or invalidated by their attack. The Act also imposes sentences of up to 10 years or life imprisonment, respectively, on persons causing the serious bodily injury or death of another person during the course of such an offense.

The law had been passed because Kathleen Marquardt and her citizen group Putting People First made it happen. Representative Charlie Stenholm (R-Texas) sponsored H.R. 2407, "An Act for the Prevention of Crimes Against Farmers, Researchers, and Other Livestock-Related Professions," and guided it through a ferocious battle from animal rights activists. At a crucial moment when it looked like the bill would be bottled up to die in the House Judiciary Committee, Kathleen got on the phone and the fax and urged her thousands of supporters to personally call the Speaker of the House, who had the power to release the bill for a floor vote, which it would clearly pass. Speaker Tom Foley (D-Washington) suddenly found his office deluged with calls from supporters of the bill demanding that it be released for fair consideration by the House. The calls stacked up one on the other until no open lines were left. The phones to the Speaker's office were all busy. The Speaker of the House of Representatives could not even call Charlie Stenholm to tell him he got the message. Speaker Foley had to walk down the hall to Rep. Stenholm's office and personally tell him, "Call off the dogs, Charlie. You can have your bill."[20]

The law instructed the Department of Justice and the Department of Agriculture to jointly conduct a study on the extent and effects of domestic and international terrorism on enterprises using animals for food or fiber production, agriculture, research, or testing, and report the results to Congress within a year.

The Justice Department, in its letter of transmittal to President of the Senate Al Gore and Speaker of the House Tom Foley, dealt with the problem of officially defining terrorism head-on: "As the Federal Bureau of Investigation has categorized only a few animal rights-related incidents as acts of domestic terrorism, for purposes of this report the term "animal rights extremism" includes all acts of destruction or disruption perpetrated against animal enterprises or their employees."[21]

Even the government has to sidestep the FBI's narrow definition of terrorism so it can use a little common sense.

The problem of tabulating monkeywrenching incidents is much more difficult than animals rights extremism. Monkeywrenchers do not post their crimes on the World Wide Web to make counting them easier. There is no federal law against monkeywrenching comparable to The Animal Enterprise Protection Act of 1992 requiring a federal report on all known incidents. It is true, however, that the U.S. Forest Service ordered a sabotage survey, conducted by Ben Hull, special agent for Wenatchee National Forest, which found during an 18-month period in 1987-88:

- 219 serious acts of vandalism to Forest Service or contractors' property, amounting to $4.5 million damage.
- 42 letters received threatening sabotage or vandalism if the Forest Service did not prevent or stop certain logging activities.
- 32 demonstrations that temporarily halted logging or road building, resulting in $201,000 in losses.
- Three fourths of this activity was in the Pacific Northwest and Northern Rockies.

One major Forest Service region refused to cooperate, making the national total seriously under-reported.[22]

Tree spiking was made a federal felony in 1988, but the law did not require a report to Congress accounting for all known incidents.[23]

Very few monkeywrenching crimes are reported. Reported crimes are seldom claimed by any perpetrator. Because of the hit and run tactics used and the cell structure of the monkeywrenchers, many monkeywrenching crimes yield no suspects and remain forever unsolved. For these reasons, complete and accurate tabulation is impossible.

A different problem emerges with eco-protests that result in criminal arrests: While they rarely involve felonies or massive destruction of property, they always incur public costs in extra law enforcement and usually involve personal injury in lost wages of workers forcibly restrained from working, and the social cost of lost domestic product. These losses are presently considered externalities, seldom accounted for as a social cost of radical environmentalism. Even obtaining estimates of extra law enforcement costs from government agencies is difficult because city, county, state and federal agencies may all send officers to handle a single protest and some do not itemize cost-per-incident. Where costs over and above normal budgets are separated out by incident, I have included accountings from various government agencies.

There is also the difference between non-violent protest and obstruction or interference. While most environmental radicals characterize their protests as "peaceful" and "non-violent," in fact their protests usually involve obstruction or interference, which is imposing physical impediments preventing persons from going where they ordinarily have a right to go. Obstruction is not a peaceful act. Obstruction is an act of physical coercion, an act of violence against another, regardless how passively performed. Even the threat of committing an unlawful act with the purpose of restricting another's freedom of action to his detriment, such as threatening to trespass in order to stop logging operations, is criminal coercion. I have not listed protest demonstrations that merely made moral argument without obstruction, no matter how noisy or radical, for a simple reason: there are no arrest records because there are no arrests because it is not illegal; it is political expression protected by the First Amendment.

When an environmental protest uses obstruction, trespass or other unlawful coercion—or in some instances becomes so large and unruly it poses a de facto threat to public safety—it becomes civil disobedience. Civil disobedience is "lawbreaking employed to demonstrate the injustice or unfairness of a particular law and indulged in deliberately to focus attention on the allegedly undesirable law."[24] Although there is a tradition of civil disobedience in America, seminally expressed in Henry David Thoreau's *On The Duty of Civil Disobedience*, there is no right to civil disobedience. No one has the right to break the law. I have thus included a range of civil disobedience eco-protests which resulted in criminal arrests.

Finally, the human cost of the violent agenda to save nature in terms of fear, chronic anxiety and intimidation caused by environmentalists cannot be calculated. The ferocious rage with which radical environmentalists hold their beliefs is impossible to comprehend by one who has not personally experienced it. Literally hundreds of eco-terror victims I contacted refused to speak to me for fear of the consequences. I have honored their requests to keep some incidents secret. Some agreed to provide information on incidents but requested anonymity. Where other verification was available, such as law enforcement or news reports, I have included a few incidents without naming victims.

The ecotage crimes listed below fit one or more of the following criteria:

● Alleged perpetrators were arrested, tried and convicted.

● A civil suit against alleged perpetrators was successfully prosecuted.

● The type of crime was described in an environmental instruction manual readily available to the public.

● The particular crime fit the modus operandi found in an environmental instruction manual.

● The target was on a hit list published in an environmental publication.

● The crime stopped, hindered or intimidated a resource extraction or conversion industry or government agency in locations where such activity was in controversy.

● Law enforcement checked and found no other persons of interest or probable suspects such as disgruntled former employees, operators performing self-inflicted damage for an insurance scam, competitors, former spouses, mentally or emotionally disturbed associates or other known enemies of the target.

● Suspects with clear ecotage motives or backgrounds had been identified by law enforcement, even if proof sufficient for prosecution was not available.

Because the publication of instruction manuals bears materially on the evolution of the violent agenda to save nature, I have included the publication of numerous instruction manuals in this list, primarily as milestones. In the United States, publishing instructions for committing crimes is protected speech under the First Amendment, and is not a crime and cannot be criminalized.

In addition, authors or publishers of such instruction manuals have not been successfully prosecuted for conspiracy in a criminal act committed by another who followed their published instructions unless they personally furnished the instructions to the perpetrator along with other resources such as money to commit the crime. In Canada and the United Kindgom most of these instruction manuals are not protected by free speech laws and are outlawed under incitement-to-violence laws.

In the list that follows, there are many officially unsolved crimes. I have included them for several reasons. Some, as will be obvious, were claimed in contacts with the media. I have included others because law enforcement officers I interviewed had strong reason to believe a known ecotage suspect committed the crime but lacked evidence sufficient to prosecute. I have omitted more than two-hundred suspicious incidents about which law enforcement entertained reasonable doubts.

Although it is annoying to the reader, I have included the sources of the entries at the end of each description to avoid the greater annoyance of continually popping back and forth from text to footnotes.

Prior to 1970

Various acts of vandalism were committed for ideological nature protection reasons before the environmental ideology was well formulated or publicly recognized.

1958. Las Vegas, New Mexico. Edward Abbey and co-workers on a Taos newspaper sawed down a dozen billboards owned by the Melody Sign Company. Abbey's father, Paul Revere Abbey, was an anarchist and member of the radical International Workers of the World ("Wobblies"). Abbey wrote his master's thesis at the University of New Mexico on "Anarchism and the Morality of Violence." *Edward Abbey: A Voice in the Wilderness*, transcript of documentary film sound track, Canyon Productions, 1993.

1959. Flagstaff, Arizona. Marc Gaede and friends cut down billboards on Route 180 between Flagstaff and Grand Canyon. Susan Zakin, *Coyotes and Town Dogs: Earth First! and the Environmental Movement*, Viking, New York, 1993, p. 41ff.

1970

4/21/70. Miami, Florida. In a guerrilla theater-style break-in during the first Earth Day celebrations, five proto-ecoteurs calling themselves "Eco-Commando Force '70" unlawfully entered six sewage treatment plants and tossed in yellow dye to prove that treated waste lingered in ocean-bound canals longer than government claimed. They did the same in the toilets of the Howard Johnson Motor Lodge and the Airport Crossways Inn, which had their own small sewage treatment plants. The group performed two media stunts in 1970, one on July 4, posting 800 "Polluted, Keep Out" signs on Florida beaches, and another on October 22, setting adrift in an ocean sewer outfall 700 bottles containing "This is where Miami sewage goes" cards addressed to the Governor of Florida and the Miami News. The group then disappeared. Like many early ecotage events, this series was only minimally unlawful and did not damage property or coerce workers, it only harmed the reputation of its bureaucratic targets. Sam Love and David Obst, editors, *Ecotage!*, Pocket Books, New York, February, 1972, pp. 160-170.

1970. Black Mesa, Arizona. For two years, a solo ecoteur known as the Arizona Phantom vandalized heavy equipment owned by Peabody Coal Company. The vandal did not gain public attention. Susan Zakin, *Coyotes and Town Dogs*, p. 41ff.

1970. Eugene, Oregon. First edition of *The Cultivator's Handbook of Marijuana* published. Signaled the beginning of synergism between marijuana growers and eco-radicals to restrict or eliminate access to national forests, each with their separate motives, one to prevent detection of illicit marijuana plantations, the other to save nature. Bill Drake, *The Cultivator's Handbook of Marijuana*, Agrarian Reform Company, P.O. Box 2447, Eugene, Oregon, Oregon 97404, 1970 (92 pages). Latest edition published by Wingbow Press, Oakland, California, February, 1987.

Summer, 1970. Kane County, Illinois. A solo ecoteur calling himself "The Fox" sabotaged steel mill drains, sealed off factory chimneys, dumped dead fish in corporate offices, and similar unlawful actions "in defense of nature." The first unequivocally criminal acts clearly guided by an ideological goal to save nature that gained public attention. Columnist Mike Royko of the Chicago Daily News praised "The Fox" in four columns in 1970 and 1971. See p. 195.

1971

Summer, 1971. Tucson, Arizona. The Eco-Raiders, a self-styled group of five young men, begin two-year campaign to protest "urban sprawl," first by sawing down highway billboards, graduating to nearly two million dollars worth of sabotage to equipment and homes under construction on the northwestern and eastern perimeters of Tucson. The first organized criminal enterprise clearly guided by an ideological goal to save nature that gained strong media attention. See p. 193ff.

1971. Luton, England. Ronnie Lee forms a local chapter of the Hunt Saboteurs north of London to interfere with fox hunts.

1972

February, 1972. Washington, D. C. While revising a 1970 book, *Earth Tool Kit*, the group Environmental Action sponsored a nationwide contest for sabotage ideas with results published in the handbook *Ecotage!* Ecotage was defined as "the branch of tactical biology that deals with the relationship between living organisms and their technology. It usually refers to tactics which can be exercised without injury to life systems." Detailed instructions on sabotaging equipment, destroying billboards, removing road survey stakes, unfurling banners, plugging industrial waste and sewer pipes, delivering sludge and dead animals to corporate offices, and other now-familiar actions. Its description of the ecotage movement presaged Earth First: "The movement's strength is that it is not formally organized and it cannot be stopped by elimination of key leaders. Though not rigidly structured, it is unified by a philosophy of respect for life." This was the earliest ecoterror instruction manual and prototype for Dave Foreman's similar 1985 manual, *EcoDefense*. Sam Love and David Obst, editors, with a foreword by Robert Townsend, *Ecotage!*, Pocket Books, New York, February, 1972. 186 pages.

Weekly, 1972. Tucson, Arizona. Eco-raiders commit acts of vandalism against new homes and construction equipment at the rate of approximately one per week.

1972. Luton, England. Ronnie Lee and Cliff Goodman form the violent Band of Mercy, dissatisfied with Hunt Saboteur non-violence.

1973

1973. England. Lee and Goodman commit two acts of arson at a Hoechst pharmaceutical plant, causing £46,000 damage.

9/19/73. Tucson, Arizona. The Eco-Raiders are arrested for smashing homes under construction, "decommissioning" bulldozers and other crimes, perpetrating nearly $2 million in damages to protest urban sprawl. Four of the five were convicted and served jail sentences. "3 Men Enter Guilty Pleas On Eco-Raider Charges," *Arizona Daily Star*, October 25, 1973, by Art Arguedas, p. 1A. See also "What is the sound of one billboard falling?" *Berkeley Barb*, Nov. 8-14, 1974, by Tom Miller, pp. 9-12.

1974

1974. Tucson, Arizona. Writer Edward Abbey drafts *The Monkey Wrench Gang*. The action in the book was inspired by the exploits of the Eco-Raiders, and based on Abbey's own monkeywrenching, some of his actions dating back to the 1950s. The characters were based on a few of his closest friends, highly fictionalized: Doug "George Hayduke" Peacock and Ken "Seldom Seen Smith" Sleight; "Doc Sarvis" was a combination of Al Sarvis, Jack Loeffler and John DePuy. "Bonnie Abzugg" was based on Ingrid Eisenstadter and others.

1975
September, 1975. Edward Abbey's novel *The Monkey Wrench Gang* published. The story's four main characters, three men and a woman, Doc Sarvis, Seldom Seen Smith, George Washington Hayduke and Bonnie Abbzug, travel across the Southwest pulling survey stakes, incinerating billboards, pouring sugar in the crankcases of bulldozers, blowing up a mining train and blasting a highway bridge, with the final target of demolishing Glen Canyon dam. It read like a comic book. Five years later, the book became the model for Earth First in method, mythology and symbolism and an icon for all radical environmentalists. Abbey became the personal mentor of Earth First, speaking at numerous events. Edward Abbey, *The Monkey Wrench Gang*, J. B. Lippincott, Philadelphia, 1975.

1976
1976. England. Animal rights activist Ronnie Lee, upon his release from prison, forms the Animal Liberation Front, claiming damage to animal enterprises worth over a quarter-million pounds sterling in its first year.

1977
5/29/77. Honolulu, Hawaii. Two dolphins released from marine laboratory at the University of Hawaii by a group calling itself the Undersea Railroad. First documented animal rights incident in the United States. U. S. Department of Justice, *Report to Congress on Animal Enterprise Terrorism,* Washington, D.C., September 2, 1993, p. 17.

1978
January 1978. Publication of *Marijuana Grower's Guide* by Mel Frank and Ed Rosenthal (Red Eye Press, Los Angeles) with explicit references to "guerrilla plantations" of marijuana on national forest land. "Clearings in forests have always been popular places to plant because they offer security from detection." "Guerrilla growers often use the same techniques as home gardeners. But the soil that they start with is sometimes marginal." Increasing amounts of federal land are occupied and controlled by marijuana growers. Protection of the marijuana crop from normal forestry practices becomes a major problem for illicit growers, especially protection from forestry brush control with herbicides sprayed by helicopter. The link between anti-herbicide activism and guerrilla marijuana plantations solidifies. "North Counties' politics of pot - Life amid the marijuana farms takes some getting used to," *Oakland Tribune*, Sunday, April 27, 1980, by Reggie Major, p. 1A.

6/18/78. Josephine County, Oregon. Paula Downing, leading anti-herbicide protester, arrested after 75-mile-an-hour chase for having 77 marijuana plants in her automobile. On July 7, 1978 she was charged by the Josephine County Prosecutor with criminal activity in drugs: "Did knowingly and unlawfully transport marijuana, a narcotic drug." She entered a guilty plea to a lesser charge and the felony charge was dismissed.[25] Downing attended protests against the Bureau of Land Management brush control spraying program. Downing was co-founder, with her husband Art, of the Headwaters Association. Both were members of the Southern Oregon Citizens Against Toxic Sprays.

1979

9/27/79. Grants Pass, Oregon. State officials said some of the people protesting the spraying of herbicide 2,4-D on Bureau of Land Management land may have been more worried about saving marijuana than possible health hazards. Josephine County deputy sheriffs dug up 8 marijuana plants in an area scheduled to be sprayed at the bottom of a ridge where protesters were camped. Deputies in neighboring Jackson County dug up 375 pot plants worth $135,000 near an area scheduled for spraying. "Herbicide foes concerned about health or marijuana losses?" *Seattle Times*, Thursday, September 27, 1979, Northwest Digest compiled from news services, p. A18.

1980

4/1-7/80. Pinacate Desert, Sonora, Mexico. Earth First formed by Dave Foreman, Ron Kezar, Bart Koehler, Howie Wolke and Mike Roselle on week-long drive. Foreman and Koehler had worked for the Wilderness Society. Kezar, a Sierra Club member, had worked as a seasonal National Park Service employee. Wolke was Wyoming representative of Friends of the Earth. Roselle, a generation younger than the others, had been active in radical left-wing groups including the Yippies and Zippies, the only co-founder without wilderness and mainstream environmental movement experience. All were convinced that the earth was in imminent danger of destruction by the technological order. All were convinced that working within the system could not save the earth. All were convinced that traditional society and industrial civilization had to be eliminated. See p. 211.

5/4/80. Siskiyou National Forest, near Takilma, Oregon. Fourteen Forest Service herbicide crew members spraying 2,4-D for brush control on newly planted forest and two sheriff's deputies are attacked by about 125 protesters armed with knives and clubs. Spray rig backpacks forcibly removed from crew and hoses cut. A rock was thrown into the pump of the Forest Service tanker. Numerous protesters attempted to slash government vehicle tires. Sheriff's deputies helped crew retreat, but protesters had blocked road with rocks and logs. Attack resumed

with increased force, protesters throwing rocks and garbage at Forest
Service crew. Forest Service Ranger Jim Schelhaas shouted surrender
and signed a document agreeing to stop the brush control program for
that season. United States General Accounting Office Report CED-82-
48, "Illegal and Unauthorized Activities On Public Lands—A Problem
With Serious Implications," p. 19. See also "EcoTerrorism," *Reason*,
February 1983, by Ron Arnold, pp. 33-34.[26]

1980. Colorado. 3.2 miles of power lines downed after line supports were
sawn through, costing the Colorado Ute Electric Association $270,000
in repair bills. "Environmental radicalism backed - Local response
said favorable," *Portland Oregonian*, January 23, 1983, by John Hayes,
p. C2.

1981

2/2/81. Near Missoula, Montana. The timber-truss Franklin bridge owned
by Montana Power Company cut with chainsaws, doused with diesel
fuel and torched, eliminating sole vehicular access to boundary of Rattlesnake Wilderness and National Recreation Area. "Upper Rattlesnake
bridge burned - Motorcycle access blocked," *Missoulian*, February 3,
1981, by Steve Smith, p. 1.

5/30/81. Near Toldeo, Oregon. A helicopter leased by Publisher's Paper
Company to apply brush control herbicides on a commercial Douglas
fir plantation firebombed and destroyed. $180,000 damage. Two
masked women held a news conference with Coast News Service to
claim responsibility on behalf of the "People's Brigade for a Healthy
Genetic Future," saying, "We sabotaged poison-spreading machines as
an act of self-defense." Two years later, anti-herbicide activist Carol
Van Strum wrote previously undisclosed details of the incident in her
Sierra Club book, *A Bitter Fog*, sparking a new investigation of Van
Strum and Sierra Club Books by Lincoln County chief deputy district
attorney Ulys Stapleton. "Herbicide foes burn helicopter," *Seattle Times*,
June 3, 1981, by The Associated Press, p. B1. "Law takes eager inter-
est in book on copter arson," *Salem Statesman-Review*, Wednesday,
February 23, 1983, by Dan Postrel, p. B1.

7/3/81. Near Moab, Utah. Vandals toppled a Utah Power and Light transmission tower carrying 345,000 volt power lines seven miles south of
Earth First's second annual Round River Rendezvous. Foreman denied Earth First had any part in the vandalism, blaming it on the victims or "Free-lance anti-environmental yahoos." The case was never
solved. Martha F. Lee, *Earth First! Environmental Apocalypse*, Syracuse University Press, 1995, Syracuse, pp. 52-53.

1982

3/10/82. Washington, D. C. General Accounting Office report, "Illegal
and Unauthorized Activities On Public Lands—A Problem With Seri-

ous Implications," issued. The GAO report, Appendix II: Marijuana Cultivation on Federal Lands, pages 6-10, cited numerous instances of violence, concluding, "Marijuana cultivation threatens public and employee safety and hinders management." The cultivation was so extensive and the profit so high, it had attracted criminal elements and biker gangs to steal the crops, causing planters to post guards and promoting violent confrontations. The report warned that law enforcement was incapable of handling the problem. Appendix IV: Trespass on Federal Lands Affects the Environment, Visitors, and Employees, under the heading, Herbicide Protest Activities, reported that numerous protests had stopped federal programs. "FS officials have charged protestors with violating 18 U.S.C. 111, which prohibits opposing, impeding, intimidating, or interfering with Federal officers engaged in performing official duties. Officials at one of the BLM's district offices said that the office had received bomb threats and forest arson threats when its intention was to spray herbicides for timber management reasons." Some projects blocked by protesters were abandoned. CED-82-48.

4/mid/82. Ranger Station in Plumas National Forest, California. During public meeting on herbicides, protester Doug Wellborne shouted at Forest Service Ranger Dewey Riscioni, "You like 2,4-D so much, you son of a bitch, let's see how you like this!" and squirted a liquid from a plastic hand-pump container into Riscioni's eyes three times and fled. Riscioni was not injured. Wellborne was convicted of assault, fined $200 and placed on two years' probation. "EcoTerrorism," *Reason*, pp. 33-34.

5/31/82. Near Dunsmuir, British Columbia, Canada. Four 500-kilovolt electrical power transformers dynamited, causing $6 million damage. A person called radio station CKNW to claim responsibility for the blast on behalf of "37 anti-herbicide protesters." A 2-page letter claiming responsibility went to 18 media and citizen group offices in British Columbia, Saskatchewan, Quebec, England and Oregon from a group called Direct Action, which was later discovered to be a group of five Vancouver-area radicals, Anne Hansen, Gerald Hannah, Brent Taylor, Juliet Belmas and Douglas Stewart, who were captured, prosecuted, convicted and imprisoned. "Herbicide foes say they set off B.C. blast," *Seattle Times*, Tuesday, June 1, 1982, Northwest Section from wire services, p. B5. See also, "Plot to sabotage military base found in British Columbia, police claim," *Bellevue Journal-American*, Friday, January 28, 1983, by The Associated Press, p. A7.

6/82. Teton County, Wyoming. Survey stakes belonging to Getty Oil Company along 2.5 miles of surveyed roads leading to a wilderness oil drilling site removed. Surveying instruments thrown into nearby creek. Direct damages, $5,000. "EcoTerrorism," *Reason*, p. 33.

6/20/82. Sacramento, California. The Sacramento Bee runs a five-part series by Jim McClung titled, "The New Lawless," on the epidemic of violence resulting from near-complete takeover of Northern California's federal lands by marijuana growers. Federal officials said they had lost control of some 100,000 acres of the Shasta-Trinity National Forest to machine-gun wielding guerrilla marijuana growers, in addition to suffering restriction of management on more than a million acres in Humboldt and southern Trinity Counties. McClung wrote, "Marijuana farmers are placing shotgun shell booby traps around the perimeter of their gardens, the shells rigged to fire when trip wires are disturbed. They also place bear traps on paths leading to the fields. Some maintain roving guard dogs and employ armed guards to protect their crops from armed robbers." Hiking and hunting become unsafe for the general public. As the price of "California homegrown" increases, so does the level of violence growers are willing to use to protect their crop. The original counterculture growers who had allied themselves with (and sometimes were) anti-herbicide protesters now faced competing growers as diverse as ex-convict street thugs and business executives, all willing to use violence. Local ranchers and loggers pay "protection" to avoid grassfires and logging equipment vandalism. Law enforcement avoids dangerous areas. All elements become hardened, reflected later in some environmental radicals. "The New Lawless," "Violence Flourishes Unchecked Where Cash And Marijuana Are The Common Denominators," "Lawlessness Worst in Denny Area - Forester Admits Authorities Can't Control Wilds of Trinity," "Outlaw Pot Growers Outgun, Outrun Lawmen," "Following The Marijuana Connection," *Sacramento Bee*, Sunday, June 20 to Thursday, June 24, 1982, by Jim McClung, all page A1 stories.

Mid/82. Undisclosed location in Olympic National Forest, Washington. Three pieces of logging equipment, a tower yarder, skidder and log loader owned by a family company that requested anonymity were dynamited and destroyed. Special Agent Robert E. Hausken of the U. S. Bureau of Alcohol, Tobacco and Firearms provided a note left on the ruins, reading, "We are acting on behalf of LIFE. The trees you have cut down were not dead—you have killed living beings. We will NOT allow this. We are incapaciting [sic] your death machines with our actions. ... Take a moment to FEEL the Earth around you, the beauty and power. Look at the worldwide destruction and pollution being done to our environment, our Earth, our home. Is the killing worth the false wealth you get???" It was signed, "PEOPLE OF THE EARTH." Author's files.

1983

5/12/83. Galice, Oregon. Earth First co-founder Dave Foreman arrested for blocking a road at Bald Mountain construction site in first major series of Earth First civil disobedience anti-logging protests in the Pacific Northwest. He was later found guilty and sentenced. See p. 226.

6/5/83. Near Summit, Oregon. Front-end log loader owned by Gassner Logging of Philomath, Oregon, torched by anti-herbicide vandals on Starker Forest Inc. lands. $100,000 damage wiped out Richard Gassner's family logging company. "No Spray. You spray, you pay, You kill, we will. PBFPF & the Swamp Fox," written on nearby shed. About 65 acres of replanted Starker forest land had been sprayed to control brush the day before. "Torching follows spraying," *Corvallis Gazette-Times*, Tuesday, June 7, 1983, by George Wisner and Steve Jones, p. A1.

6/7/83. Galice, Oregon. Eight Earth Firsters arrested for second-degree criminal trespass at Bald Mountain Road construction site for blocking Plumley Inc. construction workers from building road to logging site. Twenty-six Earth Firsters were arrested for criminal trespass in the Siskiyou National Forest area protest in two months, including Jim Goodwin; Mark M. Kelz, 36; Mike H. Perkins, 42; Christy A. Dunn, 33; Beth A. Peterson, 32; Sally E. Clements, 26; Louis H. Gold, 45 a former professor of politics at Oberlin College. The U.S. Forest Service shut down the road construction, leaving felled timber from the right-of-way clearing on the ground. "Siskiyou protest peaceful," *Corvallis Gazette-Times*, Tuesday, June 7, 1983, by The Associated Press, p. B1.

6/16/83. Logging site 40 miles east of Sweet Home, Oregon. Log yarding tower belonging to family-owned Clear Lumber Company of Sweet Home torched and destroyed. $187,000 damage. The Mid-Santiam Wilderness Committee had opposed the sale of the site for logging, said Forest Service official William Porter. No suspects. "More logging equipment torched - Fire in Sweet Home area destroys skyline yarder," *Corvallis Gazette-Times*, Thursday, June 16, 1983, by George Wisner, p 1A.

1984

Summer, 1984. Near Corvallis, Oregon. Earth First co-founder Michael Lee Roselle and others using the name Bonnie Abbzug Feminist Garden Club after Edward Abbey's *Monkey Wrench Gang* character spike trees in timber sales in the Middle Santiam and Hardesty Mountain. Susan Zakin, *Coyotes and Town Dogs*, p. 260.

1985

5/20/85. Sweet Home, Oregon. Earth Firster Michael J. Jakubal, 22, of Wenatchee, Washington, using the alias, "Doug Fir," employed his rock climbing skills to climb and sit a tree to protest the logging of old-growth timber in the Willamette National Forest. He was cited for criminal trespass by Linn County sheriff's deputies around 7:30 p.m., several hours after six other protesters were arrested on similar charges in the Pyramid Creek timber sale area, 40 miles east of Sweet Home.

"Logging Protester Arrested - 'Doug Fir' Descends From Lofty Perch." *The Washington Post*, May 22, 1985, by United Press International, p. A13.

6/16/85. Near Jackson, Wyoming. Howie Wolke, co-founder of Earth First, arrested for pulling up nearly five miles of survey stakes from a road for a Chevron Oil Company exploration operation in the Bridger-Teton National Forest. Wolke was convicted and forced to pay Chevron $2,554 in damages, along with a $750 fine, and to serve a six month jail term. "Monkey-wrenching around; Earth First!", *The Nation*, May 2, 1987, vol. 244, by Jamie Malanowski, p. 568.

1985. Dave Foreman's *Ecodefense: A Field Guide to Monkeywrenching* self-published. Nine chapters of instructions on subjects ranging from tree spiking to road sabotage, from disabling equipment to disrupting predator trapping, from jamming locks and making smoke bombs to propaganda, writing untraceable letters and evading capture. Foreman was listed as "editor" with Bill Haywood, pseudonym. Some think Mike Roselle was "Haywood," others think Foreman worked alone. Roselle later repudiated the book as Foreman's and not Earth First's. Dave Foreman and Bill Haywood, *Ecodefense: A Field Guide to Monkeywrenching*, Ned Ludd Books, Tucson, Arizona, 1985.

1986

1986. Near Salem, Oregon. After members of Earth First and a decoupling group, the Cathedral Forest Action Group, disrupted a lumbering operation, first by sitting-in and delaying the logging and then by blockading the road to keep trucks from hauling the trees out, Willamette Industries sued the individuals involved and won $13,000 in damages.

3/86. Montana. Monkeywrenchers destroyed a small firm's logging equipment and left a banner reading, "Earth First!" Martha F. Lee, *Earth First! Environmental Apocalypse*, p. 90.

5/86. Palo Verde, Arizona. Earth First monkeywrenchers cut the electrical power lines leading to the Palo Verde nuclear plant. Dean Kuipers, "Razing Arizona," *Spin*, September 1989, p. 34.

5/9/86. Corvallis, Oregon. Earth Firsters Michael J. Jakubal and Mitchell A. Friedman arrested with writer Susan Zakin about 1 a.m. near the city shops after a billboard belonging to 3m National Advertising of Eugene was cut down. Benton County Sheriff's Department Case Number 86-6642. Both Jakubal and Friedman pleaded guilty to a felony charge of first-degree criminal mischief. Benton County Circuit Court Judge Robert S. Gardner sentenced Jakubal to 90 days in jail and ordered him to pay $2,000 in restitution, 60 hours of community service and placed him on 2 years probation. On September 25 he sentenced Friedman to pay $600 restitution, 40 hours of community service and placed him on 2 years probation. The district attorney dropped similar charges against Zakin, 28, of San Francisco, who was a freelance reporter for New Age magazine in Brighton, Massachusetts. A notebook seized from her referred to "eco-commandos" who were out to eliminate cigarette ads. Zakin would go on to write a laudatory 1993 history of Earth First, *Coyotes and Town Dogs: Earth First! and the Radical Environmental Movement.* "'Commandos' suspected in sign cutting," *Corvallis Gazette-Times*, Saturday, May 10, 1986, by George Wisner, p. A1. "Billboard slayer enters guilty plea," *Corvallis Gazette-Times*, Wednesday, June 11, 1986, by George Wisner, p. B1. "Billboard destroyer gets 90 days in jail," *Corvallis Gazette-Times*, Wednesday, June 18, 1986, no byline, p. B1. "Judge sentences billboard slayer," *Corvallis Gazette-Times*, Tuesday, September 30, 1986, by George Wisner, p. B4.

7/8/86. Yellowstone National Park. Nineteen Earth Firsters arrested in protest, including Dave Foreman and Mike Roselle. "The Round River Rendezvous: A Newcomer's Perspective, *Earth First!* 6, no. 7, Lughnasadh/August 1, 1986, by Randall T. Restless [pseudonym of Phil Knight], p. 17.

10/86 through 1990. Garfield County, Utah. Garfield County Commissioner Louise Liston provided a list of vandalism costing the county more than a half million dollars in damage on the Burr Trail road connecting the town of Boulder with Bullfrog Marina at Lake Powell. The damage ranged from abrasives in road equipment crankcases and cabins burned to watering holes poisoned, fences cut and cattle shot. The county commissioners received anonymous letters from Earth Firsters claiming responsibility for the sabotage. A nearby mine sustained $175,000 in damage when abrasives were poured into the oil tanks of all of the mine's equipment. "Southern Utah County Documents Sabotage," *Blue Ribbon Magazine*, Pocatello, Idaho, August 1990, p. 9.

11/9/86. Reykjavik, Iceland. Sea Shepherd Conservation Society saboteurs Rodney Coronado and David Howitt sink two whaling vessels and wreck a whale-meat processing plant, causing $2 million in damage. Iceland did not succeed in extraditing either Coronado from the United States or Howitt from the United Kingdom. They were never prosecuted for the crime. "'Saboteur' Describes Sinking of 2 Whalers, Does Not Confess; Says Activist 'Team' Scuttled Vessels," *Arizona Republic*, Saturday, November 15, 1986, by Knight-Ridder, p. A1.

1986. Greenpeace activists "plugged 25 factory sewer pipes this past year," says Greenpeace chief Richard Grossman. "Pranks and protests over environment turn tough," *U. S. News & World Report*, January 13, 1986, vol. 100, by Ronald A. Taylor, p. 70.

1987

6/22/87. Ferry County, Washington. Three log skidders and a log loader owned by Floyd Roush and Jerry Harrington of Republic, Washington, sustained $20,000 in damage from slashed tires, smashed diesel injector pumps, sand in the fuel tanks, smashed gauges, a slashed seat cushion and cut fuel lines. "A chronology of vandalism," *Bellingham Herald*, Monday, September 18, 1989, by Leo Mullen, p. A4.

7/6/87. Ferry County, Washington. Backhoe-excavator belonging to Poe Asphalt Co., of Clarkston, Washington, working on road construction near Republic, Washington, had dirt poured in fuel tank, alternator damaged and oil filters smashed. "A chronology of vandalism," *Bellingham Herald*, Monday, September 18, 1989, by Leo Mullen, p. A4.

7/26/87. Siskiyou National Forest, Oregon. Blockade of logging roads. Earth First co-founder Mike Roselle jailed for criminal trespass in protest. "Terrorists For Nature Proclaim Earth First!" *Chicago Tribune*, Sunday, August 2, 1987, by James Coates, p. 21A.

7/23/87. Curry County, Oregon. Six Earth Firsters chained themselves to logging equipment on the Sapphire timber sale and temporarily shut down operation of Huffman and Wright Logging. Hap Huffman, vice president, took protesters to court and in 1988 a jury awarded his company $25,000 in punitive damages and $5,714 in compensatory damages. He spent $70,000 in court costs; so far, he has collected nothing. "Jury awards $30,000 to logging company," *Eugene Register-Guard*, Thursday, November 10, 1988, by The Associated Press, p. 10C.

10/5/87. Near Flagstaff, Arizona. Earth Firsters Mark Davis and Peg Millett used a propane-and-oxygen cutting torch to cut through bolts which anchored twelve pylons supporting the main cable chair life at the Fairfield Snowbowl ski resort. Davis sent a letter of demands to stop the resort's expansion into sacred Indian territory, signed Evan Mecham Eco Terrorist International Conspiracy (EMETIC). Second Superseding Indictment, *United States v. Davis, et al.*, No. CR-89-192-PHX, p. 5.

10/23/87. Mount Rushmore National Memorial. Earth First co-founder Mike Roselle and four others arrested for attempting to unfurl a Greenpeace banner over the carved face of George Washington to draw attention to acid rain. Roselle was sentenced to four months in jail. "Arrests cut short acid rain protest," *Rapid City Journal*, Friday, October 23, 1987, by Hugh O'Gara, p. 1. See also, "Activists Climb Rushmore," *Sioux Falls Argus Leader*, Friday, October 23, 1987, by The Associated Press, p. 1.

1988

5/17/88. Mount Baker-Snoqualmie National Forest, Washington. Front-end loader owned by Summit Timber Co. of Darrington, Washington smashed at a logging site near the 36-mile marker on Mount Baker Highway. No damage estimate. "A chronology of vandalism," *Bellingham Herald*, Monday, September 18, 1989, by Leo Mullen, p. A4.

6/6/88. Mount Baker-Snoqualmie National Forest, Washington. Bulldozer and excavator owned by Summit Timber Co. of Darrington, Washington had all windows and gauges smashed, all security panels and guards broken, dirt poured in fuel tanks, reservoirs, radiator and engine openings, battery destroyed, all wiring cut, all grease and hydraulic lines cut, control levers broken off, control valves stolen, grease guns and wrenches stolen, near same site as 5/17/88 attack. In addition white spray paint messages on wreckage stated, "We Never Forget," and "Sabotage." Damage, $40,000. "A chronology of vandalism," *Bellingham Herald*, Monday, September 18, 1989, by Leo Mullen, p. A4.

7/5/88. Okanogan National Forest, Washington. Twenty-four Earth Firsters arrested for blocking access to Okanogan National Forest Building for one day, in protest of timber sales said to affect lynx habitat. Arrested were: Bradd Mitchell Schulke of Seattle; David Fleak Potter of Seattle; John Craig Lilburn of Missoula, Montana; Michael Joseph Robinson of Santa Cruz, California; Lincoln Warren Kern of Seattle; Camalla Juanita Moore of Seattle; Kirsten Lee Pourroy of Bellingham; Michael Phillip Peterson of Republic; George William Callies of Seattle; Karen Louise Coulter of Seattle; Tracy Lynne Katelman of Berkeley, California; Joanne Dittersdorf of Bellingham; Elizabeth Jane Fries of Bellingham; Thomas Grey of Bellingham; Peter Jay Galvin of Portland; Philip Randall Knight of Bozeman, Montana; David Eugene Helm of Ferndale; Kurt Stein Newman of Bayside, California; Lyn "Lee" Georges Dessaux of Santa Cruz, California; Todd Douglas Schulke of Seattle; Gregory Joseph Wingard of Kent; Steven Gary Paulson of Lenore, Idaho; Mitchell Alan Friedman of Bellingham; and Kimberly Dawn Reinking of Berkeley, California. Some in this group have gone on to leadership positions in foundation-funded environmental groups.

7/15/88. Whatcom County, Washington. Two miles of road stakes uprooted on timber sale. Replacement cost, over $10,000. Sabotage attributed to Earth Firsters by Arnie Masoner, administrative officer of Mount Baker-Snoqualmie National Forest. "War of woods: Logging terrorism," *Bellingham Herald*, Sunday, September 17, 1989, by Leo Mullen, p. A1.

8/10/88. Skagit County, Washington. A bulldozer owned by Reece Brothers Logging Co. of Darrington sustained $4,000 damage while parked off Cascade Pass Road from crowbar bashing, rocks and dirt in radiator, cut oil, fuel and spark plug lines, bent control levers, water in oil tank, and torn seat. "A chronology of vandalism," *Bellingham Herald*, Monday, September 18, 1989, by Leo Mullen, p. A4.

8/15/88. Peshastin, Washington. 20 saw blades worth nearly $20,000 destroyed at the W. I. Mill when they struck spikes in logs. "War of woods: Logging terrorism," *Bellingham Herald*, Monday, September 18, 1989, by Leo Mullen, p. A1.

8/20/88. Snap Timber Sale, Mount Baker-Snoqualmie National Forest, Washington. Acme, Washington-based logging operator John Harkness discovers more than 40 10-inch nails driven into trees as his chainsaw crew arrives for work. "Ecowars: Fighting over the forest," *Bellingham Herald*, Tuesday, September 19, 1989, by Leo Mullen, p. A4.

8/20/88. Burton Peak, Washington. Dozers belonging to Louisiana-Pacific Corporation hit. Damage at $10,000. "Ecowars: Fighting over the forest," *Bellingham Herald*, Tuesday, September 19, 1989, by Leo Mullen, p. A4.

9/25/88. North of Grand Canyon, Arizona. Earth Firsters Mark Davis, Peg Millett, Marc Baker and Ilse Asplund used a saw to partially cut through twenty-nine wooden power poles which supported electrical lines serving the Canyon Uranium mine causing two poles to topple with an ensuing power outage. Davis sent a letter to a Flagstaff newspaper claiming responsibility on behalf of EMETIC. Second Superseding Indictment, *United States v. Davis, et al.*, No. CR-89-192-PHX, p. 7.

10/9/88. Near Bellingham, Washington. A front-end loader owned by Clauson Lime Co. of Maple Falls sustained $850 damage from smashed windows and lights and stolen citizens band radio and antenna. "A chronology of vandalism," *Bellingham Herald*, Monday, September 18, 1989, by Leo Mullen, p. A4.

10/24/88. Bald Mountain, Skagit County, Washington. A logging loader owned by Joe Zender & Sons Logging Co. of Deming sustained $300 in damage from vandals emptying fire extinguisher into engine, oil and hydraulic tanks. A battery and 4 floodlights were stolen. "A chronology of vandalism," *Bellingham Herald*, Monday, September 18, 1989, by Leo Mullen, p. A4.

11/18/88. Lewis County, Washington. The diesel engine, door hinges, dash and gauges smashed on a logging yarder owned by Gardin Logging, Inc, of Winlock, Washington, were destroyed causing $20,000 loss. 250 feet of wire rope, 6 gallons of chainsaw bar oil and 48 tubes of grease worth $650 were stolen. "War of woods: Logging terrorism," *Bellingham Herald*, Monday, September 18, 1989, by Leo Mullen, p. A1.

11/29/88. Medford, Oregon. Gregory Forest Products. Bandsaw explodes after hitting spiked tree as group of aides from U. S. Senate Appropriations Committee walk through sawmill. Nobody was hurt. Six logs in the mill were found to contain long nail-like spikes. The logs were salvaged from the Silver Fir fire in the North Kalmiopsis Roadless Area of the Siskiyou National Forest. "Lawmakers' aides see saw hit tree spike," *Grants Pass Daily Courier*, December 16, 1988, by The Associated Press, p. 1.

12/6/88. Near Mineral, Washington. A log yarder and log loader owned by Packwood Lumber Sales of Packwood, Washington were sabotaged at a Lewis County logging site at a loss of $72,150. Transmission covers were taken off, filled with rocks and dirt and replaced, gears were filled with dirt and rocks and gear covers discarded. 800 gallons of diesel fuel and hydraulic fluid were drained onto the ground. Instrument wires were cut and windows smashed. Spray painted on the equipment in Forest Service yellow were phrases: "Stop Eco Rape," "In Jesus' Name" and "Jesus Lives." "Tool cache may be evidence of destruction," *Bellingham Herald*, Wednesday, September 20, 1989, by Leo Mullen, p. A1.

12/6/88. Near Morton, Washington. A log loader owned by Ron J. Wells Logging Co. of Olympia parked on Weyerhaeuser Company land sustained $16,000 damage from mud and rocks dumped into the air filter, radiator, fuel tank and hydraulic tank. Law enforcement officers believe the same person or persons vandalized both the Packwood and Wells equipment. "A sampler of the sabotage," *Bellingham Herald*, Monday, September 18, 1989, by Leo Mullen, p. A4.

12/8/88. Strawberry Point, Washington. A backhoe and front-end loader owned by One Way Construction of Sedro-Wooley sustained about $12,000 damage from mud in the radiator, gas tank, sand in the crankcase and filter system. Evidence of the sabotage was cleaned up so that workers would start the equipment, worsening the damage. "A sampler of the sabotage," *Bellingham Herald*, Monday, September 18, 1989, by Leo Mullen, p. A4.

12/21/88. Matlock, Washington. Ten pieces of heavy equipment vandalized. Three bulldozers, three log skidders, two log loaders, a road grader and a front-end loader were sabotaged with sand and gravel in oil filters, radiators and fuel tanks. "A sampler of the sabotage," *Bellingham Herald*, Monday, September 18, 1989, by Leo Mullen, p. A4.

12/88. Wenatchee National Forest, Wenatchee, Washington. Survey by Special Agent Ben Hull for all national forests found during 18-month period in 1987-1988: 219 serious acts of vandalism to Forest Service or contractors' property, amounting to $4.5 million damage. 42 letters received threatening sabotage or vandalism if the Forest Service did not prevent or stop certain logging activities. 32 demonstrations temporarily halted logging or road building, resulting in $201,000 in losses. "Forest Saboteurs Drive Suspicion, Fear Into Logging - Awareness Is On The Rise," *Seattle Post-Intelligencer*, Tuesday, February 20, 1990, by The Associated Press, p. B2.

1989

2/early/89. Bellingham, Washington. Excavator owned by Henifin & Associates, Inc., sustained $7,000 damage at night after work at Sunset Square Shopping Center. Gravel had been placed in the gasoline tank and radiator. "A sampler of the sabotage," *Bellingham Herald*, Monday, September 18, 1989, by Leo Mullen, p. A4.

2/13/89. Near Bellingham, Washington. Five pieces of heavy equipment owned by Janicki Logging Co., Inc. of Sedro-Wooley, Washington sustained more than $187,000 damage while on Georgia-Pacific Corp. logging site. A log loader, $100,000; a tower and yarder, $2,000; a compressor, $30,000; an air drill, $15,000; a bulldozer, $35,000. Sand in crankcase, oil filters, hydraulic and radiator systems of loader and tower. Compressor's 4 tires were punctured and gauges broken. Hoses, wires and levers cut on all equipment. "A sampler of the sabotage," *Bellingham Herald*, Monday, September 18, 1989, by Leo Mullen, p. A4.

2/16/89. Skagit County, Washington. A log skidder owned by Kalen Parrish Logging Co. sustained $100 damage when vandals bent the steering lever and broke off the lever to lower the blade. "A sampler of the sabotage," *Bellingham Herald*, Tuesday, September 19, 1989, by Leo Mullen, p. A4.

3/10/89. Lake Whatcom, near Bellingham, Washington. Two bulldozers owned by the Oeser Co. of Bellingham, sustained $7,000 damage. Vandals used bolt-cutter to cut off 6 padlocks on fuel tanks and engine compartments, smashed gauges, poured hydrochloric acid into radiators and fuel tanks and pulled off and threw away air cleaners. Gasoline was poured over seat upholstery of both bulldozers and set afire. Paper towels were wadded up and stuck into parts of engines and ignited. "A sampler of the sabotage," *Bellingham Herald*, Tuesday, September 19, 1989, by Leo Mullen, p. A4.

3/89. Transylvania County, North Carolina. A stack of cut logs was spiked, according to U.S. Forest Service report. "Loggers See Timber Protest In Vandalism," *Charlotte Observer*, Wednesday, February 5, 1992, by The Associated Press, p. 9C.

3/29/89. Tucson, Arizona. Earth First co-founder Dave Foreman gave Mark Davis $580 to buy thermite grenades to sabotage high voltage electrical transmission towers and lines at three nuclear facilities. On March 31, 1989, Davis mailed a certified check for $500 from the funds supplied by Foreman to pay for 50 thermite grenades to hit the nuclear plants. Second Superseding Indictment, *United States v. Davis, et al.*, No. CR-89-192-PHX, p. 9.

3/29/89. Near Powell, Idaho. Earth Firsters John P. Blount, Jeffrey C. Fairchild, Arvid E. Hartley and Neil K. McLain spiked the Post Office Creek Timber Sale on the Clearwater National Forest near Powell, Idaho. "Two Plead Guilty To Spiking Trees To Stop Sales In Idaho," *Portland Oregonian*, Saturday, June 5, 1993, from correspondent and wire reports, p. B8.

4/3/89. Tucson, Arizona. Earth First co-founder Dave Foreman and four other Earth Firsters arrested at the University of Arizona during a protest against the University's plan to build up to seven telescopes on Mount Graham. The arrests came after about 50 protesters had gathered in front of the UA Administration Building. The five went inside against orders and were arrested on misdemeanor charges of interference with a public institution. "5 Arrested In Protest At UA Over Building of Mountain Scopes," *Arizona Republic*, Tuesday, April 4, 1989, by Gene Varn, p. C20.

4/15/89. Republic, Washington. Spiked logs from a Colville National Forest timber sale cost the Vaagen Bros. Lumber Inc. sawmill $57,000 in saw damage and lost productivity. "Forest Saboteurs Drive Suspicion, Fear Into Logging - Awareness Is On The Rise," *Seattle Post-Intelligencer*, Tuesday, February 20, 1990, by Associated Press, p. B2.

4/19/89. Near Snoqualmie Pass, Kittitas County, Washington. A log loader on a carrier, a log skidder and a bulldozer owned by Swiss Skyline Logging Co. of Ellensburg sustained $240,000 damage in a massive arson. The log skidder and bulldozer were singed but not destroyed. Painted over the char were the messages, "Save the trees," and "Sorry boys, no logging today." The modus operandi in this case was not characteristic of ecoterrorists, and it was investigated by the county sheriff for other possible perpetrators, including the owners, but no charges were ever brought. Law enforcement doubts its authenticity as ecoterror, but at least one newspaper accepted it. "A sampler of the sabotage," *Bellingham Herald*, Tuesday, September 19, 1989, by Leo Mullen, p. A4.

4/89. Quilcene Ranger District, Olympic National Forest, Washington. A logging yarder owned by Boulton & Yarr Logging Co. of Quilcene sustained $20,000 damage during three separate attacks in March and April. A crowbar was driven through the radiator, gasoline and hydraulic lines were cut, hydraulic fluid was director into brake shoes to destroy them, and 10 pounds of sugar was poured into the gasoline tank. "A sampler of the sabotage," *Bellingham Herald*, Tuesday, September 19, 1989, by Leo Mullen, p. A4.

5/1/89. Granite Falls, Washington. Eight tires on eight logging trucks owned by Alpine H&S Co., Inc. of Bellingham and independent truckers working for Alpine sustained $4,000 damage when they were punctured by caltrops, spiked devices welded together out of concrete reinforcement bar, twenty of which had been set on a private road near the town. "A sampler of the sabotage," *Bellingham Herald*, Tuesday, September 19, 1989, by Leo Mullen, p. A4.

5/8/89. Thurston County, Washington. Three pieces of heavy equipment owned by North Fork Timber Co. of Chehalis, Washington and kept behind locked gates sustained $16,000 damage when they were attacked at a logging site owned by Scott Paper Co. Sand and rocks were put into the water, hydraulic, diesel and oil tanks of a log loader, skidder and tower. "A sampler of the sabotage," *Bellingham Herald*, Tuesday, September 19, 1989, by Leo Mullen, p. A4.

5/8/89. Near Bellingham, Washington. Four pieces of logging equipment owned by Blockley Logging Co. of Deming, Washington sustained $16,000 damage. Dried beans were put in the radiators, where they swelled up and turned to mush, clogging the system. Dirt was put in the engine oil and some of the transmissions. "A sampler of the sabotage," *Bellingham Herald*, Tuesday, September 19, 1989, by Leo Mullen, p. A4.

5/30/89. Near Salome, Arizona. Earth Firsters Mark Davis, Peg Millett, and Marc Baker acting under the decoupling name of EMETIC attempted to destroy power pole number 40-1 of the Central Arizona Project, a water project targeted as a test run for a nuclear plant hit. They were captured by the FBI, Davis and Baker on the spot, Millett the next day in Prescott. Second Superseding Indictment, *United States v. Davis, et al.*, No. CR-89-192-PHX, p. 14.

5/31/89. Tucson, Arizona. Earth First co-founder Dave Foreman arrested by the FBI at his home in bed for conspiring with Mark Davis, Peg Millett, Marc Baker and Ilse Asplund to damage nuclear facility power lines. "'Meek' Desert Rat Denies He's Saboteur," *Arizona Republic*, Wednesday, June 7, 1989, by Randy Collier, p. A1.

Mid-1989. Elko, County, Nevada. Jim Connelley, rancher and president of Nevada Cattlemen's Association said someone drained oil from his four-wheel tractor, resulting in $1,800 in damage. Connelley said the mechanic who fixed his tractor reported doing similar repairs for six other ranchers this year. Other Elko County ranchers report herd bulls castrated, rendering them useless; fence and water pipelines cut, troughs and water tanks overturned, windmills decommissioned, steel dropped into well casings and other types of harassment. Animal Liberation Front spokeswoman Margo Tannenbaum of San Bernardino, California said the ALF's goal is "the elimination of the livestock industry." Extreme vegetarians called vegans do not believe in eating meat, dairy products, eggs or cheese and do not wear leather or wool. "Animal Activists Blamed in Wide Ranch Sabotage," *Los Angeles Times*, Sunday, November 19, 1989, by Charles Hillinger; Mark A. Stein, p. A1.

6/21/89. Near Cle Elum, Washington. A warehouse owned by the Wenatchee National Forest was destroyed along with five trucks inside by an arson fire at a loss of nearly $900,000. The arsonist entered through a back window, spread a flammable liquid throughout the building and ignited it. Seattle-based Earth Firsters were suspected. "A sampler of the sabotage," *Bellingham Herald*, Tuesday, September 19, 1989, by Leo Mullen, p. A4.

7/2/89. Okanogan County, Washington. A road grader belonging to Gary Will Logging Co. of Loomis was doused with a flammable liquid and set ablaze at a site in the Okanogan National Forest. The loss was put at $200,000. Earlier, people claiming to be Earth Firsters called the national forest office threatening reprisals if anyone logged a timber sale in Okanogan County. "A sampler of the sabotage," *Bellingham Herald*, Tuesday, September 19, 1989, by Leo Mullen, p. A4.

7/27/89. Near Bolton, North Carolina. The equipment of an entire logging operation owned by second-generation contract logger Marvin "Bobby" Goodson of Jacksonville, North Carolina, was moved by vandals into a circle around a fuel truck, doused in fuel and torched at the Green Swamp timber sale on land owned by Federal Paper Board Company, Inc. Two notes had been left on the site prior to the arson, one stating, "Get the fuck out of here. You don't belong here." FBI investigated. A track-mounted log forwarder survived, possibly because the arsonist(s) only knew how to operate rubber-tired vehicles. $750,000 loss partially compensated after insurer conducted a harrowing 18-month investigation that cleared Goodson of self-inflicted damage, but impugned his reputation. Goodson left the logging business a year later. Telephone interview with Connie and Bobby Goodson, October 29, 1996.

8/12/89. Traders Cove, 30 miles north of Ketchikan, Alaska. A workman felling trees narrowly missed hitting with his chainsaw an 8-inch spike driven into a tree. Three trees were spiked. Forest Service officer Paul McIntosh closed the site for investigation, then re-opened for logging. "Dangerous Tree Spiking Hits Tongass," *Anchorage Daily News*, Saturday, August 12, 1989, by The Associated Press, p B1.

8/17/89. Seven miles west of Leavenworth, Washington. Six pieces of logging equipment belonging to M&W Logging/Trucking of Ellensburg, Washington, vandalized and work items stolen. Never reported in the media. Property loss notice, Washington Contract Loggers Association Insurance Agency.

9/4/89. Libby, Montana. Vandals filled the engine of a bulldozer owned by Vincent Logging Company with dirt, causing about $4,000 damage. Sgt. Gary Stratemeyer of the Lincoln County Sheriff's Department announced that four separate incidents of vandalism had been reported in the past three weeks, including slashed hydraulic lines and dirt dumped in equipment engines. Damage was $200,000. "Logging opponents eyed as vandalism continues," *Great Falls Tribune*, Tuesday, September 12, 1989, by The Associated Press, p. 1A.

9/4/89. Bellingham, Washington. Ball bearing shot through four windows of Trillium Corporation, a development company, one of more than 20 unlawful acts against the firm, which had been targeted on a hit list in the July 1988 *Washington Earth First Newsletter*.

10/16/89. Stayton, Oregon. A spike in a log from the Sullivan West Timber Sale in the Willamette National Forest hit a headrig bandsaw in the Stuckart Lumber Company mill, causing over $1,000 damage. In December, a spiked log from the same timber sale hit a bandsaw in a Bohemia Company mill in Eugene, Oregon. Not reported to the media. Letter dated January 19, 1990 from Arden W. Corey, Forest Service representative, to Debbie Miley.

12/28/89. Jackson County, Illinois. Shawnee National Forest officials confirm tree spiking in the Fairview area slated for logging by East Perry Lumber Company of Frohna, Missouri. Earth Firsters camped out on the site for 80 days in summer and fall protest. "Shawnee Forest timber sale area sabotaged," *Southern Illinoisian*, January 5, 1990, by Phil Brinkman, p. 1A.

1990

2/23/90. Yaquina Head, Oregon. In early February vandals poured sugar in the gas tank of a logging tractor belonging to logger Quinn Murk. Someone mailed Murk a death threat. A protester spit at him in a confrontation. On this Friday he found his 4-year-old beagle, Charlie, dead of a bullet wound in front of his ear. "Eco-Terrorism - 'Environmental Bozos' Bite Worse Than Bark," *Washington Times*, Monday, July 16, 1990, by Valerie Richardson, p. A1.

2/20/90. Near Brevard, Transylvania County, North Carolina. Vandals broke through a gate near Pink Beds where equipment belonging to T&S Hardwoods of Sylva was stored, started a bulldozer and used it to push a rubber-tired log skidder over the side of a mountain, trying unsuccessfully to push a TMY-70 yarder after it. Damage, $10,000. "Trees Spiked," Asheville Citizen, Tuesday, March 13, 1990, by Clarke Morrison, p. 1A. See also "Earth First! - Environmental Group's Literature Advocates Controversial Tactics," *Asheville Citizen*, Friday, March 16, 1990, by Paul Clark, p. 11A.

3/9/90. Pisgah National Forest, North Carolina. Log skidder's hydraulic hoses cut and tires slashed. Cut logs spiked in timber sale. Anonymous letter sent to an Asheville TV station and postmarked Chicago warned of spikes. Writer displayed detailed knowledge of Pisgah area's forest history, also provided map of spiked area. "Logs Spiked In Protest Of Pisgah Clear-Cutting," *Charlotte Observer*, Wednesday, March 14, 1990, by Bruce Henderson, p. 1A. See also, "Tree Spikers Deserve Law's Harshest Penalty" [editorial], *Asheville Citizen*, Thursday, March 15, 1990, p. 10A.

3/17/90. West Yellowstone, Montana. John Craig Lilburn, Lyn "Lee" Georges Dessaux and several others arrested for harassing buffalo hunters in a Fund for Animals protest of a government-approved hunt. See 7/16/90. The arresting officer, warden Dave Etzwiler, charged them with harassment, a misdemeanor, in the Justice Court of Gallatin County in Bozeman, Case Number DC92-70. Lilburn was tried and found guilty, then appealed to the District Court, Case Number CR90-0964, which upheld the guilty verdict. Lilburn then appealed to the State Supreme Court, Case Number 95-331, which remanded the case for retrial by the District Court, where the jury found Lilburn guilty on March 28, 1995. Lilburn filed another appeal to the State Supreme Court, but later filed to dismiss his own appeal. The court dismissed it on November 14, 1995.

3/17/90. Escalante River, Utah. Rancher Art Lyman of Boulder, Utah, had 15 cows and 6 calves shot to death and two cabins burned. "$13,500 In Rewards In Cattle Slaughter," *Garfield County News*, Thursday, March 29, 1990, no byline, p. 1.

4/22 and 23/90. Freedom, California. After Earth Day celebrations, Earth Firsters calling themselves the Earth Night Action Group made two consecutive hits, sawing first through two of wooden power poles and then toppling a steel transmission tower belonging to Pacific Gas & Electric Company, causing a massive failure that cut off electricity at 1:37 a.m. to 100,000 Santa Cruz County residents for 10 to 18 hours. The area was still recovering from the devastation caused by the massive October 1989 earthquake. Rosina Mazzei of Santa Cruz, a victim of Lou Gehrig's Disease, nearly died when the outage cut off her respirator and her emergency power pack began to fail. Firefighters had to use hand respirators for hours before two registered nurses took over. The vandals sent a letter to the Bay City New Service and the Associated Press in San Francisco taking responsibility. The FBI investigated the incident as officially recognized domestic terrorism with ties to Earth First. Karen Debraal, a local Earth First contact, endorsed the action. Darryl Cherney, Earth First spokesman, also endorsed the action. Even locals sympathetic to Earth First deplored the hit as having no point. "Earth Day power outage - Power poles cut; cops investigate," *Santa Cruz County Sentinel*, April 23, 1990, by Steve Perez, p. A1. "Outages cut woman's lifeline," *Santa Cruz County Sentinel*, April 23, 1990, by Maria Guara, p. A1. "Group claims responsibility - Letters say 'sabotage' directed at PG&E," *Santa Cruz County Sentinel*, April 25, 1990, by John Robinson, p. A1.

4/23/90. Frohna, Missouri. East Perry Lumber Company receives anonymous pasted-letter note postmarked April 20, 1990, from Jonesboro, Illinois, stating "Stay out of Fairview - Earth First!" National Hardwood Lumber Association *Newsletter*, May, 1990.

5/7/90. Mobile, Arizona. Eighteen protesters, including three Greenpeace organizers, were arrested after disrupting a hearing on the Ensco toxic waste incinerator under construction near Mobile. Charges were dropped against 11, but five men and two women were charged with disturbing a public hearing. The Mobile hearing erupted into chaos when 400 people arrived at the Mobile Elementary School cafeteria, which is designed to hold 100, and the protesters began clapping and shouting. Deputies used stun guns to subdue protesters resisting arrest. "'Ensco Seven' Vow To Put State, Sheriff On Trial," *Arizona Republic*, Thursday, September 27, 1990, by Jon Sidener, p. B1.

5/90. Near North Bend, Washington. Earth Firsters form a decoupling group calling itself the Cedar River Action Group to protest the Sugar Bear Timber Sale on the Mount Baker-Snoqualmie National Forest in the Cedar River Watershed that supplies water to the city of Seattle. Trees were spiked and protesters provoked fistfights with loggers in front of TV cameras. Dennis Marshall, a logger from Enumclaw, Washington, won the logging contract and had to install expensive security systems to guard his equipment. Cedar River Action Group confrontations at Sugar Bear cost the Forest Service and King County police about $26,000 for additional security. "Security Gantlet Thrown Up

To Discourage Logging Foes," *Seattle Times*, Monday, August 6, 1990, by Louis T. Corsaletti, p. D1.

5/30/90. Seattle, Washington. Federal Protective Service officers arrested environmentalists occupying the Forest Service offices in downtown Seattle at noon. The protesters entered the Holyoke Building, First Avenue and Spring Street, about 10:30 a.m. and chained themselves to furniture. Michael N. Christensen (alias Asanté Riverwind), Felicia Staub, Ken Richards, Tim Dimock and four others identified themselves as members of Cedar River Action Group, protesting planned logging of old-growth timber in the Cedar River watershed. The officers said the protesters would probably be cited for disobeying federal officers. The offense has a maximum penalty of $5,000 or 30 days in jail. A Forest Service spokesman said the demonstration would not stop the contract logging of a 73-acre parcel of land called Sugar Bear. "Old-growth protest," *Seattle Times*, Wednesday, May 30, 1990, no byline, p. E1. See also "Cedar R. logging contract protested - Officials proceeding on planned harvest," *Seattle Times*, Thursday, May 31, 1990, by Linda Keene, p. F6.

July, 1990. Libby, Montana. Bruce Vincent, after speaking out on television against environmental extremists, began receiving threats against his four children. A bulldozer at his company, Vincent Logging, was sabotaged. "Eco-Terrorism - 'Environmental Bozos' Bite Worse Than Bark," *Washington Times*, Monday, July 16, 1990, by Valerie Richardson, p. A1.

7/16/90. Gallatin County, Montana. Earth Firster Lyn "Lee" Georges Dessaux arrested on two counts of misdemeanor assault, one for multiple stabbings of Dan R. Jacobs of Kalispell with a ski pole, the other for multiple stabbings of Hal Slemmer of Billings with a ski pole, in a Fund for Animals protest on March 17 to disrupt the buffalo hunt near West Yellowstone. Dessaux became so violent that the animal rights video camera operator documenting the scene stopped taping and screamed at Dessaux to stop. Dessaux was found guilty on both counts February 8, 1991 in a jury trial in Justice Court in Bozeman (Case Number CR90-0963). Justice of the Peace Scott Wyckman sentenced Dessaux to 6 months in Gallatin County Jail, suspended to 45 days. Dessaux was released after 28 days on condition he perform community service with the Humane Society in lieu of paying court debts, but the shelter rejected his help. *Sentence* in the case of *The State of Montana v. Lyn Georges Dessaux*, Case Number CR90-0963, February 8, 1991. "Shelter Rejects Bison Activist - Humane Society Won't Take His Help," *Bozeman Daily Chronicle*, Saturday, March 9, 1991, by The Associated Press, p. 1A.

7/31/90. Syracuse, New York. Twelve Earth Firsters tried to halt Onondaga County health officials from killing off mosquitoes in Cicero Swamp believed to carry eastern equine encephalitis, a disease that caused the death of a 7-year-old Camillus, New York boy in 1983. Earth Firster Andy Molloy of Syracuse said using pesticides in the swamp, about 7 miles north of the city, could kill off other insects and disrupt the bottom of the food chain. The Earth Firsters said they would hide in the swamp so applicators could not use the insect killer. "Earth First! To Hide In Swamp Protest," *Albany Times Union*, Tuesday, July 31, 1990, by The Associated Press, p. B6.

9/3/90. Schroon Lake, New York. Warrensburg, New York, Supervisor Maynard Baker slugs Earth First activist Jeff Elliott of New Hampshire in a confrontation at the entrance to Crane Pond Road. Earth Firsters came into the Adirondack area to protest the state's decision to leave the road open to customary vehicular traffic. Baker and a group of locals protested the protest. A film crew had been told that Baker and his faction would become violent, stayed for half an hour before Baker responded to Earth First insults and spitting with a flurry of fists. As soon as Elliott got up and produced a bloody nose for the camera, the crew departed and the edited footage was used in a *CBS News 60 Minutes* segment to prove the wise use movement is violent.[27] Other media didn't buy it. Provocation to violence grew into a planned tactic. "Earth First! Up Against One Hero," *Albany Times Union*, Sunday, September 9, 1990, by Fred LeBrun, p. B1.

October, 1990. Graham County, North Carolina. The Cheoah Ranger District received a letter saying trees in the Cheoah Bald timber sale had been spiked. A few days later a second letter arrived saying the trees weren't spiked. Two months later, some trees were spiked and equipment damaged in the northern part of Macon County at a Forest Service timber sale. "Loggers See Timber Protest In Vandalism," *Charlotte Observer*, Wednesday, February 5, 1992, by The Associated Press, p. 9C.

10/20/90. Near Ely, Minnesota. The Minnesota Department of Natural Resources (DNR) finds spikes in 5 large old white pines and a smaller aspen in the so-called Kawishiwi Triangle area, where the Minnesota Department of Natural Resources sold logging rights on 150 acres of state-owned land to Hedstrom Lumber Co. of Grand Marais, Minnesota. An anonymous letter sent in early October to the company, the DNR and the news media threatened the spiking. "Metal Spikes Found In Trees To Be Logged - Anonymous Letter Warned Of Act Near Ely," *Star Tribune*, Sunday, October 21, 1990, by Dean Rebuffoni, p. B1.

10/28/90. Big Reed Pond, Maine. After an August phone threat to Stephen Schley, president of timber firm Pingree Associates, Inc., spikes were found in trees on their land and that of International Paper Company. Schley was told that 400 spikes had been placed. Loggers from Seven Islands Land Company using metal detectors found more than 100 spiked trees in logging area. Earth Firsters Jamie Sayen, founder of Earth First affiliate Preserve Appalachian Wilderness, and Jeff Elliott of North Stratford, New Hampshire, served with criminal trespass notices. Elliott had a month earlier provoked a fistfight for CBS News cameras in New York. "Maine land manager confirms tree spiking near virgin forest," *Waterville Morning Sentinel*, October 28, 1990, by the Associated Press, p. 3. "Ecological saboteurs threaten foresters in old Maine forest," *Maine Sunday Telegram*, Sunday, November 18, 1990, by Tux Turkel, p. 1A. "Maine logging battle turns dangerous," *Boston Sunday Globe*, Sunday, December 16, 1990, by Denise Goodman, p. 57. "Tree spiking is an act of terrorism," [editorial], *Portland Press Herald*, November 20, 1990, p. 6A. "Spike the Ecotage" [editorial], *Lewiston Sun-Journal*, November 20, 1990, p. 6A.

12/90. Washington County, Maryland. Hardwood sawmiller Walter H. Weaber Sons, Inc, of Lebanon, Pennsylvania, received phone calls that trees in a timber sale had been spiked. Weaber confirmed that most trees had 16-penny nails driven in about 6 feet up. Fellers marked trees and screened logs with hand-held metal detectors. A suspect was identified. *Pulpwood Highlights*, American Pulpwood Association, April 1991.

12/90. Marys Peak, Oregon. Loggers find ceramic spikes in trees near Corvallis, Oregon. Ceramic spikes cannot be found with metal detectors and are much more dangerous than metal spikes. Forest Service offers $15,000 for information leading to conviction. Oregon Lands Coalition, *Network News*, 1/4/91.

12/7/90. Near Ukiah, California. Feller buncher worth $700,000 belonging to Okerstrom Logging & Trucking of Willits caught fire on Daugherty Creek while working on Louisiana-Pacific land, the probable result of hydraulic fluid sabotaged with abrasives. The same feller buncher had been burned under similar circumstances on April 3, 1990, but was thought to be an accident. "Fire called accident," *Santa Rosa Press Democrat*, April 5, 1990, p. B1. Prior to that, Okerstrom's Barko-500 log loader was burned under suspicious circumstances. The Sheriff's department was called in to investigate because of escalating tensions between radical environmentalists and loggers. An Okerstrom heel boom loader worth $200,000 was also torched at a later time. Steve Okerstrom, owner of the family company, received numerous threats demanding he stop logging and had vandalism at his home, including graffiti saying "We'll do to you what you do to the trees" sprayed on walls, cars, company vehicles and sidewalks. Several Earth Firsters including Mike Jakubal bragged they had hit the feller buncher, which Earth Firsters protested with banners during its appearance at the Redwood Logging Conference in Ukiah, but no charges were ever brought. "Logging-site fire near Ukiah," *Santa Rosa Press Democrat*, Saturday, December 8, 1990, by Mike Geniella, p. B1.

1991

2/14/91. Lake Crescent, Washington. Vandals torched the Fairholm Ranger Station and the Soleduck Entrance to Olympic National Park. Logger group posts $7,500 reward for capture and conviction of the vandals. News Release, Washington Commercial Forest Action Committee, Forks, Washington, February 26, 1991.

3/27/91. Franklin, North Carolina. Cook Brothers Lumber Company circular saws damaged when logs with concrete spikes harvested on the Nantahala National Forest in December hit blades. $1,000 damage. A skidder at the logging site had a tire flattened. A note at the site said, "These trees spiked. Green Peace." It was not clear whether the vandals had any connection to the international organization Greenpeace. "Spiked Trees Found," *Asheville Citizen*, Thursday, March 28, 1991, by Bob Scott, p. 1A.

6/13/91. Murphysboro, Illinois. Three Earth Firsters, Rene Cook of Murphysboro, John Wallace of Waterloo and Thomas Herb of Carbondale were convicted on misdemeanor charges of disorderly conduct, obstruction and criminal trespass. "Three convicted in timber protest," *Southern Illinoisian*, June 13, 1991, by Phil Brinkman, p. 1A.

June 1991. Grass Valley, California. Publication of *A Declaration of War: Killing People to Save Animals & the Environment*, by Screaming Wolf [pseudonym of Sidney and Tanya Singer], Patrick Henry Press. First overt recommendation to kill people to save nature. The book stirred controversy within the radical environmental movement. Some claimed it was published by anti-environment activists to discredit the movement, a standard "defenses of innocence" and "shifting attention" tactic. The actual authors and publishers were Sidney and Tanya Singer of the Good Shepherd Foundation, long-time animal rights activists. The publishers, now in Canada, claimed that the book had been sent to them anonymously on a computer diskette. The book is outlawed in Canada and the United Kingdom under incitement to violence laws.

9/30/91. Graham County, North Carolina. Forest Service finds 50 trees spiked with 20-penny nails in the Grassy Gap-Wesser timber sale. Southern Lumber Manufacturers Association, *Management Update*, 10/1/91.

8/27/91. Shawnee National Forest, Illinois. Twenty-five Earth Firsters protesting the Fairview timber sale were arrested. A protester, Jan Wilder-Thomas, faced assault charges for slapping a deputy. One of those arrested, Chris van Daalan, co-director of Save America's Forests, claimed his hand was broken when he was run over by a logging truck belonging to East Perry Lumber Company. Authorities at the scene said van Daalan was hit by a small log he threw in the path of an oncoming pickup truck. "Forest protest leads to arrests - Man injured trying to stop logging work," *Southeast Missourian* (Cape Girardeau, Missouri), Thursday, August 27, 1991, by Mark Bliss, p. 1.

8/30/91. Chicago, Illinois. Chicago Tribune editorial commented, "While most protesters at Shawnee pursued legal or non-violent confrontation, some engaged in tree spiking—a despicable practice meant to injure loggers wielding chain saws. This warping of values, the exalting of trees over people, serves only to lose the cause, no matter how valid the questions." "Upholding the policy at Shawnee" [editorial], *Chicago Tribune*, Friday August 30, 1991, p. 22.

10/3/91. Cheoah Bald in Nantahala National Forest, North Carolina. More than 300 trees spiked, at least the third incident of tree spiking in Western North Carolina. An unsigned September 5 letter, postmarked in Charlotte, warned the U.S. Forest Service office in Graham County of the Cheoah spikings. The letter told the service to stop the sales of six timber tracts in the area, about 90 miles west of Asheville. "Tree-Spiking Is Biggest Attack In N. C. Mountains," *Charlotte Observer*, Thursday, October 3, 1991, by Bruce Henderson, p. 1A.

11/20/91. Applegate Valley, Oregon. Three trucks, a bulldozer and a roadgrader belonging to Monte Walker, Inc. were vandalized on Boise Cascade land with mud in crankcases, transmissions and radiators. Hoses were cut, filters pried off, tires slashed and truck dashboard ripped out. Loss of more than $50,000. "Vandals Hit Equipment At Applegate Valley Logging Site," *Portland Oregonian*, Wednesday, November 20, 1991, by Roy Scarbrough, p. B2.

1992

2/3/92. Macon County, North Carolina. Hilton Cabe and Hennessee Hardwood Corporation loggers found their equipment in the Buck Creek and Rich Mountain areas of Macon County sabotaged. Damage to the equipment was estimated at $15,000. Tires were punctured, electrical wires on dozers were cut, caltrops were in the road, and hydraulic and air lines were cut. Hennessee officials said the damage would not be covered by insurance. U.S. Forest Service and county officers reported recovery of some physical evidence, including a footprint and tire prints. "Loggers See Timber Protest In Vandalism," *Charlotte Observer*, Wednesday, February 5, 1992, by The Associated Press, p. 9C.

4/8/92. Walbran watershed, Vancouver Island, British Columbia, Canada. Logger Ernie Calverley's chainsaw hits spike in tree, saw jumps back, nearly hitting him in the face. MacMillan Bloedel, the firm logging the site, halted logging to search the site with metal detectors, found 28 more. Royal Canadian Mounted Police labeled the incident "an act of terrorism." Vancouver *Sun*, 4/9/92 and *Globe and Mail*, 4/9/92.

4/24/92. Walbran watershed, Vancouver Island, British Columbia, Canada. 100 trees found with 6-inch metal spikes with points sticking out in holes drilled into trees. Holes are disguised with bark. Such spikes will fly out at high speed when hit by a chainsaw, creating "missiles in silos waiting for launch," said Constable Dan Chasic of the Lake Cowichan Royal Canadian Mounted Police. Vancouver *Province*, 4/24/92.

5/8/92. Weld, Maine. Earth Firsters trenched a road and erected a barricade of stones and dead trees to stop the harvest of 11,600 cords of wood by Timberlands, Inc. of Dixfield from Mount Blue State Park. "Earth First! protests Mount Blue harvesting," *Kennebec Journal*, Friday, May 8, 1992, by The Associated Press, p. 9.

6/8/92. Oakland, Maryland. Wood Products, Inc. harvested trees from the Hopemont timber sale near Terra Alta in Preston County, West Virginia sabotaged with 3/8-inch railroad-style metal spikes with the heads clipped off, which entered their sawmill and destroyed 2 bandsaws, causing $5,000 damage. Company offers $5,000 reward for information about the incident. News release from Wood Products, Inc., John M. Forman, contact, dated June 8, 1992.

6/11/92. La Vérendrye wildlife preserve, Quebec, Canada. Gatineau Logging finds spikes in trees slated for logging. *Ottawa Citizen*, 6/11/92.

8/13/92. Cove/Mallard, Idaho. Three Earth First activists pleaded guilty to criminal trespass and spent 9 days in jail for locking themselves to road-construction equipment in Nez Perce National Forest. Tree-spiking discovered in old-growth timber stand nearby. The three had been charged with resisting arrest, obstructing justice and injury to a vehicle. Each was fined $100 and sentenced to 60 days in jail. "Earth First! Activists Get Out of Jail - Trio Had Chained Selves To Road Construction Gear In Idaho; Spiked Tree Found Nearby," *Rocky Mountain News*, Thursday August 13, 1992, by Associated Press, p. 1.

8/24/92. Winooski, Vermont. Tom Carney, an Earth Firster from Schenectady, New York was arrested for criminal trespass after trying to disrupt a hydroelectric dam construction project. "Schenectady Man Arrested At Protest," *Albany Times Union*, Tuesday, August 25, 1992, by The Associated Press, p. B6. See also "Timber war brews - Shawnee forest management plan becomes question of use or abuse," *Chicago Tribune*, Sunday, August 30, 1992, by Hugh Dellios, Section 2, p. 1.

10/20/92. Long Beach, California. Six environmental activists, including Earth Firsters Jim Flynn of Portland, Ore., and Jake Kreilick (using his alias Jake Jagoff) of Missoula, Montana, protesting logging of rain forests, handcuffed themselves to cranes on the Sammi Superstars, a South Korean freighter docked in Long Beach and for several hours kept longshoremen from unloading plywood from Indonesia. The six were from Greenpeace, Earth First and the Rainforest Action Network (founded and led by Earth Firsters) and boarded the ship at about 7 a.m. They were charged with misdemeanor trespassing. The maximum penalty is a six-month jail sentence and $500 fine. Two others dangled from ropes on the side of the ship beside a banner that read, "Stop tropical timber imports." Illustrated cooperation between Earth First and Greenpeace. "Environmental Activists Chain Selves to Cranes of Lumber Ship Protest: Unloading of plywood from Indonesian rain forests is blocked temporarily. Demonstrators agree to leave ship and be cited for misdemeanor trespassing.," *Los Angeles Times*, Wednesday, October 21, 1992, by Rick Holguin and Maria L. La Ganga, p. B1.

11/21/92. Pagosa Springs, Colorado. For the second time in 16 months, loggers discovered foot-long, 5/8-inch-diameter spikes in at least 48 trees at a timber sale site in southwestern Colorado's San Juan National Forest. "Trees Spiked At Sand Bench Site," *Rocky Mountain News*, Saturday, November 21, 1992, by Associated Press, p. 26.

12/17/92. Bangor, Maine. Bangor Daily News editorial announces that an unsigned letter dropped off at the switchboard claimed that Earth First had driven spikes into 200 trees at Mount Blue State Park in western Maine. The News printed the story, but commented, "Of course, this is exactly what Earth First! wants: press coverage of its terroristic activities whether or not the group actually bothers to go out into the woods.... It's worth wondering whether the group really is Earth First! or Self-Promotion First!" "Something First!" [editorial], *Bangor Daily News*, Thursday, December 17, 1992, p. 10.

1992. Flathead National Forest, Idaho. The Forest Service ran up $250,000 law enforcement bills patrolling the remote 78,000 acres of the Cove / Mallard area beset by Earth First protesters. "Flathead Forest cops predict more Earth First! protests," *Whitefish [Montana] Daily Inter Lake*, October 18, 1993, by Ben Long, p. 1.

1993

5/2/93. Near Flagstaff, Arizona. Dump truck belonging to High Desert Investment Company of Flagstaff destroyed. Letter left behind said, "This letter is to inform you that your 1977 dump truck was not destroyed by young vandals, it was monkeywrenched." It was signed, "Coconino Clyde and his merry band of eco-warriors." Copy provided by Coconino County Sheriff's Department.

6/4/93. Missoula, Montana. Arvid E. Hartley and Neil K. McLain pleaded guilty in federal court to misdemeanor charges of spiking trees in Idaho. They agreed to testify against three others accused in a tree-spiking March 29, 1989. Accused were John Blount, Jeffrey C. Fairchild and Daniel A. LaCrosse. Hartley and McLain both admitted that they put metal spikes in trees with the intent to hinder a timber sale in the Post Office Creek area of the Clearwater National Forest near Powell, Idaho. "Two Plead Guilty To Spiking Trees To Stop Sales In Idaho," *Portland Oregonian*, Saturday, June 5, 1993, from correspondent and wire reports, p. B8.

7/18/93. Cove/Mallard Area, Idaho. Vandals cut hydraulic and fuel hoses, dumped dirt in fuel and oil tanks, smashed instrument consoles on logging equipment owned by Highland Enterprises, Inc. of Grangeville, Idaho, causing over $60,000 damage. During the weeks before and after the incident, Forest Service and Idaho County lawmen arrested more than 50 Earth Firsters who built barricades, climbed trees, and locked themselves to equipment protesting logging in the area. "Earth First! making enemies in Idaho anti-logging effort," *Lewiston Morning News Tri-*

bune, Tuesday, September 7, 1993, by The Associated Press, p. B13.

8/11/93. Spokane, Washington. Earth Firsters John P. Blount, 32, of Masonville, Colorado, and Jeffrey C. Fairchild, 27, Ashland, Wisconsin, were sentenced for convictions on two counts of tree spiking the Post Office Creek timber sale in Idaho in March 1989, two counts of destruction of federal property and two counts of conspiracy. Blount was handed a 17-month jail term and fined $1,000. Fairchild was given 60 days in jail and a $1,000 fine. Both were ordered to pay a one-quarter share of the $19,639 in damage to spiked trees. A third defendant, Daniel A. LaCrosse, 36, Salem, New Hampshire, was charged with conspiracy to spike trees and conspiracy to destroy government property. Two other defendants, Arvid Hartley and Neil McClain were both sentenced to 90 days home detention and ordered to pay their share of the $19,639. "Clearwater Forest Trial Opens In Idaho Tree-Spiking Case - Lawyer Denies Intent To Harm Anyone," *Lewiston Tribune*, June 8, 1993, by The Associated Press, p. 5A. See also, "U.S. tree-spiker sentences 'surprisingly' stiff," *Vancouver Sun* (British Columbia), August 17, 1993, by Neal Hall, p. C1.

8/27/93. Nez Perce National Forest, Idaho. Seven Earth First activists convicted of criminal trespass for violating a U.S. Forest Service closure of the Cove-Mallard area. Those found guilty were: Jacob Lawrence Bear, 24; Lawrence Alan Juniper, 44; Michael Richard Vernon, 43; Michele E. Pflam, 24; Beatrix A. Jenness; Peter J. Leusch; and Megan E. McNalley. Pflam received the biggest fines as Boyle imposed $250 for violating the closure and another $250 for interfering with authorities by chaining herself with a bicycle lock July 15 to the rear axle of a Forest Service vehicle. "Earth First Activists Convicted Of Violating Forest Closure," *Portland Oregonian*, Friday, August 27, 1993, by The Associated Press, p. D6.

10/15/93. Near Essex, Montana. A skidder belonging to Bruch Logging Company of Kalispell, Montana, had dirt put in gas tank and air compressor ripped off, a crawler tractor had its filters smashed and transmission damaged, and a boom crane had its tires slash, gasoline poured on the engine and set fire at the Challenge blowdown timber sale on the Flathead National Forest. $50,000 damage. A log deck by the road had spray painted on it, "Sale spiked. Fuck you very much." "Spike me," was sprayed on the crawler. Earth Firster Ronald J. Constable was convicted in federal court of spiking trees at the site after walking two miles past a locked gate, but was not charged in the sabotage because it is not a federal crime. "Logging site vandalized; trees spiked," *Whitefish [Montana] Daily Inter Lake*, October 16, 1993, p. 1. See also "Tree-spiking conviction a first," *Helena Independent Record*, Saturday, October 5, 1996, by Mark Goldstein, p. 1.

10/31/93. Reno, Nevada. A bomb in a briefcase or satchel blasted a 3-foot hole through the roof of the federal Bureau of Land Management office in an explosion heard for at least two miles. Bomb experts with the FBI, the Bureau of Alcohol, Tobacco and Firearms and other agencies searched the debris for clues in the bombing, which took place early Sunday. No one took responsibility. A month earlier, the *Earth First Journal*, Mabon, September-October, 1993, on page 34, published a section in which the Earth Liberation Front of Germany called for an "International Earth Night" on Halloween as part of an "International Action Week," October 31 through November 6, discussing government policy, urging property damage as an effective tactic for change, and recommending that no credit should be taken for ELF actions in order to thwart law enforcement. "Nation Datelines," *San Francisco Examiner*, Monday, November 1, 1993, compiled from Examiner wire reports, p. A9.

October. Moscow, Idaho. Despite two years of protests that cost taxpayers $400,000 in the Cove / Mallard area, Earth First activists failed to alter Forest Service management, said Nez Perce National Forest Supervisor Michael King. "Forester says Idaho protests haven't helped," *The Associated Press* report, October 18, 1993.

1993. Captain Paul Watson's book *Earthforce! An Earth Warrior's Guide To Strategy* published. An instruction manual in nine situations including civil disobedience, infiltration, and "striking illegally and with great destruction." Watson stressed the prohibition against injuring or killing any living being. Chaco Press, La Canada, California.

1994

1/11/94. Safford, Arizona. Saboteurs did $20,000 damage by putting an abrasive grinding compound under the valve covers of a snow blower and a front-end loader parked on the access road to the Mount Graham observatory, Graham County Sheriff Richard Mack said. The diesel loader was heavily damaged when it was started, leading crews to check the snow blower and discover the abrasive. The damage recalls sabotage methods recommended by the radical environmental group Earth First!, some of whose members opposed the project, officials said. "Equipment Sabotaged Near Mt. Graham Observatory," *Arizona Republic*, Tuesday, January 11, 1994, by The Associated Press, p. B2.

March 1994. Olympia, Washington. Allan Wirkkala Logging, $8,000 damage. Earth First claims responsibility. Washington Contract Logging Association Insurance Loss Report.

March 1994. Quinault, Washington. Tobin Logging, $10,000 damage. Earth First claims responsibility. Washington Contract Logging Association Insurance Loss Report.

April 1, 1994. Kalispell, Montana. District Court Judge Michael Keedy sentenced two brothers, Earth Firsters Daniel Sean Carter and Michael Thomas Carter, for spiking trees, cutting down billboards and vandalizing logging equipment. Daniel pleaded guilty to 2 of 4 counts of felony criminal mischief, received nine years in prison, suspended, ordered to pay $5,884 in restitution and 200 hours community service. Two charges of cutting down signs were dismissed. Michael pleaded guilty to 3 of 12 counts of felony criminal mischief, was given 19 years prison with all but 90 days in county jail suspended, ordered to pay $34,473 in restitution and 200 hours community service. Other charges of cutting down billboards were dismissed. Michael was also convicted of vandalizing equipment belonging to Schellinger Construction Company. The crimes were committed between December 1989 and July 1991. The brothers worked as carpenters. "Carter brothers to pay $40,000 for spiking trees," *Hungry Horse News*, Thursday, April 7, 1994, by Becky Shay, p. 25.

April 1994. Snoqualmie Pass, Washington. Bill Burgess Logging, arson fire, $50,000 loss. Earth First claims responsibility. Washington Contract Logging Association Insurance Loss Report.

7/4/94. Near Nathrop, Colorado. Perpetrator(s) using a .270-caliber rifle shot and killed six head of cattle, four cows and two yearlings, belonging to Frank C. McMurry on a private grazing allotment in the San Isabel National Forest. Sheriff's investigators stated there was reason to believe the shooter was an anti-cattle activist. McMurry is a Chafee County Commissioner vocally opposed to the Clinton administration's anti-rangeland proposals, leading investigators to believe he was deliberately targeted. A spent bullet was sent to the Colorado Bureau of Investigation Crime Lab. "Sharpshooter kills six head," *Western Livestock Journal*, July 11, 1994, no author, p. 1.

7/27/94. Near Olympia, Washington. Log loader valued at more than $200,000 belonging to Dave Littlejohn Logging Company, torched at about 4 p.m. See 7/31/94.

7/31/94. Near Olympia, Washington. Log skidder, two fire trucks and a bulldozer belonging to Littlejohn torched about 2:45 p.m. Damage, over $80,000. Final loss to Littlejohn, over $300,000. On August 9, the Washington Contract Loggers Association's answering machine tape held a message spoken by a computer-generated robotic voice: "The recent destruction of logging equipment was retaliation by the Earth Firsters to protect the planet earth from logging." "Two fires: A logging representative says the action looks typical of radical environmental groups," *The Olympian*, Tuesday, August 2, 1994, by Brad Shannon, p. 1A.

8/30/94. Santa Fe, New Mexico. Paving equipment worth $700,000 was sabotaged at the Las Campanas residential development by at least two sophisticated operatives who used front end loaders to tip over heavy paving equipment. A $500,000 16-ton paving machine was rolled over on its back and destroyed; a $100,000 steel-wheel roller was flipped on its side and destroyed; the cab of a $25,000 water truck was smashed, rendering it useless; a pneumatic rubber-tired machine had dirt dumped into its engine. Undersheriff Ray Sisneros said he had possible suspect information, but not enough to prosecute. "Vandals Destroy Paving Machines," *Albuquerque Journal*, August 30, 1994, p. B1. Part of the article was reprinted as "Road Equipment Jujitsu" *in Earth First!*, Mabon / September, 1994, p. 32.

9/3, 15 and 29/94. Near Mount Abraham, Maine. $40,000 worth of logging equipment belonging to Jack Frost, owner of the J. W. Frost Company in Anson, was destroyed, including a crane that was hit and repaired, then hit again. "Vandals hit loggers in Franklin County; $5,000 reward set," *Waterville Morning Sentinel*, Monday, October 24, 1994, by Betty Jespersen, p. 1.

9/6/94. Vancouver, British Columbia, Canada. Ranchers receive threats from Earth Firsters. A Vancouver-based Earth Firster using the pseudonym George Hayduke, after the character in Edward Abbey's *The Monkey Wrench Gang*, tells a reporter, "There's a war against the environment—we're soldiers in that war. We're going to cost the ranchers money. We're going to hurt them. We're going to punish them." The reporter, unaware that "Hayduke" was a pseudonym, wrote, "Mr. Hayduke's band of eco-warriors has begun a covert campaign of sabotage and intimidation against 10 ranchers he said are using a chemical called Compound 1080 to rid themselves of predators." Compound 1080 is applied only by government conservation officers when predators threaten livestock. Fence cutting was reported by ranchers. "Ecogroup terrorists: Ranchers - Earth First! at 'war,'" *The Province*, Tuesday, September 6, 1994, by Jason Proctor, p. A6.

10/13/94. Sandy River Plantation, Maine. Vandals broke down gates to a wood yard, destroying equipment belonging to nine subcontractors working for Randy Cousineau of Strong. Seven skidders were smashed, fuel lines cut, gauges destroyed. A skidder belonging to Virgil Toothaker of Livermore Falls was burned. Windows were broken out of two dinner shacks and five-gallon cans of motor oil were poured on the floor. Total damage, $25,000. That same night Jack Frost was hit again, with tires slashed and a rock rake damaged. "Vandals hit loggers in Franklin County; $5,000 reward set," *Waterville Morning Sentinel*, Monday, October 24, 1994, by Betty Jespersen, p. 1.

10/15/94. Franklin County, Maine, on Route 26. Log skidder belonging to one-man contract logger David Boynton of Kingfield was torched and destroyed. "Vandals hit loggers in Franklin County; $5,000 reward set," *Waterville Morning Sentinel*, Monday, October 24, 1994, by Betty Jespersen, p. 1.

10/28/94. Vancouver, B. C., Canada. A self-proclaimed environmental group threatened to kill government lawyers and poison courthouses. A group calling itself the David Organization made the threats in a letter to Chief Justice Allan McEachern. The letter said "unless the courts do something to save the environment it will kill us all," reported Bob Wright, Vancouver regional Crown counsel. "It talks about executions of Crown counsel. They talked about poisoning the air system, that kind of stuff." "B. C. Tightens Court Security," *Seattle Times*, Friday, October 28, 1994, by Seattle Times staff, p. B2.

1995

1/4/95. Shaftesbury, Vermont. Anonymous postcard addressed to Danny Fryar, Catron County Manager, Horse Springs, New Mexico, angrily alleged that ranchers received subsidies in the form of lower grazing fees, saying it was criminal. The card said, "I'll welcome the opportunity to confront you bastards and blow your fucking heads off and that goes for Zeno Kiehne, Betty Hyatt, James Catron, Kit Laney, Brut Stone, Richard Manning, Frank Nagol, Ed Cramer, Howard Hutchinson [all county officials, association employees or ranching advocates] and the rest of you sons of bitches." Fryar gave the card to the Catron County Sheriff, who gave it to the FBI for investigation. At press time Fryar had not heard from the FBI. Telephone Interview with Danny Fryar, October 15, 1996.

3/19-24/95. Churchill County, Nevada. A cow camp in the West Lee Canyon of the Stillwater Mountain Range near Fallon, Nevada was destroyed by arson. A bunkhouse and cookhouse owned by the Kent family in West Lee Canyon were burned to the ground. The arsonist(s) attempted to burn down the nearby corrals, and water facilities at the ranch sustained heavy damage. Bureau of Land Management signs along the road to the ranch were also destroyed by apparent shotgun blasts. "Sheriff's Department is searching for arsonist," *Lahontan Valley News and Fallon Eagle Standard*, Wednesday, April 5, 1995, p. 1.

4/12/95. Stevenson, Washington. Earth Firsters vandalize golf course at Skamania Lodge, where the Northwest Forestry Association was meeting. Rock salt etched "Stumps suck," "corporate scum," and "EF," initials of Earth First, on the putting greens of the Skamania Lodge golf course. Inside the lodge, vandals detonated a stink bomb that evacuated the building. The Skamania County sheriff's department could not identify suspects. Earth First Journal for April, 1995 published instructions on how to sabotage golf courses, concentrating on vandalizing irrigation systems. "Vandals Mar Meeting Site of Timber Executives," *Portland Oregonian*, Wednesday, April 12, 1995, by Peter Sleeth and Joan Laatz, p. C1.

4/14/95. Camas, Washington. The James River Paper Plant suffered $250,000 damage when individual(s) tripped the mill's power to its steam boilers and dumped lignin in the plant's sewer recovery system. The boilers were within 10 to 15 minutes of exploding when plant maintenance workers discovered the sabotage. Had the boilers exploded, a number of plant employees would have been killed or seriously injured. Camas Police Department.

4/15/95. Near Deming, New Mexico. Thirty-one cows and calves were shot and killed in two separate incidents. The first incident occurred between 6:30 p.m., Friday, April 14 and 3:00 p.m. Saturday, April 15, 1995, when one or more persons entered Tom Kelley's Tres Lomitas Ranch and killed 20 head of cattle with a high velocity rifle. Each cow was killed with a single shot at relatively close range. All shell casings were picked up and removed by the shooter(s). During this incident, a water storage tank supplying water to four separate pastures was emptied by removing a pipe fitting. In addition, the windmill used to fill the storage tank was disabled by breaking the sucker rod. The bolts on the legs of the windmill were removed. The vandalism was discovered before the mill blew down. Interview with Tom Kelley, Reno, Nevada, May 11, 1996.

4/20/95. Aptos, California. Three log trucks belonging to General Lumber Co. blown up with crude pipe bombs. The trucks were parked in a shed on Fern Flat Road in Aptos. Investigators checked whether the trucks' owner, Andrew Siino, had any known enemies or other problems, but found none. Rod Composti of Aptos was scheduled to start logging the following Monday on Fern Flat Road and planned to lease some of the owner's trucks to haul logs. The Sheriff's Department, working with the federal Bureau of Alcohol, Tobacco and Firearms said it was a case of eco-terrorism because of long-standing opposition to logging in the area. Neighbors who had opposed logging on Fern Flat Road ridiculed the idea. "Police Probe Bombing of 3 Logging Trucks - Aptos Neighbors Scoff at Suspicion of Environmentalists," *San Francisco Examiner*, Thursday, April 20, 1995, by Jane Kay, p. A4.

June 1995. Salem, Oregon. Associated Oregon Loggers issues a Member Alert, warning that a major private timberland owner reported explosives wired into lock boxes on their forest gates. "When you try to unlock the gate the charge goes off along with your hand." An AOL member also warned of finding syringes in gate lock boxes. Members were cautioned to examine all forest gate lock boxes before attempting to open them, to notify law enforcement if explosives were discovered and not to attempt opening a wired gate.

9/3/95. Government Flats, California. A three-year-old Brahma-cross cow belonging to Alan Flournoy was shot 15 times in the right side with a small caliber handgun about 16 miles west of Paskenta, California. The animal died a day or more later. "Cow used for target practice," *Red Bluff Daily News*, September 9, 1995, by Marsha Dorgan, p. 1.

9/17/95. Carlotta, California. 264 Earth First protesters arrested at demonstration against Sierra Pacific Industries proposed logging of second-growth trees in the Elk River area, adjacent to Headwater Forest owned by Pacific Lumber Company. Humboldt County Sheriff Dennis Lewis said one officer was knocked down and pepper sprayed his assailants. Eight Earth Firsters were arrested at the site the day before. 125 officers were called to the scene, including California Highway Patrol and deputies from other counties, funded by mutual aid agreements. Humboldt County provided food and lodging for officers from outside the area. Two Earth Firsters were charged with criminal trespass. "Anti-logging demonstrations end, Earth First curbs protests; sheriff arrests 2 on trespassing counts," *Eureka Times-Standard*, Tuesday, September 19, 1995, by Mary Lane, p. A6.

9/20/95. Carlotta, California. Law enforcement agencies tally the cost of protests of second-growth logging in the forest owned by Sierra-Pacific Industries adjacent to the privately-owned Headwaters Forest, driving Humboldt County toward high budget deficits. "Agencies tally costs of protests," *Eureka Times-Standard*, Wednesday, September 20, 1995, by Mary Lane, p. 1.

9/28/95. Arcata, California. Earth First protesters blocked Western Timber Services Inc., a forestry consulting firm in downtown Arcata that designed the logging plan for Sierra Pacific's Elk River second-growth timber. Earth Firster vandals spray painted 22 locations, including historic buildings that cannot be repainted, businesses, and statues in the public park. "Graffiti mars Arcata protest," *Eureka Times-Standard*, September 28, 1995, by Mary Lane, p. A3.

September and October, 1995. Williams, Oregon. 219 Earth Firsters and allied protesters arrested for criminal trespass in forcible work stoppage of Sugarloaf Timber Sale operated by Boise Cacade crews. 30 tree spikes were discovered after a logger broke his chainsaw on a 16-inch spike. "15 arrested in Sugarloaf timber sale protest," *Portland Oregonian*, Tuesday, September 12, 1995, by Eric Gorski, p. A1. See also "Guerrilla war looms in woods," *Portland Oregonian*, Wednesday, February 7, 1996, by Dana Tims and Peter D. Sleeth, p. B1.

10/15/95. Douglas County, Oregon. Bulldozer and excavator owned by Hull-Oakes Lumber Company of Roseburg, Oregon, had crankcases filled with abrasives, causing engines to seize up when started by workers at the Roman Dunn timber sale. Lt. Bob Urban of the Douglas County Sheriff's Department set damage at $50,000. Incident came immediately after Earth First journal ran an article on how to vandalize equipment using abrasives. "Equipment is sabotaged at logging site," *Seattle Post-Intelligencer*, Friday, October 20, 1995, by The Associated Press, p. 8B.

12/11/95. North of Ellensburg, Washington. Bulldozer belonging to William A. Hosmer was started during the night and driven into a creek with engine stuck running wide open for about 8 hours at Cook Canyon logging site. Same night equipment of T&R Logging, working nearby, was severely damaged. Never reported in media. Property loss notice, Washington Contract Loggers Association Insurance Agency.

1996

1/13/96. Eureka, California. 35 Earth Firsters arrested for obstructing Sierra Pacific second-growth logging on the Elk River reached a plea agreement with the District Attorney's office in court for criminal trespass under a "little bit more serious" penal code section than simple trespass, resulting in stiffer fines. "Earth First members reach plea agreement," *Eureka Times-Standard*, January 13, 1996, p. 1.

2/7/96. Oakridge, Oregon. Environmental activists plan hit-run tactics to halt logging of old growth in the forests on the west side of Oregon's Cascade Range, blockading roads with boulder barriers and 15-foot stakes. "Guerrilla War Looms In Woods," *Portland Oregonian*, Wednesday, February 7, 1996, by Dana Tims and Peter D. Sleeth, p. B1.

2/14/96. Southeast of Deming, New Mexico. Eleven cattle were killed on a ranch near Tres Lomitas with a semiautomatic SKS style weapon. Luna County Sheriff's Department report.

2/18/96. Keno, Oregon. Pacific Power's John C. Boyle dam was damaged by firebombs set off in the control room. Anti-dam saboteurs were suspected. "John Boyle Powerhouse turbines spinning again," *Klamath Falls Herald and News*, Wednesday, February 21, 1996, by Todd Kepple, p. 1. See also "Dam sabotage probe leads to tight security," *Klamath Falls Herald and News*, by Katy Moeller, February 26, 1996.

2/23/96. Near Olympia, Washington. A historic 80-year-old barn sitting on 10 acres of private property within Capitol Forest was destroyed by arson. The barn was owned, restored and used by the Tacoma Trail Cruisers Motorcycle Club. Arson investigators believe that individual(s) set the blaze at about 1:30 a.m. A concrete pump house was also spray painted with the slogans "Stop Destroying The Rest Of The Wildland" and "ORV = DEATH." ORV is thought to mean off road vehicle. Thurston County Sheriff's Department report.

4/3/96. Chester, California. U. S. Forest Service cancels salvage sale of fire-scorched timber in Lassen National Forest because of environmentalist protest, costing Tehama County taxpayers $85,000 in federal payments-in-lieu-of-taxes. Pressure called "economic terrorism" by local business community. "Taxpayers out $85,000 from sale," *Siskiyou Daily News* (Yreka, California), Wednesday, April 3, 1996, by The Associated Press, p. 1.

6/28/96. Lyons, Oregon. Twelve-inch steel and ceramic tree spikes in logs from the Santiam Canyon timber sale hit the blades of a veneer lathe and cost the Freres Lumber Co. several thousand dollars, but no one was injured. Millworkers found three spikes while peeling logs at the Freres veneer plant east of Salem. "Tree spikes found at controversial old-growth sale," *Corvallis Gazette-Times*, Friday, July 19, 1996, by The Associated Press, p. 1.

10/3/96. Helena, Montana. Earth Firster Ronald J. Constable, 27, was sentenced in federal court to one year in a federal penitentiary, $200 in restitution, one year of supervised probation and 500 hours of community service for tree spiking in an October 1993 incident near Essex, Montana (see 1993). Constable was permanently prohibited from entering any state or federal public lands. In June 1996 he became the first person to be convicted under a 1988 federal law against tree spiking. An undercover agent found Earth First literature in his possession. Constable admitted affiliation with Earth First. "Tree-spiking conviction a first," *Helena Independent Record*, Saturday, October 5, 1996, by Mark Goldstein, p. 1A.

10/23/96. Canyon City, Oregon. The Grant County sheriff arrested Dr. Patrick Shipsey, a John Day, Oregon, physician and chief sponsor of a grazing reform ballot measure, for killing 11 cows belonging to Robert Sproul, 85, by shooting them behind the head, using a Finnish 6 mm Sako target rifle. Shipsey was arraigned on 11 felony counts of criminal mischief. Each count carries a maximum penalty of five years in jail and a $100,000 fine. "11 Cows Slain, Range Reformer Held," *Portland Oregonian*, Wednesday, October 23, 1996, by Hal Bernton and Richard Cockle, p. A1.

10/30/96. Grangeville, Idaho. Jury awards logger Don Blewett more than $1 million in damages from Earth Firsters. After nearly 11 hours of deliberation, a jury of eight women and four men sided with Blewett against 12 Earth First defendants for damages his company suffered during the 1993 protest of the Cove-Mallard area of the Nez Perce National Forest. The jury awarded Blewett $150,000 in compensatory damages and $999,999 in punitive damages. "Highland Enterprises bulldozes Earth First!; Jury awards Blewett more than $1 million in suit over Cove-Mallard protest in 1993," *Lewiston Tribune*, Thursday, October 31, 1996, by Kathy Hedberg, p. 1A.

1996. Region Six, U.S. Forest Service (Northwestern United States). Eight timber sale protests (Rocky Brook, Enola Hill, China Left, Sugarloaf, First and Last, Red 90, Warner Creek, and Reed) incur extra federal law enforcement costs above budget in the amount of at least $1,010,931. FOIA Request 96-122-R6, U.S. Forest Service, October 9, 1996.

This is only the tip of the iceberg. The patterns that emerge from this barebones chronological chart are revealing. The non-government targets on this list, with very few exceptions, are small businesses—family logging contractors, road builders and family-run sawmills that cannot afford to pursue criminals. The scarcity of multinational corporate targets is striking, given the radical environmentalists' rhetoric of shutting down the Exxons, Norandas and Mitsubishis of the world.

9:30 A.M. WEDNESDAY, MARCH 6, 1996   *Bellevue, Washington*
"I THINK YOU'D BETTER CALL THE BOMB SQUAD," said Alan Gottlieb's secretary as I walked in.

"Where's the problem package?" I asked, slipping off my jacket.

"On your desk. On the corner."

I approached it warily. The box was wrapped in brown paper and taped. It had stamps, too many of them. I didn't recognize the return address. It was postmarked in a state where I was expecting nothing from any colleague. It was addressed in handwritten block letters to "Wise Use," not to me or to the proper name of my organization, although the street address was correct. All the wrong signals.

I gingerly picked it up. It weighed almost nothing.

"That's weird," I said aloud.

I called the Postal Inspector number we had been given and described the package, particularly its light weight.

"It could be a plastic or paper explosive. Don't open it. Put it in the proper location and call the bomb squad. I'll contact the FBI."

I took the package to the open picnic-table area we had pre-selected on the advice of Jim Bordenet and alerted the supervisors in our Liberty Park complex that we had a possible bomb. They were trained and knew what to do. There was no panic.

Within minutes the heavy bomb squad truck entered our parking lot and positioned itself between our building and the suspicious package. Squad Captain Bill Baker and his crew visually examined the package, nodding their heads.

Within a few minutes the FBI agent called from Seattle and instructed the bomb squad not to dispose of the package until she arrived, which could take up to 45 minutes in bad traffic.

The crew ran preliminary tests on the package, then slowly and deliberately set up their mobile x-ray device. They carefully took polaroids from four angles.

Captain Baker showed me one of the x-ray photos.

"Looks like your mysterious pen pal sent you a computer diskette. We probably just erased it with the x-rays."

"No explosive?"

"Not that we can see."

FBI Special Agent Patricia Jannette arrived and had a private discussion with Captain Baker. They opened the package.

It contained a computer diskette and a small note. It was from an anonymous source within The Nature Conservancy.

The enclosed disk contains some of the TNC files in my office. Your organization needs these TNC manuals and protocols; they are brazen statements of policy that TNC denies in public. The Bioreserve Manual is full of wonderful quotable quotes about how to make money off the Feds and how to set up land deals with the government before TNC has to spend any money. Unfortunately, the office copies of these manuals are battered, so scanning has made some errors. Also, it was not possible to scan all the tables and drawings because this would have taken a great deal of time and disk space, which would have been noticed. Good Luck.

I laughed. It couldn't have had a better reception committee.

I called off the alert at our complex and thanked the bomb squad for their patience with the false alarm.

"You did the right thing. It had all the earmarks," said Captain Baker.

Special Agent Jannette and I walked up to my office and I put the diskette in my computer. Its files were intact. I now had access to the operating manuals of The Nature Conservancy. So did the FBI.

"You know," I said to Patty Jannette, "we get stuff like this from inside the environmental movement all the time. We never gave a thought to anonymous packages until Gil got killed."

Special Agent Jannette had served on the Unabomb Task Force. She knew what we'd gone through. She'd gone through it too.

"It's a hard way to learn caution," she said.

We fell to talking about the Unabomber. I pushed the probable connection to animal rights or radical environmentalists. I hoped they would find him soon.

She expressed optimism about certain leads they were following that she was not at liberty to discuss.

I smiled skeptically. "You guys never give up, do you?"

"No," she said. "We don't."

As we spoke, other FBI agents huddled in the snow above a little cabin in Montana.

Chapter Four Footnotes

[1] Left-wing groups studied included the May 19 Communist Organization, composed mainly of leftovers from the 1960s Students for a Democratic Society, Weather Underground and Black Panthers; the United Freedom Front; the Provisional Party of Communists; and FALN (Armed Forces of National Liberation), one of ten Puerto Rican groups that claimed responsibility for bombings and assassinations in the early 1980s, and others. These groups are strongly committed to the destruction of American imperialism and capitalism. Leaders of several of these organizations were trained in Castro's Cuba and followed Carlos Marighella's "Handbook of Urban Guerrilla Warfare," which is available to readers in *For the Liberation of Brazil*, Penguin Books, Baltimore, 1971.

[2] The Christian Identity Movement is based on an anti-Semitic, anti-black belief that Aryans, not Jews, are God's chosen people and that America is the promised land, reserved for the Aryan people of God. Sharing that belief are leaders of: the Aryan Nations; the Covenant, Sword and Arm of the Lord; Ku Klux Klan and the Sheriff's Posse Comitatus, among others. Right-wing terrorist groups all justify the use of terrorism as a prelude to war—the Armageddon, which will establish Christ's kingdom. See Bruce Hoffman, *The Contrasting Ethical Foundations of Terrorism in the 1980s*, The Rand Corporation, Santa Monica, California, 1988.

[3] Roselle's words were quoted as, "We don't need Foreman in Earth First! if he's going to be an unrepentant right-wing thug." "Earth First! Co-founder Quits - Is Unhappy With Group's New Focus," *Arizona Republic*, Wednesday, August 15, 1990, by Sam Negri, with materials from the Associated Press, p. B1.

[4] Brent L. Smith, *Terrorism in America: Pipe Bombs and Pipe Dreams*, State University of New York Press, Albany, 1994, p. 125.

[5] *Ibid.*, p. 32.

[6] FBI Terrorist Research and Analytical Center, *Terrorism in the United States: 1994*, Washington, D.C., U.S. Department of Justice, 1995, p. 24.

[7] Office of the Attorney General, *The Attorney General's Guidelines on General Crimes, Racketeering Enterprise and Domestic Security/Terrorism Investigations*, Washington, DC, March 7, 1983.

[8] *Guidelines*, p. 13. Earth First's rhetoric could place it in this category.

[9] *Guidelines*, p. 16.

[10] *Guidelines*, p. 13.

[11] Brent L. Smith, *Terrorism in America*, p. 27.

[12] Today's official definition of terrorism is designed to protect civil rights guaranteeing social and political expression as much as to allow the FBI investigative flexibility. It resulted from criticism of FBI misconduct uncovered during the Senate Watergate hearings in 1973 and the 1975 Senate Judiciary Committee investigation that found the FBI's COINTELPRO activities to be illegal. COINTELPRO, the FBI's abusive Counterintelligence Program of domestic spying, agents provocateurs, faked documents and disinformation campaigns, was formally discarded. Congress and the public rejected the role of the FBI as "political cops" and forced the dismantling of FBI domestic intelligence units. In August 1976 FBI Director Clarence Kelley moved investigations of terrorist organizations to the General Investigative Division from the Intelligence Division, where critics

said respect for the rule of law "had been nonexistent." Whether these changes cured the problem remains a matter of dispute. See Tony Poveda, *Lawlessness and Reform: The FBI in Transition*, Brooks/Cole, Pacific Grove, California, 1990.

[13] Interviews with several field law enforcement agents in each of three federal agencies have brought up mention that unnamed "higher-ups" have stopped specific investigations of ecoterrorist groups and suspects in progress, usually under the rubric of avoiding illegal domestic intelligence operations.

[14] *Keene Report*, May 16, 1995, by David Keene, Keene & Associates, Alexandria, Virginia. I verified the report with Keene by telephone December 19, 1995. His source was present at the meeting, and expressed willingness to repeat it in a court of law or congressional hearing.

[15] Association of Forest Service Employees for Environmental Ethics: Town Creek Foundation gave $10,000 in 1992; Ruth Mott Fund gave $20,000 in 1992; Nathan Cummings Foundation gave $40,000 in 1992; HKH Foundation gave $15,000 in 1992; Mary Reynolds Babcock Foundation gave $25,000 in 1992; W. Alton Jones Foundation gave $100,000 in 1992. Public Employees for Environmental Responsibility: Educational Foundation of America (the Prentice-Hall fortune) gave $40,000 for an anti-BLM study in 1993; Bullitt Foundation gave $25,000 in 1993.

[16] The URL is http://envirolink.org/ALF/doa/nadoa92.html.

[17] The URL is http://envirolink.org/ALF/contact.html.

[18] U. S. Department of Justice, Criminal Division, *Report to Congress on the Extent and Effects of Domestic and International Terrorists on Animal Enterprises,* Washington, D.C., by Scott E. Hendley and Steve Weglian, Office of Policy and Management Analysis, September 2, 1993, 32 pages.

[19] Public Law 102-346, August 26, 1992; codified as 18 U.S.C. § 43.

[20] Telephone interview with Kathleen Marquardt, October 8, 1996. See also Kathleen Marquardt, *AnimalScam: The Beastly Abuse of Human Rights*, with Herbert M. Levine and Mark Larochelle, Regnery Gateway, Washington, D.C., 1993, p. 135. The House bill was H.R. 2407, Prevention of crimes against farmers, researchers, and other livestock-related professions. The Senate bill was S. 544, Protection of animal research facilities from illegal acts. A report was filed in the House by the Committee on Agriculture, H. Rept. 102-498.

[21] U. S. Department of Justice, Office of Legislative Affairs, *Letter of Transmittal*, by Sheila F. Anthony, Assistant Attorney General, and Eugene Branstool, Assistant Secretary, Marketing and Inspection Services, U. S. Department of Agriculture, Washington, D.C., September 2, 1993, 2 pages. Identical letters were sent to the President of the Senate and the Speaker of the House.

[22] Telephone interview with Ben Hull, October 7, 1996.

[23] United States Code, Title 18 - Crimes and Criminal Procedure, Part I - Crimes, Chapter 91 - Public Lands, Section 1864 - Hazardous or injurious devices on Federal lands.

[24] Henry Campbell Black, *Black's Law Dictionary*, Fifth Edition, West Publishing, St. Paul, 1979, p. 223.

[25] District Attorney's Information for violation of ORS 167.207, *State of Oregon v. Frances Paula Downing*, Criminal Activity in Drugs, Filed July 7, 1978,

Circuit Court of the State of Oregon for Josephine County, Vol. 117, page 2022.

[26] *Reason* magazine, in the person of then-editor Marty Zupan, thoroughly fact-checked and verified all referenced incidents.

[27] "Clean Air, Clean Water, Dirty Fight," *CBS News 60 Minutes*, Leslie Stahl, September 20, 1992.

Chapter Five
RADICALS

Noon, Saturday, March 4, 1995   *Eugene, Oregon*
The five of us waited for our orders to come up at Glenwood's, a noisy, crowded little café near the University of Oregon campus. Janet and I were getting to know the radical environmentalists who had asked me to share a time slot at the 1995 Public Interest Environmental Law Conference late in the afternoon: Tim Hermach, executive director of the Native Forest Council; Michael Donnelly, president of Friends of the Breitenbush Cascades; and Jeffrey St. Clair, editor of Wild Forest Review.

I had already encountered Hermach. A New York radio talk show host, Doug Henwood, had seen my profile in a Greenpeace booklet on "anti-environment" organizations and invited me to talk about forests as seen by the wise use movement.[1] Who would I be willing to accept as a debate partner?

I told Henwood I'd debate anyone, but warned him that the usual environmental spokespeople only mouthed platitudes about saving nature and not hurting the economy—long on appealing rhetoric, short on talk about the underlying intent to dismantle industrial civilization.

They had a predictable patter: they would first complain bitterly against the terrible damage logging does, while denying that their bans would wreak havoc upon local economies. When I listed the names of the mills they had already shut down, and enumerated the job losses they had

165

already caused county by county, they would then say the unemployment figures showed no such slump.

When I pointed out that once the loggers' unemployment benefits run out, they are no longer counted as unemployed, no longer part of the labor force, but they are still there suffering, now invisibly, then the environmentalist would go into mourning for the fallen trees, characterizing loggers as desecrators of a cathedral.

When I named particular forests that were timber gardens and not cathedrals, properly designated as commercial wood sources for perpetual sustained yield, they decried the bad methods loggers used.

When I described good methods and how logging was only the beginning, with foresters carefully planning for the future, with crews hand-planting nursery-grown seedlings from genetically diverse wild parent trees, with brush control to assure their growth, with decades of thinning operations to improve their health, with vigilant disease control and fire suppression to let them mature, then harvest again in an endless cycle, they complained that industrial tree farms were monocultures and ecologically sterile.

When I named particular tree farms I had visited and described their rich tree species mix and abundant wildlife, they'd come up with another excuse, never getting to the real bottom line: they just didn't like logging, period.

And it wasn't always their ostensible justification to save the last remnants of the wild forest—there was a lot of unspoken ideology behind that: Logging, for example, provided raw materials for thousands of products and that promoted consumerism and consumerism was a Bad Thing.

Logging was but one expression of an immense industrial infrastructure of logging companies, of equipment manufacturers that supplied logging companies, of parts manufacturers that supplied the brakes and hydraulics and electrical components for logging equipment, of the sales forces and maintenance shops that made the equipment available and kept it working, of the educational facilities that taught technology on every level from machine shop welding to physical chemistry that created new materials for industrial civilization, and industrial civilization was a Bad Thing.

But environmental spokespeople would never say it. If they did, people would realize that environmentalism has some unpleasant neo-Malthusian consequences of its own. And that revelation was politically inexpedient.

If Henwood could find someone with a little honest fire in the belly, we could have a real debate, not the usual shallow eco-joust. He found Tim Hermach.

I had known of Tim only through his New York Times and Washington Post press clips decrying all timber cutting on federal lands. He seemed to be a blunt, even brassy grassroots leader. A debate with an outspoken adversary couldn't be as unproductive as dancing the customary Sierra Club shuffle. So I told Henwood okay.

When we got together on Henwood's *Behind the News*, I was pleasantly surprised by Hermach's passionate argument for Zero Cut on the national forests. He wanted no trees cut and offered no excuses. He was totally outrageous. Damn the politics. Damn the pretenses. I argued that my constituency saw good logging and perpetual forestry as a benefit both to people and to the environment, while Hermach argued that his constituency didn't want better logging, they wanted no logging on public land. Period. And no "environmentally damaging" logging even on private lands, too. If that eliminated some destructive corporations from our ranks, so be it.

We filled half an hour with our technical and moral reasons why we felt the way we did, but the clear choice between our two opposite intentions never got blurred by temporizing or obfuscation. It was refreshing to find such a straightforward opponent.

The next day I called him and said I had enjoyed the debate. I admired his honesty. And respected his daring to be honest, even though we didn't agree about anything. But as the conversation drew on, it became apparent that we *did* agree on something: neither of us felt any great love for the Sierra Club. We had different reasons, of course. I disliked their manipulation of a friendly public to incrementally destroy industrial civilization. Hermach disliked their sell-out to politics, trading away many trees in return for a promise not to log a few.

At the end of our conversation, Hermach noted that he was less than thrilled about the big-money foundation takeover of much of the environmental movement—a takeover that my organization had meticulously documented. Because I had discovered that the foundations' control was carefully hidden, I was surprised that Tim even knew about it. As it turned out, he knew about it better than I did, and from intensely personal experience.

My organization, the Center for the Defense of Free Enterprise, had performed a detailed financial analysis of the top sixty environmental groups, including their annual multi-million-dollar revenues, six-figure executive salaries, huge foundation grants and extensive investment portfolios. We had discovered the ominous phrase, "grant-driven" to describe environmental groups that no longer controlled their own destiny, but did what their donors demanded, and the donors were a mix of multinational corporations, progressive foundations and governments. Each of these forces had their own agenda, and each used environmental groups for their own purposes—and even created environmental groups when existing ones were not sufficiently competent or compliant. We published our findings about the top dozen, and called it simply *Getting Rich.* It annoyed everyone, because its subtitle exposed everyone: *The Environmental Movement's Income, Salary, Contributor and Investment Patterns, With an Analysis of Land Trust Transfers of Private Land to Government Ownership.*

Someone from Hermach's Native Forest Council had just a few weeks earlier called and obtained my permission to reproduce *Getting Rich* for their own purposes. I had assumed it was some environmentalist trying to refute my findings, and didn't give it any further thought. I had intended the report as a warning to workers and businesses that big money interests were using environmental groups to achieve anti-competitive advantages—nationalizing private land as nature preserves so competitors couldn't buy and develop it; jacking up the costs of environmental regulation to the point that only big money corporations could afford to comply. The Native Forest Council, I found, wanted to use the report to warn environmentalists against being taken over by big money interests.

That convergence of our conflicting interests fascinated us both. Hermach suggested that since Native Forest Council director Victor Rozek would soon be in Seattle to present testimony in a court case, we should take that opportunity to get together. Perhaps for dinner. Rozek could use our conversation as the basis for a more extensive interview that might end up in the pages of their graphic newsletter, *Forest Voice*.

Victor Rozek and Cassie Daggett met Janet and me a few weeks later in the restaurant of a hotel near Seattle's federal courthouse.[2] Victor had prepared, he later wrote, to confront "an angry red-neck burdened with all the stereotypic implications of that office." Whatever he expected, the four of us spent a pleasant evening over a fine dinner and a bottle or two of good Merlot.

Rozek had done his homework. He immediately posed the question other interviewers never asked: Why? Why did I think environmentalism was wrong? What did I mean by it? How did I arrive at such a controversial opinion?

We spent hours threading through the maze of events that brought me to my stance as a defender of loggers, miners, fishermen, property owners and others who found themselves scorned as wretched and materialistic destroyers. We advanced, episode by episode, through my volunteer association with the Sierra Club during the 1960s, my election as a trustee of the Alpine Lakes Protection Society (ALPS), my career as a film-maker and magazine writer covering the logging, mining, ranching, fishing and other resource industries. We paddled through my appointment as a non-profit organization executive in 1984 and my realization that a grassroots movement of resource people was gestating across America. We explored the turning point in 1988 when my organization sponsored a conference that gave a name to what they were doing: the wise use movement. Rozek even asked about the theoretical underpinnings of the movement and I gave him my eclectic catalog of theoretical thinkers and their iconoclastic literature that had given me the intellectual ammunition to help the movement grow and gain political clout.[3]

Rozek's erudite and thoughtful rendering of that exchange soon appeared in *Forest Voice*. I couldn't have asked for a more accurate and thorough account.[4]

Thus, I didn't hesitate to accept when I got the invitation to join Hermach's friends, Michael Donnelly and Jeffrey St. Clair, in their talk at the annual environmental law conference in Eugene—even though I didn't know them and even though it was only five days before the event. I wouldn't need any preparation to deal with their subject: "Foundation / Corporate Control Over Environmental Organizations." And I was fascinated that these radical environmentalists wanted to reveal what their less-radical colleagues wanted to cover up. They were either very brave or very foolhardy.

I didn't learn which until our lunch order came up at Glenwood's.

We crammed ourselves into a booth, Janet next to Hermach, facing St. Clair, Donnelly and me. We ate our sandwiches as Michael Donnelly gave me his background: he operated a small bottled-water company from a spring in the Cascade Mountains near Mount Hood and opposed industrial logging for practical as well as moral reasons—he'd been harmed by upstream activities of the U.S. Forest Service that promoted flooding of his spring and resulting loss of business.

Jeffrey St. Clair had studied literature at an Eastern university, submitting a thesis on Thomas Pynchon's dense convoluted novel, *Gravity's Rainbow*. He possessed a formidable intellect and enjoyed a gift of kaleidoscopic expression. He was a leftist of the social justice persuasion, but adhered to no prescribed dogma and formed his opinions case by case, thinking things all the way through. His commitment to forest protection came from personal belief and experience. He knew more about what was happening on the ground than any environmentalist I had met. He also had connections to some of the national icons of the left such as James Ridgeway, whose work frequently appears in The Village Voice, and Alexander Cockburn, syndicated newspaper columnist and featured writer for The Nation.

Donnelly and St. Clair then told me why they were going to expose the big money control behind environmental groups to a room full of radical environmentalists.

In 1992, it seems, St. Clair, Donnelly and Hermach, among other Oregon grassroots environmental leaders, had been invited to a fancy private reception in Atwaters, a toney Portland restaurant perched atop a rosy-gold glass-tower bank building dubbed by the locals "Big Pink," from its garish hue in the sidelong rays of the morning sun. The foundation aristocracy, Donald K. Ross of the Rockefeller Family Fund, John Peterson "Pete" Myers of the W. Alton Jones Foundation, and Hooper Brooks of the Surdna Foundation were the hosts. They had been watching the forest debate for some time. They had great plans for it. They presented a compelling message to the locals:

> The grassroots crusade to save the Ancient Forests of the Pacific Northwest from timber operations has gotten off to a wonderful start. Now it is time for the professionals to come in, take over and finish the job. The grassroots groups will henceforth be well funded and given instructions on what to do, who to do it with, and exactly how and when to do it.

Hermach, Donnelly and St. Clair were stunned. What would happen if they didn't accept? Don Ross explained that it was an offer they couldn't refuse:

> If the grassroots groups refused to cooperate, they'd never see a dime of foundation money.[5]

St. Clair, Donnelly and Hermach found the proposition less than seductive. The three grassroots leaders politely told the foundation carpetbaggers what they could do with their money and their threats.

The foundations then made good on their threats: the grassroots operators were made pariahs to most of the philanthropic community and since then have faced continual economic stress.

However, they had been pariahs from philanthropy to begin with. And economic stress is the normal condition of a grassroots organization. So, the Green Cartel had thrown them, like Bre'r Rabbit, into their native briar patch—the encounter had been educational rather than intimidating.

Now it was time to educate others. But they needed the facts in order to be convincing. That's why I was there. I knew where all the bones were buried. The fact that I was the devil incarnate to most environmentalists made their argument all the more convincing.

I was humbled by their integrity. They had experienced at first hand what I only deduced from mountains of documentary evidence. Now I understood why these radical environmentalists would want to disseminate something as critical of the environmental movement as my Center's study—they had begun work on a similar document of their own and I had merely saved them a lot of spadework.

They briefed me on the session we would give: Hermach wasn't actually on the program, but an associate, Chad Hanson, would unobtrusively moderate the panel. Donnelly and St. Clair were the stars, and would go first, explaining the facts of foundation and corporate control of environmental organizations, while I would cap it off by describing how I located and documented the money flows and control forces.

With lunch over, we strolled to the School of Law buildings where the conference had been in session since Thursday. David Brower, former executive director of the Sierra Club, founder of Friends of the Earth, chairman of Earth Island Institute, and "the inspiration behind the environmental protection movement in the U.S.," as the brochure said, had given the opening address, a custom he had kept since the inception of the conference years earlier.

Janet and I decided to sit in on a particular session scheduled just before our late-afternoon presentation: Tarso Ramos of the Portland-based Western States Center, strategizing to expose the wise use movement as a front for big business. I had records that showed Tarso's Center itself had received $300,000 from the Florence and John Schumann Foundation for "research on community organizing for environmental issues;" and $50,000 from the Washington, D.C.-based Public Welfare Foundation "for Wise Use Public Exposure Project which opposes activities of Wise Use Movement;" annual donations of $20,000 from the W. Alton Jones Foundation to oppose the wise use movement; $15,000 from the Jessie Smith Noyes Foundation, Inc. "for research, public education and coalition building addressing Wise Use movement and need for sustainable development."[6] A wee bit grant-driven.

The room was packed, so we stepped inside the door and just stood there smiling and listening while Tarso held forth as an expert on Ron Arnold, showing his rapt listeners numerous documents about me, including Victor Rozek's new profile in Forest Voice. Although a number of audience members tried to signal him, Ramos made eye contact with me several dozen times without a glimmer of recognition.

The foundations supporting Tarso should get their money back.

Later, as we jostled through the seven hundred people converging on the auditorium, the usual gauntlet of young guns accosted me with, "You're destroying the earth," and "I just want you to know I hate you," and "You have no right to live on this planet." It evoked Dave Foreman's parting words to Earth First: "How can you be an effective activist but not be consumed by hatred?"

One particularly pinched-looking young woman approached me aggressively and said, "Are you Alexander Cockburn?" Astonished, I demurred, but asked her if Cockburn was supposed to be here, because I had read—and mostly disagreed with—his exasperatingly superb writing for years and wanted to meet him myself. The rumor, she replied, was he was to be on the same program.

The large teaching room filled. Chad Hanson, the panel moderator, introduced Michael Donnelly, who retold the foundation story: "There was a cocktail party at a Portland office tower hosted by the Rockefeller Foundation. The message was you amateurs did a pretty good job of nationalizing the issue. Now us professionals will come in and win it for you."

He told the packed hall, "Money was buying a seat at the table. The most annoying was the Forest Conference called by President Clinton in Portland in 1993. We had no say in who was representing us. The foundations did. The money took over the representation of the movement. We did not have the people there to counter the industry's many lies. We were constantly portrayed as taking food from the mouth of orphans and widows."

In Donnelly's view, the problem was a cartel led by Pew and Rockefeller family foundations.

"Not all foundations are bad. Some are funding worthy efforts," he added. "It's up to us to just say no when we're told to pull our punches for funding."

I looked out at the room, seeing a mixed crowd, all ages, more men than women, mostly students, academicians, lawyers, other guest speakers sitting in. I saw a few faces that I knew from times past. All were listening intently.

Then Jeffrey St. Clair held forth. Cockburn, unfortunately, didn't make it. As I learned from him later, nobody had bothered to tell him he was invited.

St. Clair delivered the nub of the session, turning upon this thought:

> To quote Joseph Heller: *Something Happened.* Somewhere along the line, the environmental movement disconnected with the people, rejected its political roots and pulled the plug on its vibrant tradition. It packed its bags, starched its shirts and jetted to D. C., where it became what it once despised: a risk-aversive, depersonalized, overly analytical, humorless, access-driven, intolerant, statistical, centralized, technocratic, deal-making, passionless, sterilized, direct-mailing, jockstrapped, lawyer-laden monolith to mediocrity. A monolith with feet of clay.

But, explained St. Clair, behind this transmogrification lay big money. Big foundations with vast investment portfolios in big multinational corporations. Big multinational corporations manipulating green groups for their own ends. Both taking control of the environmental movement through "grant-driven" projects dictated by the donors.

The national movement had taken multiple hits from the wise use movement because its charges "rang true," St. Clair said. "It looked elitist, highly paid, detached from working people, a firm ally of big government. Once feared as the most powerful public interest group in America, the environmental movement is now accurately perceived as a special interest group."

Then it was my turn. St. Clair introduced me: "When word leaked out that Michael Donnelly and I invited Ron Arnold to this panel, the mainstream groups freaked out. This is of course absolute hysterical nonsense. If we have any chance to succeed as a movement, we have got to demystify the opposition."

I had their attention. It wasn't my presentation: how to research the money was merely nuts-and-bolts detail. Nobody really wanted to hear it. I had been demonized by every book, every article, every authority they believed and respected. They just wanted to see the devil.

I began by pointing out that we had common cause in resisting the Green Cartel. "I'm quite surprised Rockefeller and Pew are the ones you don't like," I said.

"I'm surprised they're the ones *you* don't like," said Tim Hermach, a front row spectator.

Lest the audience get a too-friendly impression, though, I immediately made clear that my philosophy was profoundly different from theirs.

"My worldview is to preserve the project of modernity," I said.

Asked what this means and how it plays out, I went through the vast intellectual storehouse of science and technology that supports modern industrial civilization. Modernity is more than just the technology, it is also the science, the knowledge that undergirds it. And it's more than that: the spirit of adventure, curiosity, daring and achievement it excites. I acknowledged that industrialism as presently constituted is not working—it still needs to make more breakthroughs, close more loops in its systems, create more zero pollution plants for heavy industry, find materials that stretch resources, bring people into direct control.

Unlike many radical ecoactivists, who believe it is time to abandon industrialism in favor of a profoundly different system, I said that I passionately want to save it. I want to keep electricity and airplanes and highways and pharmaceuticals and cities and agriculture and written language. The project of modernity, I reminded them, created the sciences of genetics, ecology and evolution that brought them to their policies of biocentrism, biodiversity and the unmaking of industrial civilization. By destroying modernity, they would destroy their own frame of reference. The project of modernity has to do with how to make industrial civilization benign or helpful. We are learning ways to do more with less, to be more careful in supplying our needs, to think in new ways that put it all together in productive harmony. However, I admitted, "We don't know how to do that yet. But I don't want to give it up for something less. We will learn—and I don't believe that civilization is destroying the world."

By the time the question and answer session came around, I had decided this was definitely more interesting than most environmental meetings I've addressed. Some people seemed surprised that I would expose the role of big business in manipulating the environmental movement. I said that my organization's mandate was to defend free enterprise. I explained that there was a difference between corporate capitalism and free enterprise. Those who support corporate capitalism might seek monopolistic control of markets and society, but those who support free enterprise do not, I offered.

"Given what I've read about you," an audience member said, "I'm inclined to think you're playing the oldest game in the world, divide and conquer."

I personally didn't think their act was together enough for me to divide, but I made some lame excuse about coming because I was invited.

But another person asked what my motive was for this, if not to divide and conquer. So I told them what I really thought:

"The press doesn't cover you," I said, "because you're too far out. I believe, and many in the wise use movement believe, that your vision is not acceptable to most Americans. You'll end up dismantling industrial civilization. I want them to hear it from your own mouth. It's clear to me you have become the real environmental movement. For the most part, I debate environmental bureaucrats who incrementalize me to death. I'd rather get into a debate with somebody like Tim Hermach because he'll talk about the real issues. People will be able to tell the difference."

Then someone asked me whether all species would survive if I had my way and industrial civilization continued. I said no, some species will probably go extinct, but not most, and probably not many.

Another said that I had exposed the big salaries of environmental group leaders, so what was my salary at the Center? I explained that I had come aboard as a volunteer in 1984 and received no compensation. Then how do I support myself? "I write books and serve as a consultant to businesses and grassroots groups—and I give speeches on the lecture circuit." Well, then, what was my personal income? "I don't have to answer that."

Mitch Friedman, Northwest Ecosystem Alliance executive director, put in some defense for the Pew and other major foundations from which he had taken large sums of money. Without their funding, his group could not have put together large-scale wildlands preservation plans, he said. While agreeing with "a lot" of the criticism, Friedman added, "Let's not throw the baby out with the bathwater. Let's find the tools available to us and fight hard." To my surprise, he was booed.

I got an invitation from a 75-year-old West Shoshone holy man named Corbin Harney: "We want you to come out and live with us. Your ways have killed the living things today. It's getting worse. Come live with us. Will you do that?"

Why not? I'd done it before elsewhere. I had grown close to a San Carlos Apache mystic named Don Stago as a teenager, and worked as a construction laborer with the Papago near Mission San Xavier del Bac in Arizona as a young man. Hell, I'd even gone through a ceremony by a Yaqui *brujo* who healed a bullfighter friend whose *corrida* I was traveling with near the Sonoran town of Caborca in Mexico. I had also written numerous magazine articles about tribal logging enterprises such as Navajo Forest Products, and had a lot of Native American contacts in the Bureau of Indian Affairs from having written the history of James Watt's tenure as Secretary of the Interior, so I said, "I will."

I was a little skeptical about the welcome I'd get at Corbin's Nevada home: I told him in front of the audience that I'd invited one of his friends, Chief Raymond Yowell of the West Shoshone National Council, to address our Wise Use Leadership Conference in 1993, and he had to cancel because a big foundation heard about it and threatened to cut off their funding if he appeared on our program.

I went to Nevada twice in the months that followed, but both times Dee Dee Sanchez, Harney's contact person—Corbin has no phone—said he was off somewhere holding drummings for environmentalists and selling his new book, *The way it is: one water— one air— one mother earth*, (Blue Dolphin Publishing, Nevada City, California, 1995, $16.00).

An outraged young woman, who evidently held Corbin Harney in high esteem, wanted to know what my creation myth was such that I could be the way I am. I took her seriously and was about to talk about my cosmology when Chad Hanson and Jeff St. Clair indicated that another question might be more to the point.

In the middle of one of my long-winded discourses on the theoretical basis of wise use—a discussion of the philosophers who led up to Jurgen Habermas's work on communication, as I recall—one disgusted young man at the rear of the room corrected my pronunciation of Max Weber's name.

At the end, David Brower—the Archdruid, as writer John McPhee tagged him—stood up. "I want people to congratulate you for what you achieved," he said. "You are one of the people most responsible for what happened on November Eighth that we don't like." A slight exaggeration, I pleaded, and replied, "Dave, everything I know I learned from you."

To some extent it was true. Dave was 83 at the time, and he had suffered a stroke and had a pacemaker implant during the past year or two. I had met him first at a wilderness conference in 1968, when he was executive director of the Sierra Club and I was a lowly volunteer. I have studied his methods and honored his boldness and vision ever since, although our paths diverged as environmentalism became more and more radical. Though I don't agree with him about much of anything, and he can be a terrible pain in the ass, I profoundly respect his ability and his integrity and I can't stand the idea of a world without him.

As usual, Brower surprised me. He told the audience about our last meeting, the winter before, at the Vermont Law School. It was a ferocious debate on property rights in front of the television cameras. He said he had won the debate and I acknowledged never winning an argument with him. Then he told the audience about the faculty dinner that snowy evening in South Royalton where he and I were seated at opposite ends of the room—to prevent a scene, presumably. He smiled at that, and told what happened afterward: we both hung back as guests departed the restaurant, and then, when the coast was clear, we greeted each other cheerfully, went into the bar and talked over old times and new. That's not something either of us usually admits in public.

I asked him to tell what we did there.

"We closed the bar," he said, adding that our personal discussions over a few beers in the wee hours had also fallen in his favor. I admitted they had.

Neither of us mentioned the fact that accompanying him in the bar that night was Mathew Jacobson—using his alias, Buck Young—an Earth First organizer who lived in Bondville, Vermont. Young discussed the growth of the wise use movement, and I ventured that it was now sufficiently sturdy and organized that it was running on its own. Young said matter-of-factly, "No, you're what keeps it alive. If we took you out, it would fall apart."

There were many things we didn't bring up in Eugene. David Brower's comments signaled the end of the session and a young photographer named Elizabeth Feryl asked us to pose together. We agreed, even though we both thought it an odd juxtaposition that would have no conceivable use—certainly it would not be any great fundraising tool for either of us. However, I subsequently found one of Feryl's shots in Alexander Cockburn and Ken Silverstein's 1996 book, *Washington Babylon*, calling me the "Ahab of the Wise Use Movement," who "once worked for Sierra Club; spends life seeking revenge."[7]

Janet and I joined Tim Hermach and Victor Rozek the next morning for breakfast before our long drive back to Bellevue. We talked about a joint project, something the two sides could do together that would compromise neither of us. The idea came up of a co-authored point-counterpoint book. That interested us both. It should be a lively exchange of passionately held beliefs, we agreed. It would also be an object lesson to all in the debate that unbending opponents can treat each other with respect.

A day or two later I called Jeff St. Clair to ask for some help finding corporate filings on the Internet, something he had become good at. Jeff mentioned a note he had received from David Brower after the conference. Dave was furious with St. Clair for sharing the podium with the likes of Ron Arnold. Brower told him that even if it was true that the foundations and corporations controlled some major environmental groups, environmentalists shouldn't talk about it in public. I didn't mind him resenting my presence; it was his domain. An angry Archdruid I can deal with. A hypocritical Archdruid almost makes me cry.

NOON, FRIDAY, JUNE 16, 1996   *Bellevue, Washington*
ALEXANDER COCKBURN WAS AS CRUSTY AS I HAD EXPECTED HIM TO BE. And as delightfully offbeat. Raised in Ireland, he came to the United States in 1973, and since then has written for The Nation and in syndication to many newspapers. He is one of a few people I know who still admits to being a Marxist, and, even rarer, has actually read nearly everything Marx wrote. He is wild and free-roaming as his late father Claud Cockburn, a genuine Communist who used to hawk literature on London streets, and, under the pen name of James Helvick, had written the novel that inspired

one of my favorite screwball movies, the 1954 Humphrey Bogart vehicle, *Beat The Devil.*

Alexander was in Seattle for a book signing of his latest opus, *The Golden Age Is In Us*, and took the opportunity to come over to my office at the Center for the Defense of Free Enterprise across Lake Washington in Bellevue. He was very proud of the 1968 Dodge Dart GT he'd bought in 1990, and insisted on taking me down to our parking lot and lifting its hood to show me the compact engine surrounded by enough empty space for a mechanic to step into and do his work uncompressed. He had driven the remarkably preserved old car up from his California home in Petrolia, on the Mattole River near its passage through the King Range headlands into the Pacific.

I took Alex to lunch at Morgan's Lake Place near my Center, where we enjoyed an expansive conversation and a nice meal while watching the geese in Lake Bellevue swim toward us and out of sight beneath this restaurant on stilts.

Now we settled on my living room couch facing the 600-acre wood of Wilburton Hill to get down to a real discussion. But first, I had a copy of *Political Ecology*, a 1979 reader Cockburn had edited with James Ridgeway, and I wanted him to autograph it.[8] The book had been a valuable reference to me for years, getting inside the minds of the opposition. Ridgeway had already signed my battered copy when he sat on the same couch two weeks earlier with Jeff St. Clair, interviewing me for a Village Voice piece on the wise use movement and range rights, so Alex signed it cheerily, "Here's the other half!"

Now Cockburn wanted to know more about big money's takeover of the mainstream environmental movement. He had already published an article about my Center's report, *Getting Rich*, and grasped the ground plan of the big foundations.[9]

Among the most damaging items I gave him were transcripts of tapes of a three-day closed meeting of foundation bigwigs at Rosario Resort in the nearby San Juan Islands, which had been convened by a shadowy consortium of more than 160 foundations called the Environmental Grantmakers Association. Cockburn already knew of the meeting from my abbreviated account in *Getting Rich*, but now I gave him the full verbatim transcripts of four major sessions of the annual retreat.

The last session was called, "The Wise Use Movement: Threats and Opportunities." It was based upon two documents, a preliminary February 1992 study by the W. Alton Jones Foundation,[10] and, in September, a 292-page detailed report by the Boston political strategy firm, MacWilliams Cosgrove Snider, examining the wise use movement and coming to some alarming conclusions. The MCS study had been sponsored by the Wilderness Society and was called, "The Wise Use Movement: Strategic Analysis and Fifty State Review."[11] It took six months and cost over $50,000 to produce.

The W. Alton Jones Foundation of Charlottesville, Virginia gave $50,250 to the Wilderness Society in 1992 "to undertake a national review of public perceptions of environmental protection efforts." That innocuous-sounding review was the MacWilliams Cosgrove Snider study. Forbes magazine subsequently called the MCS review "The Search and Destroy Strategy Guide."[12]

The session presenters were Debra Callahan of W. Alton Jones Foundation and Judy Donald of the Washington, D.C.-based Beldon Fund. They revealed that, even though their preliminary report had characterized the wise use movement as "command and control, top-heavy corporate-funded front groups," in fact, "what we're finding is that wise use is really a local movement driven by primarily local concerns and not national issues.... And in fact, the more we dig into it, having just put together over a number of months a fifty-state fairly comprehensive survey of what's going on [the MCS Report], we have come to the conclusion that this is pretty much generally a grassroots movement, which is a problem, because it means there's no silver bullet."

I told Cockburn I had acquired these tapes because of the wise use movement's first amphibious landing. The Oregon Lands Coalition hired a boat and Chuck Cushman of the American Land Rights Association invited Janet and me to join a 17-person sign-carrying delegation to Rosario Island to protest the gathering. We sailed into the private harbor and docked in time for a lunch-break protest. The foundation executives were just filing out of their beachfront meeting hall and recognized me. They notified the resort manager, who threatened to call the cops. I told him we had rented a room at his resort and that we would not leave because we were registered guests. Impasse.

Then an EGA leader offered to let us tell our story on the grassy dockside to whomever among their attendees would listen. We agreed. While we talked to about forty of them, one of our protesters sauntered through their empty meeting room and gathered up all the litter they had left behind. Among the papers was an order form for the tapes of the whole conference, twenty-four sessions, packaged in a neat plastic binder for $125. We ordered them before the EGA prevented the commercial tape duplicator from selling any more to the public.

We found that MacWilliams Cosgrove Snider, calling wise use the most serious threat to the environmental movement that had yet appeared, recommended a multi-pronged attack: "Tar Wise Use Leaders" and "Focus public attention on ties between Wise Use and extremists." The report recommended tying us to groups such as the Lyndon LaRouche organization, Rev. Sun Myung Moon's Unification Church, the John Birch Society, Scientology, Neo-Nazis, militias or any other unpopular bunch. I recall going into the office of my colleague, Alan Gottlieb, after reading the MCS report and saying, "Hey, Alan, did you know you're a Neo-Nazi?" To which he replied, "How many Jewish Neo-Nazis are there?"

Cockburn was fascinated. He then wanted to know what my take was of David Helvarg's recent scurrilous articles in The Nation attempting to tie the wise use movement to militias and the Oklahoma City bombing.[13] Cockburn had just blasted Helvarg for his unfounded accusations in his *Beat The Devil* column in The Nation.[14]

I told Alex it was the continuation of another strategy suggested by the MacWilliams Cosgrove Snider study, to accuse wise users of perpetrating acts of violence against environmentalists.

Helvarg's articles were part of a grant-driven campaign by the Green Cartel, with the W. Alton Jones Foundation in the lead. Helvarg had, a month earlier, participated in a Washington, D.C. news conference arranged by left-wing public relations expert David Fenton of Fenton Communications. There he actually accused the wise use and property rights movements of complicity with militias in blowing up the Oklahoma City federal building by this linkage: "James Nichols, held as a material witness in the bombing of the Oklahoma City Federal Building, is a member of the Michigan Property Rights Association, founded by property rights activist Zeno Budd. Budd, in turn, was a featured speaker at a militia forum in Detroit."[15]

Cockburn had seen the news release, which contained a dozen similar accusations about bombings, threats and arsons in many parts of the country, plus Helvarg's assertion that the wise use movement was the primary recruiting source for militias.

"Therefore, the wise use movement blew up the federal building," he laughed.

I smiled and said, "Exactly, just as you wrote in your *Beat the Devil* column. But if you put all those things in the same sentence with a sufficiently accusatory tone, the impression you leave is that there's some actual connection. You can't build a criminal case on such remarks, but you can convince your supporters that you have documented something you haven't. And you can demonize a whole class of people as monsters."

I told Cockburn that the idea originally surfaced in reaction to the rising awareness of ecoterrorism, violence done by environmentalists, not to them. A workable plan to deflect the growing evidence of ecoterrorism came from Jonathan Franklin of the San Francisco, California-based Center for Investigative Reporting. In 1992 he had written a piece for the Center's magazine, Muckraker, titled, "First They Kill Your Dog," alleging that several personal attacks and arsons against environmental activists were perpetrated by the wise use movement.[16]

The Center for Investigative Reporting in 1992 received $100,000 from W. Alton Jones Foundation "for reporting on current dynamics of national environmental organizing efforts" and $105,000 from the Florence and John Schumann Foundation "for research on environmental conflicts in the West." The W. Alton Jones Foundation was behind the strategy from the beginning, as was its close ally in the Environmental

Grantmakers Association, the Schumann Foundation. They were the architects; the Center for Investigative Reporting was the first contractor.

The next was CBS News. Mike Wallace of CBS News is a long-time supporter of the Center for Investigative Reporting—he even wrote a November 18, 1991 fundraising letter for them on CBS News stationery—and the CBS News 60 Minutes crew thought Franklin had a great story idea. While Jonathan Franklin was cooking up his Muckraker article, CBS News was working in tandem on a nearly identical "wise-use-is-violent" story, starring some of the same "victims," for release in the fall, the same time Franklin's article would break.

On Friday, May 8, 1992, CBS News Producer Rebecca Peterson called me requesting an interview for a story about "community violence"—news people are rarely forthcoming about what they really have in mind—and I agreed to speak on-camera with Leslie Stahl at a suite in the Alexis Hotel in downtown Seattle late in May. I also agreed to meet Peterson and her camera crew in Austin, Texas on May 16, so she could get some field action footage at a speech I was giving to the Texas Wildlife Association.

In Austin, it became clear that CBS News wanted me to incite the crowd to violence. My message wasn't "strong" enough. I complained to Peterson that her show was a setup, and wasn't going to mention environmentalist violence—and there were plenty of arrests, convictions and prison terms to talk about, as well as sabotage instruction manuals, journals inciting to sabotage and other apparatus of long-term systematic violence, none of which existed in the wise use movement. But she said that since 60 Minutes had nailed Earth First in a negative report just a couple of years earlier, this was a balanced report.[17] Being a realist in the face of such irrefutable logic, I said okay and decided I was in for some lumps from CBS News.

I then recalled that four months earlier, Mark Brodie, another CBS News 60 Minutes producer, had asked if I would be willing to appear in a story on the wise use movement trying to explain to the public what it was all about. In response, I sent him a sizable package including 1) my event calendar through April; 2) a list of names and numbers of other wise use leaders who would probably be willing to appear; 3) a list of visual resources (films and video); 4) a stack of print resources (eight or nine books I had written or edited); 5) news clips of recent print media; and 6) a "screen treatment" script of my proposed message.

Brodie examined the package and wanted an exclusive on my story. He said it would be broadcast sometime in the fall of 1992. In the meantime, CNN's Sharon Collins had asked me to tell much the same story. CBS News couldn't guarantee that their story would actually run (it's not uncommon for a controversial figure to grant such an exclusive and then find his or her story has been spiked to keep it from the public). CNN guaranteed that their feature would air as the lead story in Collins's Network Earth program at an early date. I told CBS News I couldn't give them an exclusive and did the CNN feature instead. It never occurred to me that I had sent CBS News 60 Minutes the equivalent of my near-term battle plan.

Finally smelling the proverbial rat, I called around and found several people Stahl had already interviewed, so I figured out what the ambush was about before my appointment: my comments would come after several weeping women—Jonathan Franklin was writing for Muck-raker about the same ones—spent a lot of screen-time accusing uniden-tified wise users of beating, burning, raping and stabbing them, and, since I wouldn't recommend violence, my role was reduced to providing the lame sound-bite saying, gee, no, we don't do violence. Cut back to the weeping women showing their wounds.

When I got to the Alexis at 1:00 p.m. on May 26, and we were all wired up, I pointed out to Leslie Stahl that the so-called attacks she wasn't telling me about were incidents of random violence, some done by teen-age ruffians and others by unaffiliated hooligans. There were no arrests, no convictions, no wise use connection. It was all lies. That landed on the cutting room floor. As expected, I ended up as the weak denial sound-bite.[18]

When it was broadcast, wise use didn't come out as badly as I had expected, so I sent CBS News a rather double-edged letter saying their story had lived up to the time-honored standard of television investigative reporting. They, of course, ran that as soon as possible.

However, as it turned out, their story didn't quite live up to that standard. CBS News 60 Minutes later had to run a retraction about one of their women, Stephanie McGuire, stating that law enforcement said the woman had lied.[19] It took them sixteen months.

Helvarg's Sierra Club book, *The War Against the Greens,* and his articles in The Nation were the next bricks in the Muckraker / 60 Minutes process of myth-building.

W. Alton Jones Foundation had even created a myth-building en-vironmental group, giving $145,000 in startup grants in 1993 for the ex-clusive purpose of keeping track of the wise use movement and institu-tionalizing the MacWilliams Cosgrove Snider strategy. The group was called CLEAR, Clearinghouse for Environmental Advocacy and Research, and it was a project of a little-known outfit called the Environmental Work-ing Group, which itself was a project of San Francisco's Tides Founda-tion, a huge eco-money funneling operation. CLEAR's first release was the MCS Report, the Search and Destroy Strategy Guide. It has since spent hundreds of thousands of foundation dollars for the sole purpose of gathering and disseminating anti-wise use propaganda. Now they even have a fancy web page that tracks some 2,000 wise use groups.

Others who joined W. Alton Jones Foundation in funding a lucra-tive wise use-bashing industry among environmentalists are Changing Horizons Charitable Trust and the Winslow Foundation of New Jersey (one of whose directors is Wren Winslow Wirth, moneyed wife of former Colorado Senator Timothy E. Wirth, the Clinton administration's Undersecretary of State for Global Affairs who was helped into place by Al Gore).

Jones, Tides, Changing Horizons and Winslow spent a lot of money on a 1993 anti-wise use report, *How the Biodiversity Treaty Went Down*, which tried to figure out which of us defeated the Biodiversity Treaty in the Senate, yielding a guilt-by-association extravaganza that doesn't even mention the people who actually rallied the troops to make the telegrams and phone calls come in when they saw "biodiversity" stealing their land after reading the whole treaty, which the Senators hadn't seen.

The small army of anti-wise use experts hired by the big money included:

Chip Berlet of Political Research Associates in Cambridge, Massachusetts, whose left-wing research of right-wing groups lumps together anybody to the right of Chip Berlet, which is practically everybody. His perception tends toward All You Right-Wingers Look Alike To Me. He accuses all populists, conservatives and libertarians of scapegoating the following: federal law enforcement officers, abortion-rights supporters, gays and lesbians, welfare recipients, people of color, immigrants, and, oh, yes, environmentalists. He's one of the guys who thinks my colleague Alan Gottlieb is anti-semitic. A real expert.

Sheila O'Donnell, of Pacifica, California, professional snoop, partner with an elderly friend, Beverly Axelrod, in Ace Investigations. Her business card is the Ace of Spades with an address and a private investigator license number. O'Donnell is "The Green P.I." lauded in David Helvarg's book, *The War Against the Greens*. She was covertly working on the wise-use-is-violent project at the same time in 1992 that Jonathan Franklin of Muckraker and CBS News 60 Minutes were working on their sneak attacks. O'Donnell wrote an August 20, 1992 report titled *Common Sense Security* for environmentalists that scared its recipients into believing they were under siege by right-wing monsters. The report advised environmentalists to think about getting an unlisted phone number and to make sure their document shredder was the confetti type, not the strip type, because strips can be reconstructed with a little patience, and not to talk to the FBI without a lawyer present because they harass environmental activists (scapegoating federal law enforcement officers, oh my!). All very common sense. Well, don't *you* have a confetti document shredder?

The Green Cartel, I discovered, had Sheila O'Donnell nosing into my activities while I was writing this book. Her Ace Investigations was being paid in late 1996 to gather intelligence on me by the Tides Foundation, the W. Alton Jones Foundation and the Winslow Foundation.

Paul F. deArmond of Bellingham, Washington, former computer systems employee at Western Washington University's Bureau of Faculty Research, now occupied with attending meetings of right-wing groups to see who's there. Anybody who shows up at a particular meeting is linked to everybody else who shows up at that meeting. Then everybody who showed up at *that* meeting is linked to anybody who showed up at any *other* meeting that anyone at that meeting ever showed up for. The deArmond Connect-the-Dots Principle is reminiscent of John Guare's wonderful play, *Six Degrees of Separation*, the premise of which is that everyone in the world is connected to everybody else by only six intervening relationships; or the Internet game in which people try to name the movies connecting the actor Kevin Bacon to other show-business personalities; or the "Erdös number" which reflects a mathematician's degree of separation from legendary Hungarian mathematician Paul Erdös—those who have co-authored a paper with Erdös get the number 1, those who have co-authored a paper with a member of this group but not Erdös get the number 2 and so on. Watch out, you may be part of the wise use movement if you go to the supermarket, which will give you a Ron Arnold number of 250,000,000 (the approximate population of the United States).

All these folks have written a lot about me but never contacted me once. If that's the best intelligence team money can buy, maybe we don't need money.

The anti-wise use campaign had friends in high places, I discovered on February 24, 1994. ABC News Nightline with Ted Koppel that evening ran a report titled, "Environmental Science For Sale," produced by Jay Weiss. It was an investigation of the wise use movement, probing my activities and those of scientist Fred Singer of the Washington, D.C.-based Science and Environmental Policy Project, among others.[20]

Koppel opened this edition of Nightline with a stunning revelation: Vice President Al Gore had given him the story. Koppel explained that he and Gore had met by chance waiting for an airplane, and, over coffee, Gore urged him to investigate connections between the wise use movement and such elements as big industry, Lyndon LaRouche and the Unification Church of Rev. Sun Myung Moon.

Koppel had first covered the wise use movement almost exactly two years earlier, on February 4, 1992.[21] On that date, after a five-minute introductory segment interviewing me and a number of other wise use advocates, the program switched back to the studio and a face-off between conservative radio talk show host Rush Limbaugh and then-Senator Al Gore. Koppel was the first broadcaster to note that environmentalism was no longer a motherhood and apple pie issue, but now had serious challengers for the moral high ground.

Gore was deeply upset by the rise of wise use. By 1994 he was Vice President of the United States, and the time had come to strike back.

So, on the night of February 24, Koppel told Gore's story—but notified his viewers exactly where it had come from, a highly unusual move in a medium that normally goes to extremes protecting sources. And he sounded annoyed.

While Koppel explained that Gore's office had sent him a stack of documents, an image of the fanned-out papers filled the TV screen. If you've seen such graphics, you know that the top document is always totally illegible so that a certain amount of anonymity is preserved for the source. However, peeking out from behind the first sheet was a letterhead just beyond legibility—unless you knew what it said to begin with. I did. It said, MacWilliams Cosgrove Snider.

So—Vice President Al Gore was keeping a dossier on us, courtesy the Green Cartel: MacWilliams Cosgrove Snider, a political strategy firm, hired by The Wilderness Society, using a grant from the W. Alton Jones Foundation (the CitGo oil money) authorized by director John Peterson "Pete" Myers, who has given away hundreds of thousands of dollars to smear the wise use movement. Knowing that Al Gore has been secretly keeping tabs on me, do I need to call Psychic Hotline to know why the Winslow Foundation gave money so that Sheila O'Donnell of Ace Investigations could gather intelligence on me? Could it be because Wren Winslow Wirth is the wife of Clinton administration official Tim Wirth who was given his State Department slot with the help of Vice President Al Gore?

Vice President Gore, Koppel told his viewers, was particularly concerned about Dr. Fred Singer of the Washington, D.C.-based Science and Environmental Policy Project, well known for debunking the ozone depletion and global warming scares.

Laws have been passed against important industrial chemicals because computer models predict them to deplete ozone or cause global warming. Dr. Singer points out flaws in computer models, noting that realistic risk assessments rather than computerized guesswork or emotional scare tactics are needed for sound public policy.

Michael Oppenheimer of the Environmental Defense Fund told Koppel he was so worried about the wise use movement because, "If they can get the public to believe that ozone wasn't worth acting on, that they were led in the wrong direction by scientists, then there's no reason for the public to believe anything about any environmental issue."

What about those Moonie ties and big industry money? When asked by Nightline, Dr. Singer acknowledged having accepting free office space and science conference travel expenses in the past from the Unification Church, as well as funding from large industries. The Moon support lasted only a short time, but the industry funding continued. "Every environmental organization I know of gets funding from Exxon, Shell, Arco, Dow Chemical, and so on," said Singer. "If it doesn't taint their science, it doesn't taint my science."

Koppel evidently felt used by Gore, saying, "In fairness, though, you should know that Fred Singer taught environmental sciences at the University of Virginia, that he was the deputy administrator of the Environmental Protection Agency during the Nixon Administration, and from

1987 to 1989 was chief scientist at the U.S. Department of Transportation. You can see where this is going. If you agree with Fred Singer's views on the environment, you point to his more impressive credentials. If you don't, it's Fred Singer and the Rev. Sun Myung Moon."

Koppel noted that Dr. Singer's predictions about the low atmospheric impact of the Kuwait oil fires was accurate and the environmentalists' forecast of doom, as voiced by the late astronomer Carl Sagan, was wrong.

Koppel handled the segment about me much the same way, saying that I had once served on a local board of the American Freedom Coalition, "a political organization, which, in the past, has received substantial funding from the Rev. Sun Myung Moon." There were no allegations that my Center had received Moonie money, or that I was a follower of Moon or his church, or that some nefarious Moon-influenced plot was afoot, unlike the Green Cartel's version of the story. Somebody at ABC News had actually done some fact checking.

Then I remembered. Three months earlier, on Tuesday, November 9, 1993, ABC News producer Bob Aglow had called me on behalf of correspondent Bettina Gregory, asking for an interview for the "American Agenda" segment of World News Tonight with Peter Jennings. I had previously appeared in that segment and was treated fairly. I agreed. That Friday, November 12, Aglow and Gregory taped the interview in my office. Among other things, I gave them a stack of my Center's financial statements showing where our budget really came from: small donations from members, book sales and conferences, with less than 5% coming from foundations and corporate grants.

However, the segment never aired. But the film that Koppel used in his Nightline broadcast was the footage taken by Bob Aglow with correspondent Bettina Gregory. Someone on the Nightline staff had obtained it from the World News Tonight staff—evidently along with my financial data.

At the end of the Nightline feature, Koppel pointedly rebuked Gore's recruitment to a hatchet job, concluding, "The measure of good science is neither the politics of the scientist nor the people with whom the scientist associates. It is the immersion of hypotheses into the acid of truth. That's the hard way to do it, but it's the only way that works."

There was something odd about this edition of Nightline. Why did Koppel reveal the source of his story? And why did he take such pains at fairness that it repudiated Gore's premise? I contacted the network to see what they knew about their source. ABC News Nightline producer Jay Weiss wouldn't say why Koppel told of Gore, but neither he nor Koppel knew that the Search and Destroy Strategy Guide existed because Gore did not provide it, only a stack of anti-wise use articles and news releases provided by MacWilliams Cosgrove Snider. So I sent them a copy.

A little poking around also led to an interesting discovery: Al Gore himself took $1,000 from the Rev. Sun Myung Moon's Unification Church to address their American Leadership Conference just before accepting the vice presidential nomination. Two high ranking environmentalists had also taken $1,000 from Moon's Unification Church for speeches at a media conference: Marion Clawson of Resources for the Future and Donella Meadows, lead author of *The Limits to Growth*. What, if anything, did that mean?[22]

A little more poking around revealed that Jay Weiss was not the producer originally assigned to investigate Gore's allegations. The original producer of the "Environmental Science for Sale" segment had been 12-year ABC News veteran Tara Sonenshine. Sonenshine had started her career as a booker, the person who finds newsmakers and makes appointments for interviews. She had a Rolodex® to kill for by the time she became an assistant producer. She knew just about every newsmaker in the world when she received the promotion to full producer, including Al Gore and Tim Wirth and his rich wife Wren.

Sonenshine took Gore's story and ran with it as if she were Gore's advocate. She scripted it as a truly vicious hit piece. Her original version had painted Lyndon LaRouche operative Rogelio Maduro as a crackpot with ties to the wise use movement, the culprit who allegedly sank the Biodiversity Treaty.[23] It also crucified University of Virginia Professor Patrick Michaels—who, like Fred Singer, challenged global warming computer models—for accepting research funding from industry.[24] It took every cheap shot in the book: sinister lighting to make Professor Michaels look unsavory, industry-sponsored film footage with no context, a one-sided slam against everyone it didn't like. It was the perfect Green Cartel reprisal.

Sonenshine's show was scheduled to air early in February, but a Nightline assistant producer told me Koppel didn't like its tone and demanded changes. Sonenshine was chagrined. My source said that during an acrimonious staff meeting, Sonenshine departed. Whether she was fired or resigned depends on who you ask.

The February 8 edition of The Washington Post carried "Rumour du jour: Tara Sonenshine, editorial producer at ABC News's 'Nightline,' is headed for a policy job with national security adviser Anthony Lake. She has been with 'Nightline' for 12 years."[25]

The Washington Post reported on February 14 that Tara Sonenshine had been appointed special assistant to the president and deputy director for communications at the National Security Council, "working on longer-term projects, which some uncharitably call an effort to make NSC chief Anthony Lake more TV-genic."[26]

Did Al Gore give her that job as a weenie for doing a hatchet job on the wise use movement? Or as a getaway route when the hatchet broke?

Ten days later, "Environmental Science For Sale" was broadcast, much changed, a combination of clips from Sonenshine's hit piece and the Weiss remake.

Sonenshine lasted less than a year at NSC before going to work covering national security for Newsweek.[27]

It had been a long afternoon with Alexander Cockburn. I capped off our conversation by suggesting that what was happening to resource producers in America today reminded me of something Karl Marx had written about the elite forcibly removing the people from their property and pushing them into the urban work force. I went to the bookshelf where I keep the collected works of Marx and Engels.

"It's Chapter Twenty-Seven of *Capital*," Cockburn said. "Expropriation of the Agricultural Population From the Land."

I found the place.

Cockburn continued: "It went on for many years during the Reformation in England. It hurled the peasant into the proletariat. Did the same to a lot of monastery inmates."

"It's happening right now," I said. "And environmental regulations written by the rich and powerful are the crowbar prying them from their land."

"Tell me, Ron, did you really blow up the federal building?"

He cracked his big Irish grin. He had something up his sleeve.

Two months later I found out what.

In response to a letter to The Nation from David Helvarg complaining that Cockburn's "claim that there's no evidence linking the Wise Use/Property Rights network to the militia movement ignores a well-documented and expanding list of overlapping organizers, members, groups and materials," Alex replied:

> For years now David Helvarg has been backed by environmental groups such as the Sierra Club to investigate and smear the Wise Use movement by any means necessary. This goes back to the early 1990s when the Environmental Grantmakers Association offered a de facto bounty for material discrediting Wise Users as (a) a front for corporations or (b) part of a far-right terrorist network.

Alex then went through the 1992 Grantmakers session on wise use in which Debra Callahan concludes "that this is pretty much generally a grass-roots movement, which is a problem, because it means there are no silver bullets."

Cockburn noted that she advised, "Attack Wise Use. . . . We need to . . . talk about the Wise Use agenda. We need to expose the links between Wise Use and other extremists: the Unification Church, the John Birch Society, Lyndon LaRouche. We need to talk about the foreign influences."

Some argued back that the Wise Use line wasn't entirely off. As Barbara Dudley (then running the Veatch Foundation, now head of Greenpeace) put it, "It is true that the environmental movement has been...an upperclass, conservation, white movement....They're not wrong that we're rich. . . . We are the enemy as long as we behave in that fashion." But Dudley's caveat was rejected.

In his letter, Helvarg claims that the "wave of terrorist violence" against greens was ignored until recently by the big D.C.-based environmental organizations. To the contrary. These same organizations long ago decided that the proper way to deal with Wise Use was to pay people like Helvarg to smear it.

One rarely sees such blunt argument in print. I was impressed. Cockburn went on:

And so we have the unlovely sight of Helvarg behaving like an F.B.I. agent. He prowls across literature tables at Wise Use meetings and ties all the names on the pamphlets, letterheads and books into his "terror network." The trouble is, he never makes his case. Helvarg never comes up with the terrorist conspiracy he proclaims, because there hasn't been one.

Then Cockburn pointed out the real tragedy of the situation:

As Barbara Dudley so presciently pointed out in 1992, there was a tremendous opportunity for the green movement to reach out to the farmers, ranchers, residents of rural communities who have been shafted by big government and big business. But it was an opportunity deliberately rejected in favor of pursuing a self-declared war on the West, where ranchers and millworkers alike were seen, in the repulsive words of Andy Kerr of the Oregon Natural Resources Council, as merely "collateral damage" in the effort to purge the landscape of inconvenient humans. Who declared war on whom? Challenge a person's livelihood and place on the land, and you must expect a response.[28]

Cockburn closed his letter with a reminder. It was, after all, the environmentalist Edward Abbey who wrote in his foreword to what some have called a genuinely terrorist document, *Ecodefense* by David Foreman:

If a stranger batters your door down with an axe, threatens your family and yourself with deadly weapons, and proceeds to loot your home of whatever he wants, he is committing what is universally recognized—by law and morality—as a crime. In such a situation the householder has both the right and the obligation to defend himself, his family, and his property by whatever means is necessary. This right and this obligation is universally recognized, justified and even praised, by all civilized human communities. Self-defense against attack is one of the basic laws not only of human society but of life itself, not only of human life but of all life.[29]

Helvarg gradually faded from the wise use-bashing industry. In the interim, he complained bitterly to his friends that the Sierra Club did not promote his book enough and that he was making a meager living only by whatever derivative magazine articles he could drum up as interest in his accusations waned. A second edition of *The War Against the Greens* came out in paperback in 1996, adding a few paragraphs to beat up Alexander Cockburn.

Others came to fill Helvarg's niche, Andrew Rowell most recently, whose 1996 book *Green Backlash: The Subversion of the Environment Movement* is disappointingly just an update of Helvarg's book, whacking most of the same people with the same meat cleaver, sort of a *Son of War Against the Greens*. It even has the same cast of expert informants including supersnoop Sheila O'Donnell handing out the same old W. Alton Jones Foundation line that they're-all-Moonies-and-neoNazis out to kill everything. It was so sloppy that I found a little sticker on the inside front cover saying: "Erratum: On page 19 is stated that the Washington Post is a Moon-backed newspaper. The Washington Post is not connected to Sun Myung Moon. We regret the error."[30]

However, another book with the title *Green Backlash: The History and Politics of Environmental Opposition in the U.S.* came out in 1997 that took the wise use movement seriously—and it was written by a serious scholar, Professor Jacqueline Switzer of Southern Oregon State College. Switzer examined the growing opposition to organized environmentalism in a factual and objective manner. Her book was based on actual field work, visits and interviews with her subjects and a scholarly regard for the truth, none of which could be said for Andrew Rowell and Sheila O'Donnell, who have never contacted me once.[31]

THE POINT OF THIS LONG DIGRESSION IS THIS: Nobody I have described in this chapter is an ecoterrorist. Being a radical environmentalist does not mean that you are an ecoterrorist. It is wrong to demonize a whole class of people for the acts of a bad core of criminals. I have shown you these radicals as I know them in order to demythologize them. I passionately disagree with environmentalists, radical and mainstream, as they passionately disagree with me. But I will not let my distaste for their opinions and actions alloy my respect for them as human beings. This is a book about criminals. They are not criminals.

Chapter Five Footnotes

1 Henwood's show is broadcast each Thursday by listener-supported station WBAI in New York City. The debate with Hermach was broadcast November 3, 1994. Henwood, an economist, is also the publisher of Left Business Observer and author of *The State of the U.S.A. Atlas: The Changing Face of American Life in Maps and Pictures*, Simon and Schuster, New York, 1994. His latest book is *Wall Street*, from Verso, 1997.

2 November 17, 1994.

3 Rozek listed these items: The collected works of Karl Marx and Frederick Engels, particularly Marx's *Capital* and Engels's *The Dialectics of Nature*; Max Weber's *Economy and Society*; Emile Durkheim's *The Division of Labor* and *The Elementary Forms of Religious Life*; Sigmund Freud's *The Future of an Illusion* and *Civilization and its Discontents*; Carl Jung's *Archetypes and the Collective Unconscious*; Jurgen Habermas's *Theory of Communicative Action*; Ronald Inglehart's *The Silent Revolution*; Abraham Maslow's *Motivation and Personality*, Second Edition; Lovejoy and Boas's *Primitivism and Related Ideas in Antiquity*; Lewis Coser's *The Functions of Social Conflict*; Lewis Feuer's *Ideology and the Ideologists*; Teilhard de Chardin's *The Phenomenon of Man*; Hugh Dalziel Duncan's *Symbols In Society*; Luther Gerlach and Virginia Hine's *Lifeway Leap* and *People, Power, Change*.

4 "Who Put the 'Wise' in Wise Use?" *Forest Voice*, vol. 8, no. 1, January-February 1995, by Victor Rozek, p.12.

5 This incident was kept quiet by the foundations. A radical environmentalist publication called Cascadia Planet later recounted these events.

6 The Foundation Center, *Grants for Environmental Protection and Animal Welfare*, 1994-1995, New York, 1995, grants number 883 (p. 26), 2205 (p. 67), and 3201 (p. 93). Form 990, W. Alton Jones Foundation, 1993.

7 Alexander Cockburn and Ken Silverstein, *Washington Babylon*, Verso, New York, 1996, p. 225.

8 Alexander Cockburn and James Ridgeway, *Political Ecology: An Activist's Reader on Energy, Land, Food, Technology, Health, and the Economics and Politics of Social Change*, Times Books, New York, 1979.

9 "The Collapse of the Mainstream Greens," *CounterPunch*, Vol 1, No. 17, October 1, 1994, edited by Ken Silverstein and Alexander Cockburn, p. 1

10 "The wise use movement," released by Pete Myers and Debra Callahan, W. Alton Jones Foundation limited-distribution document, Charlottesville, Virginia, February 6, 1992.

11 "The Wise Use Movement: Strategic Analysis and Fifty State Review," MacWilliams Cosgrove Snider, Clearinghouse on Environmental Advocacy and Research, Washington, D.C., March 1993.

12 "Fighting Back," *Forbes*, July 19, 1993, by Leslie Spencer, p. 43.

13 "The anti-enviro connection (paramilitary groups and anti-environmentalists)" (Cover Story), *The Nation*, May 22, 1995 v260 n20, by David Helvarg, p. 722(2). *See also*, "Anti-enviros are getting uglier: the war on Greens," (Cover Story), *The Nation*, Nov 28, 1994 v259 n18, by David Helvarg, p. 646(4).

14 "Beat the Devil," *The Nation*, Volume 260, Number 23, June 12, 1995, by Alexander Cockburn, p. 850.

15 News release: "Militias Linked to 'Property Rights' Movement; Federal Em-

ployees Leader Asks for Hearings," Fenton Communications, May 2, 1995, contact Charles Miller or Helen Pelzman, 202-745-0707.

[16] "First They Kill Your Dog," *Muckracker: Journal of the Center for Investigative Reporting*, Fall 1992, by Jonathan Franklin, p. 1.

[17] "Earth First!", *Sixty Minutes Transcripts*, vol. 22, no. 24, March 4, 1990.

[18] "Clean Air, Clean Water, Dirty Fight," *CBS News 60 Minutes*, Leslie Stahl, September 20, 1992.

[19] "Clear Air, Clean Water, Dirty Fight—Update on Past Stories," *CBS News 60 Minutes*, Leslie Stahl, January 9, 1994. The voice-over of Leslie Stahl said, "This Florida woman, Stephanie McGuire, told us that because of her beliefs about protecting the environment, she had been slashed, burned and raped, but the State of Florida Law Enforcement Department says that's probably a lie." McGuire was also a featured "victim" in Jonathan Franklin's "First They Kill Your Dog," and David Helvarg's Sierra Club book, *The War Against the Greens*. CBS News 60 Minutes was the only one of them to run a retraction.

[20] "Environmental Science For Sale," *ABC News Nightline*, Ted Koppel, Transcript No. 3329, February 24, 1994.

[21] "The Environmental Movement's Latest Enemy," *ABC News Nightline*, Ted Koppel, Transcript No. 2792, February 4, 1992.

[22] Telephone interview with Tom Ward of the Unification Church, New York, March 10, 1994.

[23] Telephone interview with Rogelio "Roger" Maduro, Leesburg, Virginia, February 25, 1994. The actual individuals behind the anti-treaty call-in campaign were Tom McDonnell of ASI (see p. 42ff), consultant Michael Coffman, Ph.D. and Kathleen Marquardt of Putting People First.

[24] Telephone interview with Prof. Patrick Michaels, Charlottesville, Virginia, February 25, 1994.

[25] "The TV Column," *The Washington Post*, February 8, 1994, by John Carmody, p. C4.

[26] "The Federal Page - In The Loop," *The Washington Post*, February 14, 1994, by Al Kamen, p. A13.

[27] "Media Notes," *The Washington Post*, June 21, 1995, by Howard Kurtz, p. D1.

[28] "Cockburn Responds," *The Nation*, Volume 261, Number 5, August 14, 1995, p. 150.

[29] Dave Foreman and Bill Haywood [pseudonym], editors, *EcoDefense: A Field Guide to Monkeywrenching*, "Forward!" by Edward Abbey, Ned Ludd Books, 1985, p. 7.

[30] Andrew Rowell, *Green Backlash: The Subversion of the Environment Movement*, Routledge, London, 1996.

[31] Jacqueline Vaughn Switzer, *Green Backlash: The History and Politics of Environmental Opposition in the U. S.*, Lynne Rienner Publishers, Boulder, 1997.

Chapter Six
HISTORY

12:35 A.M., WEDNESDAY, SEPTEMBER 5, 1973   *Tucson, Arizona*
JOHN WALKER AND GARY BLAKE HURLED FIST-SIZED ANGULAR DESERT STONES into the newspaper's darkened windows. The plate glass cracked, then shattered. They sent rock after rock into the splitting panes, watching big dagger-shaped chunks collapse inward. Blake pulled a crowbar from within his army surplus jacket and knocked the remaining jags off like so many broken teeth. As a gesture of contempt, he smashed the adjoining glass door to splinters. The Arizona Territorial had paid the price for giving them bad press. Now perhaps the owner, E. D. Jewett, Jr., chairman of the Pima County Planning and Zoning Commission, would think twice. Walker spray painted the wall with two-foot high letters: ECO-RAIDERS.

The two dark figures quickly left North Oracle Road for a side street, shadows among shadows. There were few neighbors here to jump at their noise. No astonished eyes had spotted them in the clear night— they had waited fifteen minutes after the bright quarter moon was down. They slipped into the back seat of the waiting car and Pat Salmon drove them away from the wreckage, taking a circuitous route home to Oro Valley.

As that midnight glass shattered, so did their lives. They didn't know it yet, but it was their last action as the Eco-Raiders, suburban guerrillas, enemies of urban sprawl who had left a swath of destruction behind

193

them two years long and nearly two million dollars wide. Walker, Blake, Salmon and the two other University of Arizona students were unstoppable. They were minor folk heroes, celebrated in alternative newspapers across the country. Yet, in a few short autumn days, they would drop into a police snare already winding about them, betrayed by one of their own. When the shards that were the newspaper's windows finished their airy pirouettes and froze into a silent crust over everything in the little office, the criminal career of the first authentic ecoterror group in America was over.[1]

The Eco-Raiders did their first actions in the summer of 1971. Tucson—the sleepy little mountain-held town in the basin of the Santa Cruz and Rillito Rivers, south of the Santa Catalina Mountains, west of the Tanque Verdes and the Rincons, east of the Tucsons, north of Helmet Peak and the sparse Santa Ritas—was just beginning to grow and everybody wanted growth except the Eco-Raiders. They were still students at Canyon Del Oro High School, sneaking out of their affluent North-side homes at night to cut down encroaching billboards alongside Highway 89 where it left town as North Oracle Road on its way through the mountains to the mining towns of San Manuel, Globe and Florence, the back way into big-city Phoenix.

They had no name then and there were only four. It was just a few friends of John G. Walker, who was seventeen at the time, intelligent, moody and alienated, close-mouthed, dark-eyed, long-haired, bluejeaned, the cliché of the 1970s teenager. The whole thing was his idea. His father held a technical post at the University of Arizona that supported the family in comfort. John kept shy of both parents and his older sister, more reclusive than rebellious.

His closest friend was Gary E. Blake, also seventeen, not quite as bright, sociable enough to have a part-time job as a bank messenger, harder at the corner of the eyes than Walker but more round-faced, emphasized by center-parted shoulder-length stringy hair—he would later grow a wispy post-adolescent beard giving him the mock Jesus look of an early Grateful Dead devotee. Then there was Patrick G. Salmon, 18, very unlike Walker or Blake, outgoing, personable, more smart than intelligent, light hair only collar-length and a longer more defined patrician face with widely spaced rectangular eyes. And last, Chris Morrison, 17, from a working-class family on the way up—his father a foreman at Tucson Gas & Electric—tall and thin, short red hair and glasses, a type nobody would notice in a school hallway, a little too eager, straining to be more than he was.

Canyon Del Oro was a mixed but mostly affluent North-side high school. A teacher named Michael Rowe probably sparked John Walker's interest in ecology.[2] Walker had hiked the slopes of the Catalinas because they were literally in his back yard on North Christie Drive, just beyond the swimming pool and the guest house of his family home. He knew the desert creatures, old coyote and jackrabbit, he knew the aloof mountain sheep, he knew the red-blooming ocotillo and the thorny cholla and the paloverde tree and the lavender ironwood and the giant saguaro, and he knew how to avoid rattlesnakes and the little scorpions that were most dangerous, and what the lair of the black widow spider looked like. But Rowe introduced him to people who wrote about such things, and such new friends can change your life. Walker read Edward Abbey, whose *Desert Solitaire,* a biographical essay on living as a seasonal ranger in Utah's Arches National Monument (now Arches National Park), propelled him to the forefront of the burgeoning environmental movement. Abbey's sensitive nature writing awakened many to the love of wilderness, especially of the desert West, and John Walker became one of the awakened. Abbey's call to protect the desert also sowed a peculiar rage in the souls of those who had awakened, and John Walker harbored its seeds.

Walker kept his important papers in an old briefcase his father gave him. An article by Chicago Daily News columnist Mike Royko found its way into his briefcase, about a man from the Fox River Valley near Aurora, Illinois, who became a one-man antipollution Zorro. He blocked steel mill drainpipes, he tried to seal off their smokestacks, he put dead skunks on executives' porches, he dumped dead fish in corporate reception lobbies, and he always left a note telling why, signed, "The Fox." The Kane County sheriff spent more than a year hunting The Fox without success.

In a second column, Royko made a legend of The Fox with an account of a meeting arranged to explain why the saboteur did his deeds of ecotage. "In the finest traditions of all mystery crusaders," wrote Royko, "'The Fox,' by day, is an ordinary, soft-spoken citizen. He's approaching middle age, has a respectable job, a family, and has never before gone outside the law." He was driven to break the law, he told Royko, because he remembered what Kane County was like when he was young and saw what these companies were doing to the air and the streams. Places where he used to fish and watch ducks were now uninhabitable.

"Nothing seemed to make them stop," said the Fox. "So I decided that even if I was only one man, I'd do something. I don't believe in hurting people or in destroying things, but I do believe in stopping things that are hurting our environment. So I have been doing something. I want them to know why it is being done, so I always leave a note suggesting that they clean up their mess, and I sign it 'The Fox.' That's because of the Fox River.'"

Royko spent three columns taunting victims of The Fox about their just punishment at his hands, noting that allies had begun to work with the ecoteur, multiplying the law enforcement problem. However, accomplices brought problems to The Fox, too. Royko stopped his columns after the fourth, concerning a narrow escape: someone gave the sheriff a tip where and when The Fox would try to block a soap company drain with concrete, and a squad car searchlight caught him in its beam. He ran like hell and got away, but realized that someone he trusted had ratted on him. The danger had become too great. That was the last exploit of The Fox. But they never caught him.[3]

As it was Edward Abbey who awakened John Walker to the love of wild nature, it was The Fox that inspired him to action. Highway 89 ran barely a block east of Walker's high school, and stark ugly billboards were going up alongside it in droves, advertising new subdivisions coming soon. They were big gawky slabs mounted inelegantly on two standard telephone poles. And they were just a short walk away from school. They stirred that peculiar rage in Walker. He decided they were a target worthy of attention. He talked to Gary Blake about it, a confidant already indoctrinated in the ways of The Fox. Walker took a handsaw from his father's garage tool rack and together the two cut down their first billboard late one June evening. It was easy. It was exhilarating. They could knock out a billboard in less than an hour. Ordinary precautions could avoid unwelcome notice. Another week, another blow struck for the desert. Walker did not care who might be harmed by his deeds. Another month, and Walker began to see things more systematically.

By midsummer he and Blake sized up Pat Salmon and Chris Morrison as being concerned about the environment and the impact of development on the desert. Walker gave them the story of The Fox, and in separate guarded discussions, got the right response. He casually suggested what a great idea it would be to cut down one of those ugly billboards, like those vandals had been doing. Again, he got the right response. After separate initiations in the rites of ecotage, Walker told each of them about their little group and its mission. Morrison and Salmon became part of the billboard crusade. A classmate of Morrison's, Don James, was taken on an initiation raid in late August, but the next month school work got to be too much and he was never brought into the group.

Those who remained active acquired a four-foot two-handed saw "for those tough cleaning jobs" and the billboards continued to fall.

In December Walker approached another friend who had grown close, Mark Quinnan. Walker was a little hesitant: Quinnan's brown hair was not as long as his, and he had an after-school job at the Oro Valley Country Club restaurant. Walker gave his friend some printed material. Quinnan does not recall exactly what it was, but Walker said, "read this." He was preaching to the converted: Quinnan was in. And he had his own car, like Blake. Now the five were a band.

For the next few months, they cut down billboards in carefully planned raids, week after week, more than a dozen of them. The police were beginning to take notice. It didn't look like casual vandalism anymore.

In March of 1972, Walker found a newly published book called *Ecotage*. It was a compendium of more than a hundred clever ways to sabotage the wheels of civilization to save nature, and it reprinted Mike Royko's columns on The Fox. It also contained sections on the exploits of Florida's Eco-commandos, Michigan's Billboard Bandits and Environmental Action, the Washington, D.C.-based group that sponsored the ecotage contest which produced the clever ideas.

Walker invited his four friends to the backyard swimming pool after school for a few beers and an important discussion: It was time to broaden their field of operations. The public demand that all glass beverage containers be returnable was meeting with resistance from bottling companies. That required some action.

More seriously, urban sprawl was coming. Five-thousand new homes were going up around the edges of Tucson this year, most of them in massive developments that scraped up great chunks of desert and replaced it with tacky little boxes stuck into regimented streets and decorated with alien trees and cactus brought in from Mexico.

And they needed a name. It was time to start stringing up banners and leaving notes that explained their actions, like The Fox did. They had to sign the notes with something.

But what? They agonized over it. Finding a parallel to "The Fox" wouldn't be too hard: "The Coyote" was symbolic, The Trickster of local Indian stories. But the people buying the new houses probably knew nothing of coyotes or symbolism or Indians. When darkness fell, they moved into the guest house and agonized some more. They went through a dozen names.

But when they came out, they were the Eco-Raiders.

Walker gave the orders. He was the undisputed leader. During late March he began cruising unfinished housing developments in the family car after school, then looking again at night to see where the construction lights were. Soon he took Blake and Quinnan on their first raid against homes for sale, lobbing stones through all the windows and spray painting the words "Eco-Raiders" in big letters on brick sidewalls. There was no way to clean the graffiti from porous block, so the whole surface had to be painted at considerable expense.

An April Fools Day raid got the name "Eco-Raiders" into the newspapers for the first time: hundreds of broken non-returnable bottles and aluminum cans were dumped Saturday midnight in the entry of the Kalil Bottling Company office on South Highland Avenue, a complex that housed nearly all the soft-drink bottlers in town. The note they left said, "A little non-returnable glass: Kalil makes it Tucson's problem. We make it Kalil's problem." The Arizona Star wrote, "The note signed 'Eco-Raiders,' carried the postscript 'Buy returnable, they don't litter.'"[4]

In early May Walker sent a letter to the largest developers, including the Estes Company, Samuel Sneller and Robert King. It read: "It has come to our attention that one of the major threats to the Tucson area is urban sprawl. Even now, developers are ruining the desert by leveling large areas and packing houses in row after row, with no respect for nature. We have seen urban sprawl in Phoenix and we do not want the same here. We have decided to act in our own way to stop this spoiling of nature. – The Eco-Raiders."[5]

They put their exclamation mark on the letter by spray-painting a big Estes Company sign on East Broadway with the words: "Stop Urban Sprawl."[6] Word of mouth began to pass around.

John Walker had been drawing up the battle plan of each raid on lined notebook paper and putting it in his briefcase. He never let anyone see the plans. He located potential targets by reading local newspapers for legal notices of rezoning petitions and advertisements of new developments. He studied each job. He mapped it. He went to each site first and examined it personally, always with Blake, who began to affect the army surplus jacket and army boots that Walker wore.

Walker realized early that all of the Eco-Raiders should never go on a raid together. He watched each Eco-Raider to see what his special talents were. Quinnan and Blake had their own cars, but the others could borrow their parents' cars when necessary. Walker never mixed and matched his crew: Some he never sent on a raid together. Some he always sent together. Quinnan never did an action with Pat Salmon. Blake never did an action without Walker. Even though most of their hits lay within walking distance of their homes, Walker arranged to have the driver for the night drop them off near the attack site, tell him a primary and secondary pick-up location, and have the driver cruise different routes that would take him by the two pick-up points at fifteen-minute intervals. If the raiders missed one pick-up, they'd get the second or the third.

Walker knew the best raid hit as many houses as possible in the shortest time possible so the gang could devastate a company and then lay low for a while. He knew how to vary the tactics: hit homes for an extended time, then dump debris on the steps of City Hall once, then hang a Stop Urban Sprawl banner on a busy freeway overpass, then on to hitting developer's offices instead of homes, then back to hitting homes.

As their damage attracted increased media attention, Walker watched the newspapers to see which companies were posting guards and using dogs, then shifted to hitting the smaller builders like Marved Construction that was putting up only 42 homes around North La Cholla.[7]

Now the raids on housing developments accelerated. The tactics focused on the most expensive damage possible. When a complicated set of engineering stakes had been posted in the ground by survey teams, the Eco-Raiders came along and quietly rearranged them, whole blocks at a time. The cost of resurveying a development would easily exceed $10,000.

When road graders, backhoes, flatbed trucks and bulldozers were left unattended, the Eco-Raiders filled their tanks with sugar, cut their fuel lines, and left notes advising that it would be better if operators didn't try to start the machines. Such heavy equipment repair quickly climbs into the thousands of dollars for each machine.

When a concrete slab floor was laid with its plumbing and sewer and electrical conduits sticking up ready for the house to go up around them, the Eco-Raiders would sweep in and break off every protuberance at the floorline, hitting dozens of houses in a night. Contractors were then forced to chip out the surrounding concrete down to undamaged tubing, repair the pipes and conduits on the stub below floorline, and re-pour the concrete. When the home was later completed on the repaired foundation, the Eco-Raiders would return and cave in all the drywall with sledge hammers, smash toilets and sinks, rip up carpets with a crowbar and throw paint over everything, hitting dozens of houses a night. Insurance claims on the earlier vandalism often meant higher deductibles or no insurance at all on the later vandalism.

By the end of 1972, everybody in Tucson knew about the Eco-Raiders, and the Southern Arizona Home Builders Association was worried. An early 1973 report of 55 contractors claimed that Eco-Raider vandalism had cost about $250 per house or $180,000. In fact, the cost was vastly greater, but contractors feared to make it public lest it encourage the vandals or generate copycats. However, a slipup on another page of the report gave the sharp-eyed a more realistic clue: vandalism had been inflicted on 4,000 of the 5,000 homes built in 1972. $250 times 4,000 homes is $1 million.[8] The actual figure was probably somewhere in between. The Eco-Raiders had been busy.[9] But they themselves had no idea of the money they were costing.[10]

Walker's rage grew bolder. He sent a typewritten letter in January of 1973 to County Supervisor Ron Asta and Supervisor Joe Castillo trying to explain what builders should do and why:

> The dry climate slows the growth rate of most plants and reduces the chances of successful natural seeding. When a bulldozer clears more land than is necessary for home construction and fails to provide for untouched common areas, the barren earth is slower to recover its natural greenery. The scars remain for years, often decades. Land clearing has an adverse effect on animals, eliminating sources of food and shelter, and destroying everything from bird nests to animal burrows.
>
> Often for the sake of convenience an area will be cleared for construction, irrespective of hills and washes that provide for water runoff. When rainwater arrives to replenish the diminishing water table, it either fails to enter washes and evaporates or runs its own course causing erosion and further land damage....
>
> The worst by-product of the real estate developer's lack of environmental concern is the way in which urban residents are

separated from the beauty of the natural desert environment. Only people having a familiarity with the real desert will ever be concerned enough to halt its destruction.[11]

Walker also allowed a statewide weekly alternative newspaper, The New Times, to publish a four page spread with photographs of Eco-Raider actions—but only with Blake, and only with the two of them wearing ski masks. Walker let them publish a lengthy position paper. The whole story was reprinted in alternative newspapers across America. Walker then allowed Tom Miller of the Berkeley Barb to come along on a raid with a photographer, and let him watch the windows of a real estate office being smashed out.[12]

The pressure on county politicians grew steadily. Close to a million dollars in losses that law enforcement couldn't stop was simply intolerable. In early February of 1973, the Pima County Sheriff's Department let newspapers believe they were "on the verge of making arrests in connection with the vandalism." They weren't.[13]

As contractors grew more vigilant and security got more difficult, Walker ordered whoever had the car for the night to act as lookout several streets from the action, with a horn honk code to warn of approaching trouble. On one occasion near the intersection of Snider Road and Soldier Trail, east of the Tucson city limits, while Walker and Salmon were tearing up plastic plumbing on a dozen homesites, a sheriff's patrol cruiser pulled up behind Chris Morrison's parked lookout car. The deputy asked for a driver's license and inquired what the young man was doing sitting there alone on the edge of town in the middle of the night. Morrison gave the standard dumb response like, "I'm late from my girlfriend's house and I'm scared what my Dad's going to say," which, as always, got a "Get along home, now, son, your folks will be worried." The cops didn't have a clue.[14]

And the builders knew it. By mid-1973 the damage had mounted to nearly $2 million. Developers were frantic. One contractor, Sam Sneller, said the industry was "near a state of chaos." Local contractors were blaming the insensitivity on big development outfits that had come in from New York and Chicago. The Eco-Raiders were have a small but measurable impact on how a few contractors looked at the desert.

The Eco-Raiders turned away from hits on actual development and onto the sales and advertising apparatus downtown. They spray painted billboards with "Stop Urban Sprawl." They broke out the windows of real estate companies. They hit real estate sales offices with liquid metal in their keyholes.[15] When they got out into the developments, they focused on finished display homes for sale, simply driving by at night and breaking the windows out, realizing that prospective homebuyers would not likely be attracted to such a neighborhood. And Eco-Raider hits were damaging equipment every week.[16] Some of the Eco-Raiders, including Mark Quinnan, began to wonder if Walker hadn't lost his original focus on protecting the desert from developers—his rage seemed to be diffusing into mere destructiveness.[17]

The Sheriff was an elected official and he had to do something. He had already given the Eco-Raider case to the homicide squad, the cream of the detectives, and they were no closer to solving it than when he gave it to them six months earlier.[18]

So in early July the Sheriff created a five-man Special Problems Task Force headed up by newly made Sergeant Duane Wilson. Each man on the squad was a young Vietnam veteran, seasoned in guerrilla tactics, and each was eager to catch these thugs who snatched away the roofs over people's heads. They had but one case to investigate: the Eco-Raiders. They had no hours—someone was to be on duty at all times of the night and day to instantly respond to any call. Wilson juggled their days off and rotated shifts to avoid overtime. The county attorney's office gave them six special investigators, which the task force ignored.

Sergeant Wilson immediately sent his crew to examine all the cases of damage they could document. Together, they assembled a time-line, trying to spot regularities and changes in the vandals' modus operandi over time. Wilson quickly saw that weekends and Wednesday midnights were their preferred hit times. He tried to determine how many Eco-Raiders he was up against. He saw that the raids were originally clustered only in the northwest corner of the city, spreading later to other areas. He tried to predict where they would hit next. He saw that big projects like those of the Estes Company were hit several times a month. So he sent his crew in their unmarked cars to lay out for several nights in the Eco-Raiders' favored areas at their favored times. One unit responded to a report of noise near their stakeout point, but by the time they were able to drive to the vandalized site on unfinished streets, no vandals were to be found.

The two officers asked Sergeant Wilson to order them dirt bikes so they could give chase over the desert terrain. They got their dirt bikes. They practiced desert riding at night. They got very good at it. On Wednesday, July 11, at a stakeout both men heard the telltale clatter of vandals' tools breaking up drywall on the far side of a large development. They sped directly to the source of the noise—and found nothing but damaged homes. They concluded that the perpetrators must have a sophisticated lookout system.

In fact, the Eco-Raiders never knew about the dirt bikes—Quinnan and Walker had simply finished a short job at an Estes Company site and left for C&D Pipeline, where they smashed a load of plastic pipe, then plugged door keyholes with liquid metal at Richard A. Huff Realty Inc., 4897 East Speedway Boulevard and at Trans-Arizona Development Corporation, 1717 N. Swan Road, and then went home. They did not forget to leaves notes that said, "We are the Eco-Raiders."[19]

That was as close to catching the Eco-Raiders in the act that the task force ever came.

After a month of intensive work, the Special Problems Task Force had nothing. The Eco-Raiders were always two days ahead of them. Wilson said to his crew, "If we don't catch those bastards soon, we're gonna go nuts."

A call for Wilson came in just then, but it was nothing about the Eco-Raiders. It was an elderly friend of his family asking for some help getting rid of the chickens she had let run loose on her ranch south of town. She wanted to know if some sheriff's men would like to come out and shoot about fifty chickens for her. Wilson passed the word around and Deputy Gary Martin, one of the best men on the task force, said, "Sure." Shotguns and high-powered rifles would be too dangerous with horses and livestock nearby, so Martin, an avid gun collector, selected from his racks an air rifle he particularly liked and went killing chickens on the last weekend of August.

The rifle was still on the back seat of his unmarked car when he took it into the sheriff's fuel center to gas up Monday morning. The attendant was a summer hire Martin had come to know through his incessant chatter, a minimum wage worker who washed and filled patrol cars to help pay for college. As the young man wiped down the windows he spotted the weapon laying on the seat and said, "Hey, Deputy Martin, what's the air rifle for? You going to shoot out some windows and blame it on the Eco-Raiders?"

Martin had become accustomed to taking flak about the Eco-Raiders and passed it off as a smart-ass remark. He reported in to Sergeant Wilson and commented on the young man's wisecrack.

"Let's talk to him," Wilson said.

Deputy Martin went back and brought the young man into headquarters where Wilson questioned him.

"What's your name, son?"

"Don James."[20]

"What the hell do you know about the Eco-Raiders?"

"Me? Nothing! I just think they're doing a good job."

Wilson sensed the young man was lying. He kept at him.

After fifteen minutes of talk tag, James finally said, "Ok, I went on a couple of billboard raids with them when I was in high school."

Wilson thought he was just leading them on.

"How many went on those raids?"

"Three or four. I don't know for sure."

"What were their names?"

"I never knew. We didn't have any classes together. Except one guy, the one who asked me to go along."

"What's his name?"

"I don't remember. That was two years ago and it was just a guy

I knew in Miss Roten's humanities class."

"You're in a lot of trouble, son."

"Hey, look, it wasn't me did all that stuff. I can't afford to get into trouble. School's starting and this is my last week here."

Wilson and Martin drilled into James for half an hour with nothing further. No names. A vague memory. Martin finally took Wilson aside and told him, "Look, I really believe this kid doesn't remember, but maybe we can jog his memory. He's been talking to me all summer about buying some little red sports car he saw on a lot. Maybe we should see about getting a few hundred bucks approved and make him a paid informant. Give him some time to remember."

"Okay," said Wilson, "I'll take care of the money and you go drive him by that car lot real slow, as many times as it takes. I want him to stare at that little red sports car till his tongue hangs out. I want everything he knows."

It took a few days for Wilson to get the money and a week for James to come up with something. He finally thought of his high school year book.

Four hundred dollars bought the task force a high school picture with a name on it: Chris Morrison.

It was their break. But it gave them only one name: Don James could not identify the other Eco-Raiders from their high school pictures.

The task force quickly learned that Chris Morrison had returned as a sophomore to the University of Arizona's College of Architecture. He lived at home with his parents. They put a twenty-four-hour-a-day tail on him. He was their only lead to the others.

The Arizona Territorial, a little weekly newspaper out on North Oracle, reported an Eco-Raider attack, but Chris Morrison had been at home all that night. Wilson began to wonder if the name had been worth the four-hundred dollars. After a week of frustration with nothing on Morrison, Wilson asked the Tucson Police Department for help tailing the suspect at night with their new helicopter. The Department readily agreed and arranged a strategy meeting with the flight crew. The pilots told Wilson their basic problem: "From the altitude we have to fly at, all cars look alike at night, two white headlights approaching, two red taillights receding."

The task force would have to do something to make Morrison's car distinctive.

"What if we shot out a taillight?" Wilson asked. "Would that work?"

"One taillight will work if you tell us which one is out."

That night Deputy Martin took down his air rifle again and shot out the left taillight of the Morrison family car while it was parked at home. It put a hole through the red lens but failed to break the lamp.

The helicopter crew said, "That's even better. We'll get a red and white taillight. We can't miss it."

Sunday night, September 16, the helicopter crew easily tracked Morrison's branded car to a home on West Hardy Road where he picked up a passenger and then drove to a remote site near Saguaro National Monument. There the car stopped for a long time and the helicopter returned to base. Two unmarked task force cars made their way to the location and took turns driving past the parked target vehicle. They saw two people in the front seat, evidently absorbed in conversation. They took positions a discreet distance away. Nothing happened. Morrison's car then drove back to the West Hardy home where the passenger got out and went in. Morrison then drove home.

A quick check identified the young West Hardy resident as Patrick G. Salmon, 20, languages student at the University of Arizona.

The next morning, armed with a Justice of the Peace warrant, the task force arrested Chris Morrison and Pat Salmon separately and took them in different cars to headquarters. Morrison's father was brought along with his son. Wilson and Martin questioned Chris Morrison upstairs while two other officers questioned Salmon on a lower floor. The two Eco-Raiders were unaware of each others' presence in the building. The task force was unaware of the names of the other Eco-Raiders.

Morrison quickly told the officers, "Ok, I'll admit to what I did, but I won't tell you who the others are."

Wilson snapped back, "I don't give a shit what you won't tell us. You mean you're not going to tell us about Pat Salmon? We know who you guys are. What we don't know is which ones did which raids so we can charge everybody with the proper offenses."

Wilson let that sink in. The two officers downstairs were doing the same thing with Pat Salmon. Salmon adamantly refused to cooperate. Upstairs, things were different.

"Ok, Chris," said Sergeant Wilson after consulting with the prosecutor, "we're ready to make you a deal. You tell us which ones did what and we'll see that the county attorney gives you immunity from prosecution."

Chris's father growled, "Take the goddam deal."

A short time later Chris Morrison signed a deposition linking Pat Salmon, John Walker, Gary Blake and Mark Quinnan with 35 counts of misdemeanor vandalism. At the same time Pat Salmon was granted limited immunity even though he refused to divulge the names of his cohort: he agreed to plead guilty as an Eco-Raider to several charges of malicious mischief in Justice Court, including the Arizona Territorial attack, in which he was involved as the driver.

On Tuesday, September 18, the task force arrested John Walker and Mark Quinnan. The next day, Gary Blake, who was rooming with Walker but had been at school during the arrest, surrendered to authorities.[21]

Their cases never went to trial. On October 24, Walker, Blake and Quinnan, all 19 years old, entered a plea of guilty to 32 acts of malicious destruction of property in Justice Court and received various sentences: Walker and Blake, 6 months; Quinnan, 90 days. Salmon pleaded guilty to three counts of malicious mischief, sentence, 60 days.[22]

The Estes Company filed a civil lawsuit against all five, asking for $5,000 each in actual damages and another $10,000 each in punitive damages.[23] The suit was settled out of court and each Eco-Raider paid a lesser amount to Estes.[24]

The Eco-Raiders failed to stop or even slow the development of Tucson. Their two-year campaign did cost contractors a great deal of money and also brought greater awareness of the desert to a number of high-end developers who made a sales point of building appropriately and aesthetically to fit the environment. Many of the tract houses the Eco-Raiders hit are now gone, victims of economic evolution and changing tastes, replaced by more expensive homes with landscapes restored to some semblance of the original desert ecology, which has become trendy in parts of Tucson.

None of the Eco-Raiders surfaced again as ecoteurs, with the possible exception of Walker, the leader. John Walker took his jail time hardest of the four, hating the confinement, growing bitter and misanthropic after six months in the tank with forty or more drug dealers, car thieves and bank robbers. He was mercilessly ridiculed—"You got six months for vandalism? What a pussy." His family never came to visit him in jail, yet Quinnan's father came several times a week. After his release, Walker became a wildlife painter and wood carver of some merit, exhibiting in Tucson street fairs. Deputy Duane Wilson suspected later that a number of sabotaged mountain sheep hunts up in the Catalinas might have been Walker's work, but he could never prove it and no charges were ever brought.

Wilson came in contact with John Walker again in 1975 when the struggling artist filed a sheriff's report of a burglary at his home. Officers recovered Walker's possessions and Wilson found an aging briefcase among them. He looked inside. There, with fading copies of Mike Royko's columns on The Fox and a battered copy of *Ecotage* were the plans of every attack the Eco-Raiders had done from the beginning, all handwritten and diagrammed on ruled notebook paper. Wilson studied them all before returning them to their owner. The Eco-Raider case was history now and the plans made no difference to the law, but they explained so much that Wilson had never understood.

"He was a military genius," Wilson says. "A damn genius."

"I didn't know what they were saying back then," Wilson adds thoughtfully. "But I've lived up in Marana all my life, and now Tucson has caught up with us. It's overcrowded, overgrown. It's a mess. A damn mess. If I had to catch those kids again today, I don't know that I'd do it. Except it's my job, you know."

After their release from jail, the Eco-Raiders met each other only in chance encounters now and then, one by one, never in a group. During their brief chats they never discussed their former lives. Mark Quinnan married Gary Blake's sister, but it didn't last long. Gary Blake moved back East, somewhere in western Pennsylvania, and Pat Salmon was thought to have taken work in Saudi Arabia. Mark Quinnan saw Chris Morrison one day at the University of Arizona some time around 1976 and went to speak with him, but he ran away. John Walker was rumored to have become a hermit in the Catalinas for a while, then quietly disappeared.

It was over.

Except for the shouting: Edward Abbey's immensely influential 1975 novel *The Monkey Wrench Gang* was a fictionalized account of the Eco-Raiders.[25] Although the characters were based on Abbey's personal friends Doug "George Hayduke" Peacock and Ken "Seldom Seen Smith" Sleight[26] and Ingrid "Bonnie Abzugg" Eisenstadter and Al Sarvis and John DePuy and Jack Loeffler and others, and not the Eco-Raiders, the concept of the novel sprang from the exploits of a teenager named John Walker and his four young friends, the first authentic ecoterror group in America. Their names were obliterated by Edward Abbey's fame and Abbey did nothing to identify them. When Mark Quinnan introduced himself to Abbey once at a little bar in Tucson in the late 70s, Abbey did not recognize him and Quinnan, unassuming, did not mention the Eco-Raiders. The tribes of radical environmentalists do not acknowledge the name of John Walker because they have no memory of the name of John Walker. Yet he called the muse to their poet. He is step-father to them all.

TUESDAY, JULY 17, 1979   *Leixoes, Portugal*
ALEX PACHECO STOOD ON THE DECK OF THE *SEA SHEPHERD*. Captain Paul Watson had just told his crew they had ten minutes to decide what to do. The whaling ship Sierra stood off the harbor of Leixoes, and Watson was about to ram her.

"Look," Watson said, "I can't guarantee you're not going to get hurt, but I can guarantee you're going to jail after we ram the *Sierra*."

Ten minutes later sixteen of the crew stood on the dock. On board the *Sea Shepherd* remained only Captain Watson, chief engineer Peter Woof, third engineer Jerry Doran, and Alex Pacheco.

Watson contemplated the twenty-one-year-old Pacheco, still unsure of his goals even though he had already discovered his organizational talent after founding a campus animal rights group at Ohio State University. The *Sea Shepherd* foray was his summer job. But Watson didn't want to see him rotting in some Portuguese jail for weeks or months and missing the next school year.

"Alex," he said in his best brotherly tone, "I want you to get off the boat and do the photography to document what we do here and be a spokesperson."

Pacheco obeyed.

Watson rammed the Sierra. The Portuguese Coast Guard ordered the three pirates to sail the *Sea Shepherd* into the harbor where it was seized to pay for damages to the Sierra. The port captain charged Watson with negligence, but Watson retorted that it was not negligence, that he had hit the Sierra exactly where he intended.

The port captain said, "I see. Well, we have received no complaint from the owners of the Sierra, so you are free to go."[27]

Pacheco immediately left Portugal for England, where he stayed the rest of the summer. There he met Kim Stallwood, a young and politically savvy animal rights journal editor with the traditionalist British Union for the Abolition of Vivisection. Stallwood was far more radical than BUAV's somewhat refined membership, openly supporting the violent Animal Liberation Front with sympathetic coverage in his periodical, Liberator.[28]

And there in England Pacheco joined the Hunt Saboteurs Association, the turning point that led him the next year to become the founder of America's largest and most powerful animal rights organization, People for the Ethical Treatment of Animals.

The Hunt Saboteurs Association dated from 1962, when antivivisectionists melded with class-conscious protesters to harass fox hunters, seen as conspicuous symbols of the rich "landed gentry" torturing an innocent creature for sport.

The modus operandi of the Hunt Saboteurs, known locally as "sabs," has hardly changed since those early days: The fox bounds out of the woods doing thirty miles per hour, takes a sharp right, zips across an open road and disappears into another dense copse of trees.

The hounds come to the road a full minute later, looking everywhere and filling the air with irritated, baffled cries. Shambling after the dogs is the underground attack group, a gaggle of men and women in sweatshirts, jeans and sneakers howling and flapping their arms—and spraying citronella from plastic squirt bottles to confuse the scent of the reddish-brown quarry.

Next come half a dozen determined constables who catch a young woman and two of the men and tussle them to the ground.

Last come some thirty huntsmen straight from a Currier & Ives lithograph—black woolen coats, bowlers, cream breeches and knee-high leather boots (the master of hounds and those of adequate prestige in scarlet coats)—astride sleek, expensively groomed mounts.

The hunters will watch the two male college students and the daughter of a wealthy industrialist as police bundle them off to a waiting van where they will be arrested for breach of the peace. The three sabs will spend the rest of the day in jail. The dogs will run after the fox. The fox will make a clean getaway.

The attack of the underground Hunt Saboteur Association will be publicized by an aboveground group, the League Against Cruel Sports, which will stay just far enough away from the actual sabs to avoid prosecution. This underground / aboveground arrangement will be copied by ecoterrorist organizations in the years to come.

As the years go on, the hunters and sabs will begin to recognize each other on sight, get to know each other well, call each other by first names, and trade insults and animosity. "They look like dinosaurs on horseback, don't they? I don't see any reason why animals should suffer just so this bunch of toffs can have a bit of sport." And, "They are very hateful and nasty people. It's not really fox hunting that they hate. It's that we've got more money than they do."[29]

During all this the Hunt Sabs made a point of nonviolence: on the rare occasions when they were slugged by an angry fox hunter, they never struck back.

A nineteen-year-old man named Ronnie Lee changed all that. In 1971 he formed a Hunt Sab chapter in Luton, north of London, adhering to their nonviolent tactics despite misgivings. The next year saw his friend Cliff Goodman sustain an eye injury during a sab, and the incident convinced Lee that a different, more violent direction was called for, one that went beyond annoying the upper class to target any part of industrial civilization that, in his view, abused animals. The result was the Band of Mercy, founded in 1972 by Lee and Goodman.

In less than a year the respectable-sounding organization had committed two arsons at a Hoechst pharmaceutical plant, causing some forty-six thousand pounds sterling in damage. They got away with a string of attacks on various animal enterprises until 1975, when Lee and Goodman were apprehended as they returned to the scene of a break-in. After a sensational trial they were convicted and sent to prison. It was during this trial that supporters advanced the premise used by all ecoterrorists since: Their criminal actions were justified because they stopped actions that were more criminal.

Lee used his jail time fruitfully, drawing up an invincible battle plan: by using hit and run tactics and a diffuse cell structure, he and a few carefully chosen militants would inflict economic loss on animal enterprises. Each raid would be publicized, using horror stories of animal abuse and any other means to enlist support for their cause from the broader society. Each raid would result in less profit to plow back into animal use. As militant supporters were recruited, raids would escalate damages to the point that all animal enterprises would eventually be unable to operate. For the militants, it would take an impossible combination of utmost secrecy and broad publicity—tactical secrecy to protect the raiders, strategic media publicity to change society—but the Hunt Saboteur Association / League Against Cruel Sports underground / aboveground arrangement could be easily copied.

Lee got out of prison in 1976 and found himself revered by many as a martyr. He immediately took advantage of his recognition to form the Animal Liberation Front. It was a declaration of war on industrial society.

The first ALF hit was against the Charles River Laboratories; raiders damaged vehicles, doing several thousand pounds worth of damage. In its first year of operations, ALF inflicted a quarter-million pounds sterling in damage. Its targets included any institution in any way connected with animals. Butcher shops, furriers, animal breeders, chicken and beef farmers, fast food outlets, and horse racing tracks—all were hit. ALF smashed the windows of several Islamic halal butcher shops in Bedfordshire; smashed the windows of six shops in Banbury for displaying circus posters; planted a bomb under the car of a cancer researcher.[30]

Even the dead were not safe from Animal Liberation Front terrorism, which set out to shock the public out of its apathy about animal mistreatment. In January 1977 three ALF activists broke into the graveyard of St. Kentigern's Church in the small Lake District village of Caldbeck, Cumbria, to desecrate the grave of Robert Peel, the legendary huntsman and most English of folk heroes, who had lain there a hundred and twenty-three years. They smashed his headstone and dug up the grave. The activists, who did not bill the desecration as an ALF raid, even called the media to report they had exhumed Peel's remains and thrown them in a cesspit. The police found no evidence of this, but discovered a stuffed fox's head in the dug up grave. One of Ronnie Lee's colleagues, Mike Huskisson, and two other activists were captured and sentenced to nine months in jail for the desecration, of which Huskisson served six months in 1977.[31] Today, including the Peel incident, he has served 12 months in prison for involvement with animal rights militants. The last occasion was in 1985, following a complex case in which stolen documents were used to prosecute the Royal College of Surgeons for causing suffering to a research animal (the college was acquitted on appeal).

The cell structure of ALF grew up quickly. One anonymous activist told a reporter, "You get a call from someone you trust, about an activity which needs to be undertaken. If you trust them, you go out and do it and don't ask many questions. It's much more effective, run on a cell-based structure like that." The activist had liberated seventy chickens from a farm near Cuckfield in Sussex, nine goats from an agricultural college in Kent, and a horse and a donkey from a research establishment near Tunbridge Wells, Kent.[32] And, like The Fox, the activist was just another citizen with roots in the community, totally untraceable.

By the time Alex Pacheco arrived in late July of 1979, the ALF had grown to not quite a hundred participants.

The Washington Post Magazine provided a brief biography of Pacheco, who was born in 1958: "The son of a doctor, he'd grown up first in Mexico and then Ohio, where he graduated from high school and entered Ohio State, planning to become a Catholic priest. One summer in the mid-'70s, while visiting a friend in Toronto, Pacheco took a tour of a slaughterhouse. It was the turning point in his life, and he still speaks passionately of what he saw there: 'the stench of the blood, the excrement everywhere, the screaming of the animals.'"[33]

His visit was followed by indoctrination by "two brilliant activists," one a founder of American Vegetarians and the other an "artist, feminist, and animals rights activist."[34] He stopped eating meat and started telling everybody about what he'd seen. Most people were indifferent, but one sympathizer slipped him a copy of Peter Singer's book *Animal Liberation*, the manifesto of a militant new animal rights movement that was little known in America but very active in England.[35] The die was cast. Pacheco quickly founded a campus animal rights group and discovered he was good at it. Then came the summer that brought him to England, where animal rights activism was well-known.

After Pacheco's British sojourn, back in the United States, he transferred to George Washington University and became a political science major. He also started volunteering at a local dog pound, where he met Ingrid Newkirk, who worked there.

Ingrid Ward Newkirk was born in Surrey, England in 1949, and raised in a household full of dogs, cats, chipmunks, mongooses and exotic birds. Her father was a navigational engineer who took his family on assignments around the world, her mother a social worker wherever they went. They moved to India while she was a young girl—it was there she first remembers observing cruelty to animals.

She has lived in the United States since 1967 when her father migrated to Eglin Air Force Base in Florida. There she met a race car driver named Steve Newkirk. They married that year and moved to Poolesville, Maryland, where she began studying to become a stockbroker. She tells the story of how her budding career was derailed by a neighbor who abandoned nineteen cats, which she took to the local humane society. "It was the biggest dump I'd ever seen," Newkirk says. "Dogs cringed as you approached them. Animals sat in their own filth while workers sat on garbage cans smoking and laughing the day away." She volunteered to clean the place up and didn't stop there. She became Washington, D.C.'s first female poundmaster in 1978 and immediately halted the sale of animals to labs. Soon after, she became Director of Cruelty Investigations for the Washington Humane Society.

That was where Pacheco came in one day in 1980 to volunteer with the animals and met Newkirk. They quickly became friends. He gave her a copy of Singer's *Animal Liberation*. She was stunned by its message. Singer opposed "speciesism" and argued that animals deserved equal moral consideration and rights. Newkirk told People magazine, "Before, I simply thought that people shouldn't cause animals unnecessary pain. I had never thought that maybe they don't belong to us, that they have their own place on the planet."[36] The realization sparked in Ingrid Newkirk the peculiar rage that had touched John Walker, the quiet rage that could manifest itself as charm and wit when needed, yet underneath made you cold and methodical and absolutely ruthless.

Pacheco and Newkirk decided in her kitchen to form a new group: People for the Ethical Treatment of Animals. Pacheco incorporated it as a Delaware corporation in July, 1980, with Newkirk as co-director. PETA soon assembled a core-group of eighteen members that met in a Takoma Park, Maryland basement. PETA's first operation was picketing a poultry slaughterhouse in Washington. During that eventful year, Newkirk's new devotion to animal activism—and a string of civil disobedience arrests—strained her childless marriage to the breaking point: she and Steve Newkirk were divorced.

People for the Ethical Treatment of Animals was to become the richest, most powerful and most ruthless animal rights group in America. As Dr. Edward Taub, a scientist whose life was shattered by PETA, reflected after a protracted battle with the organization, "They're extremely dangerous and extremely malicious and they'll do anything. And it is minor to them to destroy a person."[37]

Many disgruntled former PETA members express similar terror.

FRIDAY, APRIL 4, 1980   *Interstate 10 approaching Lordsburg, New Mexico*
DAVE FOREMAN SHOUTED "EARTH FIRST!"

Howie Wolke, sitting next to Foreman in the front seat of the eastbound Volkswagen minibus and popping more cans of Budweiser for the two of them, liked it. So did Mike Roselle, who lay stoned in the back of the bus. It put all their feelings into a single tough expression.

Roselle grabbed a piece of paper and something to write with through his marijuana haze and drew a squiggly circle with a clenched fist in it, like the emblem on Dave Foreman's motorcycle helmet. He passed it up to the front of the minibus. From somewhere the words "No Compromise in Defense of Mother Earth" appeared. Nobody recalls where. It only mattered that it was a great motto for their new environmental group with the great name Dave Foreman just called out. They would turn it into a logo. The fist would have to be green. It would be very hard to compromise if your logo was a green fist. They would write a manifesto of radical anarcho-environmentalism. They would have their own newspaper. Make big wilderness proposals. Save the Earth. It was war against industrial society. They would be Edward Abbey's Monkey Wrench Gang come alive.[38]

The founding events of Earth First have become part of the folklore and mythology that attracts and holds adherents. The story of the five co-founders' desert adventure varies from one version to the next, but the spirit of rebellion against industrial civilization and the flouting of social controls runs through them all.

Susan Zakin wrote the most detailed history of Earth First, *Coyotes and Town Dogs: Earth First! and the Environmental Movement*. It is not possible to understand Earth First in its own terms without reading her book. Zakin is a friend of Dave Foreman's and had access to people and materials no one else had. She shares the goals of Earth First. She was arrested with Mike Jakubal and Mitch Friedman on a raid in 1986 and charged with felony first-degree criminal mischief for helping cut down a billboard near Corvallis, Oregon. The charges were dropped after she showed she was a freelance reporter for New Age magazine in Brighton, Massachusetts. Her book describes the event thus: "Jakubal promptly got himself arrested for sawing down a billboard with a reporter from a national magazine along to record the event." That's all you get. You should expect a whole book of similarly truncated reporting. While rich in detail in some areas, it is blank in others, especially the important process Dave Foreman went through in writing *Ecodefense*. Not a word. But what you get in personal biographies alone is worth the annoyance of many crucial gaps. Her book was paid for in part by the agenda-laden Center for Investigative Reporting.

Christopher Manes wrote a lesser but useful book on Earth First, *Green Rage: Radical Environmentalism and the Unmaking of Civilization*. It is mostly polemical, as its title suggests, but valuable for the fact that it addresses criticisms of environmentalism. Manes, for example, actually read some of my books, particularly *Ecology Wars: Environmentalism As If People Mattered*, and devoted generous space to reasoned refutations, unlike Zakin, who referred to me by simply regurgitating the MacWilliams Cosgrove Snider Search and Destroy Strategy Guide.

The truly indispensable guide is Martha F. Lee's *Earth First! Environmental Apocalypse*. It is less detailed than Zakin, but, as a scholarly account, it has the great virtue of objectivity. Lee's analytical powers are extraordinary, an added bonus in puzzling out what it all means. Lee is a professor of political science at a Canadian university and interviewed a broad spectrum of Earth Firsters in extensive field work, showing a great respect for the principles of historiography. Lee gets to the heart of things without trying to protect or glorify the leading characters. Dave Foreman reviewed Lee's manuscript, so we can assume that he agrees to its factuality, even in the numerous places where it makes him look unattractive.

All but a few accounts tell a creation story something like this: [39]

Five friends spent a week driving around Mexico's Pinacate Desert in the VW bus, camping, hiking, drinking, visiting whorehouses and talking environment. The five, which included Bart Koehler and Ron Kezar, formed a cameraderie that grew into solidarity that grew into the decision to form Earth First. Koehler and Kezar got off the minibus on its return trip to Albuquerque before the decision to form Earth First was made (in Kenneth Brower's *Harrowsmith* version Koehler was still with them at decision time). [40]

Their trip to the Pinacate would also become symbolic of the time of testing that each Earth Firster would go through before total commitment. The particular desert they chose was important. On Mexican maps the area surrounding the Pinacate appears as *El Gran Desierto de Altar.* Dave Foreman liked to quote Edward Abbey on the Pinacate: "Abbey once said Saguaro National Monument is high school, Organ Pipe Cactus National Monument is undergraduate, Cabeza Prieta is graduate school, and Pinacate is post doc as far as the Sonoran desert goes." Abbey called the Pinacate "the final test of desert rathood," where seekers may find "Mystery itself, with a capital 'M'," what Susan Zakin called "the transcendence that lies at the core of the real world."

The Pinacate was a metaphor for all nature, the trip into it a shamanic journey deeper and deeper into one's own spiritual wilderness until one reached the dangerous wild heart of existence and came back with knowledge of the ultimate truth. The fact that the Pinacate is an easy drive just across the Mexican border might have had something to do with it, too. Geologically, the Pinacates are a thirty mile wide volcanic field with many separate small volcanoes, rather than one giant volcano. The most recent volcanic activity occurred there about 1,300 years ago. The Pinacates (the name means stink bug, and that somehow never got into Earth First mythology) are mostly in a Mexican National Park, but the Arizona border cuts diagonally across the top of the volcanic field. Over 300 volcanic vents adorn the Pinacate field, looking like polka-dots in satellite photos. The highest elevation of *Cerro Pinacate*, Pinacate Peak, is about 4,000 feet, with a false summit called Carnegie Peak, a long walk from a roadside trailhead.

The five took the desert journey because each of their lives had reached a turning point with no clear future. They were at loose ends and destiny was calling, a condition which would resonate with many future Earth Firsters.

David William Foreman was thirty-two and burned out. He had returned to New Mexico in defeat from a year-long stint lobbying in Washington for the Wilderness Society. It was a repeat of his whole life: a series of promising starts that fizzled. An army brat, he began as a too-bright mama's boy moving from one base to another; he grew into an awkward church-going youth who wanted to be a preacher; his grades at the University of New Mexico were mediocre; he joined the libertarian youth group, Young Americans for Freedom, where he abandoned the liberal-conservative mainstream and shed his religious beliefs in the objectivist glow of Ayn Rand's *Atlas Shrugged*; he joined the Marine Corps and went AWOL to avoid training at Parris Island, North Carolina, was turned in by his own father, slapped in the brig and booted out with an undesirable discharge; back home he tried to fit into the cowboy West in a horseshoeing class but wasn't very deft at it; he finally managed to do a good enough job volunteering with a local environmental group to attract favorable attention from a Wilderness Society executive who brought him aboard as a state-level staffer for a few years. They sent him for lobbying seminars to Washington and then offered him a position there as the Society's Coordinator of Wilderness Affairs, liaison between headquarters and the entire field staff at $20,000 a year. He took it. During that grueling year, he lost all the issues he cared about to compromises with other interest groups and Congress—particularly RARE II, the second Roadless Area Review and Evaluation process. The Forest Service ended up recommending only 15 million acres out of 62 million acres under study for wilderness designation, a paltry amount in Foreman's estimation. Worse, internal Wilderness Society leadership struggles left him disillusioned and feeling discarded. Worst of all, when he packed his belongings in his beat-up pickup truck for the long drive home, his wife Debbie Sease wanted to keep her job with the Wilderness Society. She did not come with him. He lost it all.

Ron Kezar, at thirty-seven, was the oldest, a medical librarian Foreman had met in Texas, a Sierra Club conservation chair who had bought some land in Glenwood, New Mexico, near the Gila Wilderness Area. When his friend Dave Foreman got in a funk over Debbie, Ron took him on hikes to cheer him up. He helped Foreman plan the Pinacate trip not long after Dave traded in his dented pickup truck for the minibus. On March 29, 1980, they drove to Tucson where they met the three Wyoming guys at the Greyhound station:

Howie Wolke, a 28-year-old Jewish Brooklyn-born graduate in conservation studies from the University of New Hampshire, had drifted west and in 1974 became the Wyoming representative of the Friends of the Earth, working menial jobs around Jackson to supplement his $50 a month salary. The mentor who convinced Friends of the Earth to take him on was

Bart Koehler, thirty-two, the same age as Foreman, who had worked seven years for the Wilderness Society in Lander, Wyoming, but suffered a serious manic breakdown after pulling too many all-nighters during the intense forty-day state legislative session in 1979. He was still lithium-deficient and buzzed—he never sank into the depressive phase of the disease—when Wolke took him to the Greyhound station for the trip to the Pinacate.

Michael Lee Roselle, at twenty-four, was the youngest of the five. He had wandered across the country after running away from a poor violent alcoholic family in Los Angeles and attached himself to the Yippies at the bloody 1968 Democratic convention in Chicago. He drifted to Kentucky's hillbilly counterculture before arriving in Wyoming in the winter of 1975, where he met Howie Wolke while they were both working as busboys at a Jackson town square restaurant. Wolke thought of Roselle as a skinny six-foot-six walking left-wing bumper sticker when he took the big kid under his wing. Wolke turned him into a fire-breathing solo environmentalist. He learned to stop mumbling into the microphones at public hearings and become theatrical. Roselle eventually took to going out nights on the highways around Jackson and sawing down billboards, which was a little too much for Wolke, who didn't go along. Roselle's drug of choice was not beer, but marijuana, a generational difference. Roselle was also the only one who had not devoted a substantial fraction of his life to defending wilderness and who had no experience working with mainstream environmental groups.

They called themselves the Buckaroos. They were tough rednecks going to a tough desert. After their meeting at the Tucson bus station on Saturday, March 29 they enjoyed an all-night party with a case of Tooth Sheaf stout, a heavy Guiness-type beer, then the next morning drove southwest through the Papago Reservation, then Organ Pipe Cactus National Monument, then crossed into the Mexican state of Sonora at the dusty little border town of Sonoita where they restocked their beer supply. The Buckaroos drove around the winding rutted roads into Pinacate National Park Sunday afternoon and visited the 722-foot-deep Sykes Crater. Monday they put on their backpacks and tried for *Cerro Pinacate* itself. They walked up to Carnegie Peak, the false summit, and stayed there swigging beers except for Ron Kezar, who walked up to Pinacate Peak and signed the visitor log book. Tuesday they drove an hour south to Puerto Peñasco on the Sea of Cortez and bought two shrimp dinners each. Wednesday they zoomed 120 miles westward through *El Gran Desierto* down Sonora Route 2 and finally stopped at San Luis Rio Colorado, a marginally industrialized border town beside the Colorado River, where they toured the cantinas in the red light district. As the night progressed, each man in turn left the bar with the lady of his choice. Then they drove homeward Thursday, presumably having found "the transcendence that lies at the core of the real world," and the next day decided to start Earth First.

During all this time they talked endlessly about the failure of the mainstream environmental movement. It seduced grassroots environmentalists to Washington and into a spiral of salary and prestige that became more important than saving wilderness. The RARE II debacle had proven it ineffectual at adding new acreage to the National Wilderness Preservation System in amounts the radicals wanted, that amount being essentially all federal land. Foreman, Wolke, Kezar and Koehler now hated the political system they once worked with and respected, because Congress demanded they compromise with competing interests, and that meant they didn't completely get their way, and that meant wilderness would be "destroyed," and that was intolerable. Roselle was aware of their cause for complaint from the campaigns he had worked on with Wolke in Jackson against Getty Oil's exploration proposals on the Bridger-Teton National Forest, but he had always had the leftist's disdain for anything establishment.

The group they formed had only one entry requirement: Foreman's simple declaration, Earth First! They really didn't have their principles and creeds in place yet. Their first public act was for eight hikers to mount an imitation historical plaque in the New Mexico ghost town of Cooney in a tongue-in-cheek commemoration of Victorio, a Mescalero Apache chief who had attacked a mining camp to protect the mountains "from mining and other destructive activities of the white race," according to the Earth Firsters who trekked to the wilderness townsite for the ceremony.

One of the hikers told a reporter, "We think the Sierra Club and other groups have sold out to the system. We further believe that the enemy is not capitalism, communism or socialism. It is corporate industrialism whether it is in the United States, the Soviet Union, China or Mexico."[41] Another predicted, "We will take pure, hard-line, pro-Earth positions. No nukes, no strip mining, no pollution, no more development of our wilderness. We are concerned about people, but it's Earth First."

There is some other fun in this opening ritual: Howard Bryan, a journalist from the Albuquerque Times, believes Victorio and his Apaches actually stole gold anywhere they could find it, with no intent of protecting anything beyond a Spanish treasure cache they had stumbled upon in a cave in what is now called Victorio Peak. Robyn Wagner of the University of New Mexico wrote, "Victorio would attack wagon trains, churches, immigrants, mail coaches, and anything else that promised riches. Often he would take prisoners back to Hembrillo Basin, where he would subject them to elaborate torture before they were killed."[42] The Earth Firsters were probably well aware of Victorio's true nature.

Contrary to later assertions that Earth First has always been a "movement" and not an "organization," and that its adherents are "Earth Firsters" and not "members," it was well-organized at the beginning and adherents were called members.

And contrary to the Pinacate legend, the organization was not formed solely by "the group of five" Buckaroos. In fact, it may even have been planned prior to the trip to the desert: Susan Zakin wrote in *Coyotes and Town Dogs*, "News traveled so fast in Lander [Wyoming] that Louisa [Willcox, a reporter for High Country News] heard about Earth First! practically before the Buckaroos limped homeward." Zakin remained silent on how news traveled so fast. Regardless exactly when and how it was formed, an internal Earth First memo shows the organization's doctrine was influenced by at least four other leaders: Mike Comola, former president of the Montana Wilderness Association; Randall Gloege, former Northern Rockies representative for the Friends of the Earth; Sandy Marvinney, past editor of the Wilderness Report; and Susan Morgan, a former education coordinator for the Wilderness Society.[43] Louisa Willcox, too, became an early participant. Yet "the group of five" would get all the publicity, and Dave Foreman would become the sole charismatic leader.

The group of founders envisioned an anarchic structure with no leaders and no followers as the only effective response to the overwhelming power of the corporate state. "So what is the one kind of human organization that's really worked? The hunter/gatherer tribe, so we tried to model ourselves structurally after that."[44] The tribalist ideal of rejecting hierarchy and authority that they preached was beyond their grasp—and beyond actual tribal behavior with its chiefs, shamans, elder councils, captives, tortures, taboos, berdaches, outcasts, and executions—so Earth First began with both hierarchy and authority.

The first general meeting of Earth First, the Round River Rendezvous on July Fourth, 1980, established a central authority divided into two governing structures: the Circle of Darkness and *La Manta Mojada* ("The Wet Blanket"). The Circle of Darkness had twelve members, including the five Buckaroos and Susan Morgan, who was put in charge of the bank account, membership dues, the membership list and the newsletter. *La Manta Mojada* was made up of eight advisors to the Circle, whose names were kept secret because they all came from moderate conservation groups. *La Manta* didn't last long and didn't do anything, leaving the Circle as the sole ruling elite. They intended to keep tight central control. They didn't want any sellouts.

About sixty-five people showed up for the kick-off Round River Rendezvous at the Cross-T Ranch in DuBois, Wyoming. The name for the gathering was picked by Bart Koehler from Aldo Leopold's essay, "The Round River—A Parable," which recounted the Paul Bunyan story of a river in Wisconsin that "flowed into itself, and thus sped around and around in a never-ending circuit." Leopold's editors included the essay in the posthumous *A Sand County Almanac*: "The current is the stream of energy which flows out of the soil into plants, thence into animals, thence back into the soil in a never-ending circuit of life."[45] It sounded a lot like Yoda teaching Luke Skywalker the ways of The Force, and *The Empire Strikes Back* was a hot ticket among potential recruits in 1980. The "Rendezvous" was a reference to the sometime meetings of 19th century mountain man trappers and explorers such as Jim Bridger and Liver-Eating Johnson and Jedediah Smith and Big Anton Sepulveda and the rest. Nowadays, "Mountain Men" societies throughout the Rockies hold annual summer Rendezvous and period costume pageants, so it was a natural idea.

The first Round River Rendezvous established the organizational structure. Foreman quickly sent out the initial newsletter, which was first called "Nature More" from a famous line in Lord Byron's poem, "Childe Harold's Pilgrimmage"—*I love not man the less, but nature more*. Others objected that it was too literary, and it was quickly changed to "Earth First! Newsletter," later simply to "Earth First!"

Critics of Earth First rarely realize how intellectualized and self-consciously strategic these decisions were. The second issue of the newsletter included a check list of what supporters could do "to be active in the outfit, to help the gang," which ended with "Create the poetry of the movement. We need myth, ritual, and song."[46]

It was more important than one might think. The "redneck" leaders of Earth First were well read in biocentric philosophy from Aldo Leopold and Arne Naess to Bill Devall and George Sessions to Gary Snyder and Edward Abbey and dozens of others. Yet they counted on horror stories of depredations against nature more than philosophy for their success. Dave Foreman had also studied the Wobblies, the International Workers of the World, a radical labor organization formed in the early part of the century—Edward Abbey's father, Paul Revere Abbey, was one of the elderly remaining Wobblies. The IWW used music, art and stickers, which they called "silent agitators," to get their message across, and thereby gained political clout beyond their numbers, as Earth First wanted to do. Foreman borrowed IWW's ideas: Earth First began to market a collection of bumper stickers, T-shirts and little round "silent agitator" green-fist logo stickers he called "snake oil and trinkets." But the IWW took radicalism too far: in 1917, Wobbly leaders were arrested for conspiring to hinder the draft and encouraging desertion during World War I and given long prison sentences. Their leader, Big Bill Haywood, jumped bail and escaped to Russia, where he died ten years later. The Wobblies disdained contracts, feeling negotiations frittered away workers' energies, and instead concentrated on direct action: general strikes, obstruction, and organizing. Foreman was well aware that radical action is spurred by rage, not rationality; one might die for a slogan, but not a syllogism. Thus, even though their philosophy had exact intellectual source points, the leaders of Earth First were highly uncomfortable about identifying them.

Seattle-based Earth Firster George Draffan told an interviewer, "Most people in Earth First! are not dependent on books to explain their own views of things. I don't think it has much effect. We have a pretty simple philosophy and very simple feelings about things."[47]

It would not do for the macho redneck Buckaroos to be revealed as elitist snobs, for real rednecks do not read. The founders preferred to convey their ideas in "myth, ritual and song," in part so that adherents would perceive them as "intuitive feelings." They preferred followers to feel that the evidence for their beliefs appeared in the world around them and pointed to an environmental crisis, which would stir the necessary rage. The written word presented two problems to Earth Firsters: first, it could lead to dogma, and second, it could reduce their beliefs to just another viewpoint. Either could stifle their preference for action over thinking.

One feature of the newsletter has puzzled many: its use of odd-sounding names for the publication dates, such as Mabon and Eostar and Lughnasadh and Beltane. They are Pagan holidays, first chosen by Dave Foreman to flout mainstream tradition, but which later became a real point of contention within Earth First between those who had ideological reasons to use them and those who didn't.

The dates themselves (mostly Celtic or Teutonic words):

Samhain, October 31, popularly known as Halloween, is the Witches' New Year, and marks the beginning of each numbered volume of the Earth First periodical. Pronounced "sow-in" in Ireland, "sav-en" in Scotland, and "sam-hane" in the United States where we do not speak Gaelic.

Yule, December 21, the Winter Solstice, closely associated with Christmas in the Christian tradition.

Brigid, January 31, February Eve, the holiday of the Celtic Fire Goddess Brigid, whose threefold nature rules smithcraft, poetry/inspiration, and healing.

Eostar, March 21, Vernal Equinox, day and night are equal as Spring begins. The Germanic Goddess Ostara or Eostre (Goddess of the Dawn), after whom Easter is named, is the tutelary deity of this holiday. It is she, as herald of the sun, who announces the triumphal return of life to the earth.

Beltane, April 30, May Eve. The name "Beltane" means "Bel's Fires." In Celtic lands, cattle were driven between bonfires to bless them, and people leaped the fires for luck.

Litha, June 21, Summer Solstice or Midsummer. On this day, the noon of the year and the longest day, light and life are abundant.

Lughnasadh, July 31, August Eve, (pronounced "LOO-nah-sa") one of the Celtic fire festivals, honoring the Celtic culture-bringer and Solar God Lugh. Playwright Brian Friel recently wrote a moving drama, *Dancing at Lughnasa*, centered about the festival; it has been well received in regional theaters.

Mabon, September 21, Autumnal Equinox, or Harvest Home. This day sees light and dark in balance again, before the descent to the dark times.

Many Earth Firsters were baffled by the names, but went along with them as an oddball piece of fun.

As Professor Martha Lee has pointed out, all these founding members of Earth First were remarkably similar in age, education and background. They all shared the belief that modern society and its destruction of the natural world could only end in apocalyptic crisis—The Apocalypse Creed, I call it. But none realized that the envisioned apocalypse could have different consequences for different Earth Firsters. They would find out the hard way.

The event that put Earth First on the political map was the "cracking" of Glen Canyon Dam on March 21, the Spring Solstice, of 1981. That morning, seventy-five Earth Firsters gathered at the Colorado Bridge overlooking the dam, to stage what appeared to be an ordinary protest demonstration. Their sign waving and chanting duly occupied the dam's security agents. Edward Abbey showed up to give a speech about how beautiful Glen Canyon had been before the dam drowned it. While everybody was looking elsewhere, an ersatz Monkey Wrench Gang consisting of Dave Foreman, Howie Wolke, Louisa Willcox, Tony Moore and Bart Koehler climbed over the fence and ran to the center of the dam, where they secured and unfurled a narrow three-hundred-foot-long plastic banner that looked like a big crack down the face of the dam. The media loved it. It was the perfect news clip. It was short, snappy and sassy. It was defiant. It was funny. It was serious. It was lighthearted. It needed no sound bite. Federal law enforcement couldn't catch the culprits, who vanished into the crowd, leaving no evidence and no fingerprints on the banner. No one was arrested. Earth First was glorified. Authorities were a laughingstock. The leaders had their myth: a phony crack in a suggestive place on a real dam.

Now for their ritual: Edward Abbey told them: "Oppose. Oppose the destruction of our homeland by these alien forces from Houston, Tokyo, Manhattan, Washington, D.C. and the Pentagon. And if opposition is not enough, we must resist. And if resistance is not enough, then subvert." It was the boilerplate Abbey tag line to every speech, sometimes accompanied by "Of course, I never advocate subversive activity of any kind— except at night and if you are accompanied by your parents." Abbey had contributed $200 towards the purchase of the plastic crack.[48] He launched a petition to raze Glen Canyon Dam.

To make sure the "myth, ritual and song" all materialized, Bart Koehler assumed his identity as country-singer Johnny Sagebrush with his verses of "Were You There When They Built Glen Canyon Damn?"

It made enough press and enough word of mouth among environmentalists to get Earth First off the runway. Now they had to learn how to fly.

MAY 1981   *Silver Spring, Maryland*
PETA BEGAN ITS ATTACK AGAINST DR. EDWARD TAUB, a behavioral scientist of national standing who loved animals and loved working with them.

Dr. Taub came to the attention of PETA because Alex Pacheco found his name on a list of government research grant recipients. Taub was the closest researcher to Pacheco's Takoma Park home, just a short distance away in Silver Spring, Maryland. Taub used monkeys in research to help stroke and head injury victims regain use of paralyzed limbs.

Dr. Taub at the time was studying deafferentation, the loss of all sensation in a body part—although some control over movement returns, in practice, most affected limbs atrophy and become useless. Through research with monkeys, Dr. Taub discovered that paralysis in surgically deafferented arms and legs was only temporary, and that the monkey could thereafter be trained to use the limbs. Dr. Taub discovered the crucial fact that in humans, the loss of use was not the direct result of stroke or injury, but of frustration and inability to learn to use a limb that cannot be controlled through the instinctive sense of feel. The benefits patients have received from this discovery are beyond measure.[49]

Alex Pacheco infiltrated Dr. Taub's laboratory, posing as a student interested in his research. He seems to have been following the action plan recommended in the animal-rights manual *Love and Anger* by Richard Morgan: "Since most researchers don't think there's anything wrong with what they're doing, they might even be willing to discuss their research with you, as long as you approach them innocently."[50]

Pacheco ingratiated himself with Dr. Taub, who truly believed in his work, and volunteered to work at night. Taub gave him the keys to the place, telling his wife, "We have a wonderful new student." Pacheco secretly took pictures of conditions in the lab that he thought were "horrifying," and took them to New York animal rights groups. Cleveland Amory, who had financed the *Sea Shepherd*'s ramming of the *Sierra* in 1979, gave Pacheco money to buy a better camera and some walkie-talkies, which enabled the infiltrator to photograph inside the lab while staying in touch with a sentry posted outside to warn of any unexpected visitors.[51]

When Dr. Taub went on vacation for two weeks in August, 1981, he left the monkeys in the care of lab assistants with whom Pacheco had struck up a friendship. One day, one of the lab assistants improperly lashed an experimental monkey known as Domitian to a "chairing" device in a shocking quasi-crucified pose. While the lab assistant was out of the room, Pacheco took pictures of the setup. He later used one of the photos in a poster emblazoned with the motto: "This Is Vivisection. Don't Let Anyone Tell You Different." Thus a photo of an improper setup became the emblem of the animals rights movement against animal suffering.[52]

While Taub was on vacation, two graduate student lab assistants mysteriously failed to show up for work to clean the cages or feed the animals on certain days. John Kunz, the graduate student left in charge of the place, said, "Both of them stopped coming in. They called in with different excuses. I didn't come down on them hard. In hindsight, maybe they were taking advantage of that situation." On days when the lab was improperly staffed and in disarray through no fault of Dr. Taub, Pacheco brought in sympathetic academicians and animal rights activists, including members of the Humane Society, on unauthorized "tours" of the lab.[53]

Pacheco then obtained affidavits from his "tourists"—as a lawyer described it, he "covered himself with paper"—stating that the monkeys were living in poor and unhealthful conditions. He took the photos and affidavits to local law enforcement agencies, who agreed to obtain a search warrant to raid the lab and seize Taub's animals—but not before animal rights activists spent days building cages to house the 17 monkeys that would be seized. The night before the raid, Pacheco and Newkirk smuggled in one final witness, veterinarian Richard Weitzman, who did not agree that the animals were in any danger and said his reaction was, "Why didn't you confront the gentleman and tell him what's wrong and have him fix it?" Pacheco and Newkirk did not inform Weitzman of the next morning's raid. When Weitzman heard about the raid on the news, he said, "I knew there was something not too right about this. I felt they were people who were against the research more than anything else."

Police executed a search warrant and seized the monkeys on September 11, 1981. PETA promptly ran a large fundraising ad based on the story.[54] Readers were exhorted to "be part of a historical first" by sending money. Contributors were told "Money is urgently needed for civil legal costs, expert witnesses, other professionals, etc. for this on going project," although the only legal action pending was one brought by the State.

The prosecuting attorney, Roger Galvin, brought a seventeen-count information against Dr. Taub, of which eleven counts were dismissed at trial, five ended in acquittal, and the one remaining was overturned and dismissed upon appeal.[55]

The single residual charge was failure to provide adequate veterinary care for six of the animals. Seven veterinarians gave testimony on that count regarding the advisability of bandaging nerve-severed limbs. Five had expertise in deafferentation, and supported Taub's decision not to use bandages. Two vets with no specific expertise held that he had been negligent in omitting bandages. The court sided with the two dissenting opinions and found Taub guilty of one charge of animal cruelty. His conviction was overturned on appeal and four scientific societies also exonerated him in independent investigations.

PETA continued to raise funds with the Taub story as if he had been convicted. The rage fanned by their campaign shattered Dr. Taub's life. His laboratory had been raided and his name forever connected with the abuse of innocent creatures. Anonymous animal rights activists sent him scalding anti-Semitic hate mail. One letter said, "Bastard, too bad the Nazis didn't get you."[56] NIH terminated his grant, leaving him with no income. For the next four years he survived on his savings and his wife's salary in a relentless battle to clear his name. By 1986, he was successful enough to be hired as a professor of psychology at the University of Alabama at Birmingham. PETA organized demonstrations against him in Birmingham. He no longer does animal research, and he suspects that pressure from PETA will prevent him from ever working with animals again. The hate that PETA is able to generate is quite phenomenal.

The Silver Spring monkeys were eventually relocated to Tulane University in Louisiana, and Alex Pachecho would later take part in planning a raid to kidnap them.

PETA launched its pattern of enraging people about animal abuses with the Taub case. The sociologists James M. Jasper and Dorothy Nelkin commented, "Murky passions are shaken loose by such deep rage, and a certain number of animal rights activists turn to an invective that betrays their own misanthropic feelings."[57]

Those manipulating such emotions are not always what they seem. Prosecuting attorney Roger Galvin may not have been the impartial official he was supposed to be in the Taub case: He shortly helped to found the California-based Animal Legal Defense Fund, which began working closely with PETA.

But PETA's true importance is its relationship with the Animal Liberation Front, which migrated to North America from England shortly after PETA got started. The Animal Liberation Front is the violent underground terrorist group, while PETA is its aboveground agent to the press and the public.

While the Animal Liberation Front's own web pages count eleven ALF-related raids in North America from 1977 to late 1982, a book by Ingrid Newkirk dates the first ALF raid in the United States at Christmas, 1982, and attributes the American ALF movement to a woman known only as "Valerie." [58] Many believe "Valerie" is Newkirk.

A 1993 People magazine interview with "Valerie," wearing a wig and dark glasses, presented her as an ordinary person no one would suspect, married and the mother of a small child, doing volunteer work at a local library—she reads mysteries, watches popular television shows and attends Tupperware parties. "My friends think I'm only into my baby and gardening," she was quoted as saying.[59]

The reporters had to travel from the Washington National airport to Valerie's hideout laying in the back of a windowless van driven first by Ingrid Newkirk and then a replacement driver. The story they got follows Valerie from mid-October of 1981, when she took a leave of absence from her job. She had read about the Animal Liberation Front that had been raiding animal research laboratories in England. She made inquiries about ALF to a British animal rights group, then flew to London and was met by intermediaries. Valerie convinced them that she wasn't a spy, then was led to a pub and introduced to ALF's leader, Ronnie Lee, 41 at the time, who had started the group six years earlier. In Britain, ALF members taught her how to pick locks, disconnect alarms and obtain a fake driver's license. She arrived home in autumn 1982.

ALF's American debut came on Christmas Eve, 1982, when Valerie and two others broke into a lab at Howard University in Washington, D.C., where researchers had been using cats to study the effects of drugs on nerve transmission. They found about 30 cats, photographed them, then took them away to a sympathetic veterinarian who treated them, after which they were placed for adoption. The next day Valerie anonymously dropped the pictures at the Maryland offices of People for the Ethical Treatment of Animals. Soon afterward, PETA's director, Ingrid Newkirk, called a press conference, and ALF U.S. was on the map.

People magazine wrote, "Although PETA has no connection to ALF, it has consistently publicized evidence it has received from the group." As time went on, law enforcement officers would come to doubt that separation.

The ALF connection also proved to be a huge recruitment and fundraising bonanza for PETA, which began to grow rapidly.

On May 28, 1984, the Animal Liberation Front broke into laboratories at the University of Pennsylvania Experimental Head Injury Clinic in Philadelphia which were developing therapies for serious head injuries, such as automobile accidents, based on research using baboons—research which helped develop antitoxins that counteract the effects of stroke or trauma and greatly limit the damage.

The ALF did $60,000 worth of vandalism to computers and medical equipment and stole six years worth of research data, including sixty hours of the researchers' videotapes of the injuries and treatment. PETA immediately distributed ALF news releases attacking the research. From the sixty hours of videotape ALF stole, PETA edited down a 26 minute indictment of technicians performing useless experiments on animals that were insufficiently anesthetized. Penn officials asserted that the experiments were medically of great value and that because the baboons had to be awake during the experiment, a disassociative anesthetic was used, which only gave the appearance they had not been anesthetized. Provost Thomas Ehrlich was concerned that the edited tape, less than one percent of the total stolen, may have been doctored. PETA chairman Alex Pacheco was subpoenaed to bring the stolen tapes before a grand jury and explain how they came into PETA's possession.[60]

But the niceties of science were lost on an outraged public. PETA members staged a sit-in at National Institutes of Health on Rockville Pike in Washington, D.C.[61] The pressure led Secretary of Health and Human Services Margaret Heckler to suspend funding for the Penn project. The NIH slapped Penn with a citation for "material failure to comply with Public Health Service policy for the care and use of laboratory animals." The U. S. Department of Agriculture, which is responsible for enforcing the Animal Welfare Act, stuck Penn with a $4,000 civil penalty. Ultimately, the Head Injury Clinic was closed.[62] The Washington Post later characterized PETA and related groups as "medical vigilantes."[63]

PETA forwarded copies of the documents stolen by the Animal Liberation Front—PETA says it does not accept stolen materials for legal reasons (receiving stolen goods is a crime) but does accept copies of them—to Roger Galvin's Animal Legal Defense Fund for legal consideration. ALDF's annual report states that it was involved in defense of activists facing "criminal charges arising from the break-in at the University of Pennsylvania Head Injury Laboratory and the showing of the videotape... After the receipt of 60 hours of head injury videotapes by PETA, ALDF attorneys began a series of FOIA requests." The annual report also notes ALDF's victory in preventing grand jury testimony relating to the break-in, and in five of their attorneys overseeing "the highly successful NIH sit-in, which is credited with forcing the final closing of the head injury laboratory."[64]

6:00 A.M., THURSDAY, MAY 12, 1983   *Medford, Oregon*
DOUG PLUMLEY WAS IN NO FINE MOOD. Since the mob of scruffy protesters first jumped out from behind the bushes three weeks ago to obstruct his road building crew, it had been almost daily upset. Earth Firsters protested his Medford-based logging company in a place called Bald Mountain up in the Kalmiopsis area above the little town of Galice.[65]

A lot of strange things had been going on in the woods lately. The Oregon Forest Protection Association's manager, James B. Corlett, had sent out a memo warning loggers against interfering with marijuana plantations discovered on private land or guerrilla plantations in national forests. Corlett told of a small forest owner in Humboldt County, California, who notified the sheriff of marijuana patches he had found on his land. Shortly after, his house and barns were burned to the ground. Loggers were told to be wary of local motels when marijuana crops came to harvest, because they were booked with people from San Francisco carrying briefcases and accompanied by bodyguards and they paid local merchants in hundred dollar bills. Loggers were advised not to expect things to change. One incident told why: A young man in the Garberville area who had not worked in some time paid $220,000 cash for the ranch of a forest owner. Merchants all over northern California turned their heads because the cash flow was the best they had ever seen. Also, some tiny environmental groups began to receive large anonymous cash gifts after harvest because their work helped discourage unwanted visitors from the forest.[66]

Doug Plumley knew that a new breed of heavily armed seasonal marijuana growers from San Francisco, not the resident back-to-the-land types that grew for personal use and traded for a little cash through the Garberville connection, had virtually taken over the national forests just across the county line in Northern California—even federal law enforcement agents were afraid to go into the Shasta-Trinity National Forest and the Six Rivers National Forest. Some loggers down there were being forced to pay protection money by this new violent bunch.

Here, in the Siskiyou National Forest, which contained the Kalmiopsis Wilderness Area, things were getting bad too. Even the peace-and-love hippies down by Cave Junction and Takilma were buying guns to protect their dope plantations from takeover by the city-breed growers and theft by biker gangs.

And there was the Fishline Alliance, a shadowy Northern California ecotage group that had passed out flyers with diagrams showing how to cut the brake lines of logging crew buses and how to string fishline rigged to shotgun triggers across trails loggers would use. A crew bus near Happy Camp, California actually had its brake lines cut. And there were frightening reports from Olympic National Forest in Washington State of a whole logging operation blasted to bits. The Bureau of Alcohol, Tobacco and Firearms had quietly warned the logging industry to be alert.

Nothing like this protest had happened to Doug Plumley's company before and he didn't know what to do. Who were they? Big-time dope growers? Biker gangs? Hippie protesters? When his job boss, Johnny O'Connor, first radioed in on April 26 that four young men had showed up and said they were going to shut the operation down the next day, all he could think to say was, "Well, if they do, just wait for the law to come."

The next day Mike Roselle, who had adopted the alias Nagasaki Johnson, showed up with Steve S. Marsden, George Pedro Tama and Kevin E. Everhart, carrying an Earth First clenched-fist logo banner. They stuck their banner on the dozer blade and harassed Plumley's road construction crew. Plumley called the Josephine County Sheriff's Department. Deputy Bud McConnell came and asked the bunch to step away from the dozer. When they refused, he arrested them on disorderly conduct charges and took them into Grants Pass for booking. The court sentenced Roselle to a year's supervised probation, fined him $150 and ordered him not to go within five miles of the Bald Mountain road.[67]

Plumley couldn't keep a crew at work because protesters shut them down day after day. The Josephine County Sheriff quickly ran out of budget for such oddball crimes as disorderly conduct up on some remote logging site, so Doug Plumley had to hire two deputies out of his own payroll. The U.S. Forest Service refused to send law enforcement officers to the protest site, even though it was their timber sale being blocked. They just sat in their green vans on the other side of the mountain, secretly pleased to let the loggers take the heat. Plumley then tried to rally other independent loggers to band with him and fight this new threat to their livelihoods, but they didn't want to get involved—or attract similar attention to timber sales they were logging. And getting help from the big corporations—it wasn't even worth thinking about.

Johnny O'Connor began to see the pattern in the protests: Monday through Thursday only, so as not to spend the weekend in jail. Bring the media with you and give them a show. Find the bulldozer run by Les Moore and run in front of it so it looks to the cameras like you are being buried by a deliberately brutal operator. Pile more dirt on your body if it doesn't look good enough. When the bulldozer backs up, run after it. Most of all, humiliate the whole crew by getting television shots of the sheriff leading them to work while protesters harass them from all sides. When the protesters took a break long enough to talk to the road builders, it became clear to O'Connor that many of them were liberal-arts college students. About a quarter of the protesters had no idea what they were protesting, about a quarter were stoned, about a quarter were local back-to-the-land types who had migrated from the East and the rest were organizers and support team leaders. Somebody smart was behind this, Johnny thought. [68]

Then, early on Thursday, May 12, Dave Foreman, who now used the alias Digger, set up a roadblock on an access road ten miles from the construction site for the benefit of two television crews he had invited. He also brought Dave Willis, a mountain climber who had lost his hands and feet to frostbite, in his wheelchair, perfect camera fodder. They were going to get some real media mileage out of this one. A Grants Pass Earth Firster named Charles Thomas, alias "Chant," led a support team that swung a dead snag down from the bank on the right side of the road and obstructed the way. The slope on the left side of the road dropped away several hundred feet into a ravine, so nobody was going to drive around the log that direction.

Sheriff's Deputy John Bebb showed up at 6:00 a.m. and told Foreman and Willis to move. They refused. Bebb winched the log around to the left of the road, more or less out of the way but still a minor obstacle. Willis rolled his wheelchair to the right of the road, beside the ditch, about thirty feet in front of the log, while Foreman joined the rest of the protesters out of sight up on the ten-foot-high cutbank.

A few minutes later, cat skinner Les Moore came around the curve toward them driving the crummy, as crew buses are called; the vehicle was Plumley's six-pack pickup, hauling loader operator Dick Payne, shovel operator Leland Townsend, laborer Tim Stone and two choker setters.[69]

This is what Foreman was waiting for.

Moore passed the protesters' truck parked on the road as a bottleneck and then saw the television crews, Channel 5 from Medford and a Portland station, he couldn't see which one. There was a man in a wheelchair alone in the road near the ditch. He knew what they wanted: the humiliation shot, Deputy Bebb leading them through the roadblock.[70]

Moore thought, to hell with that. He steered to the left toward the falloff just enough to miss Willis by about fifteen feet—and the photo opportunity of a redneck running over a guy in a wheel chair. Then he cut sharply to the right, up the bank, cut a hard left around the log and right once more to straighten out on the road.

Suddenly, there was Dave Foreman in front of the pickup, stormed down from the bank.

Moore had seen him two days earlier while driving the crew to work down at Riggs Creek, where Foreman had blocked the bridge and got himself chained to the roadblock. Moore, knowing the terrain better than his tormentors, simply took a cutoff through a mining claim and came up on the back side of them without having to play their game. The television cameras only saw a bunch of loggers in a pickup truck on the far side of the bridge blowing their horn and laughing at Earth First as they drove off to work. That didn't get on the news and it made Foreman mad.

Today was get even time. Foreman grabbed the pickup's grill dead center, fully expecting Moore to stop in front of the cameras. A voice from the back seat said, "Hell, run over the son of a bitch."

Doug Plumley had told them to do nothing until the law got there. Well, the law was there.

"Nah," said Moore, "we'll just give him a little ride."

Les Moore put the six-pack in its lowest gear and drove on up the road with Foreman holding onto the grill and running backward at about two miles an hour, then three, then four miles an hour.

"He huffed and puffed a lot at first," said Moore. "He was hangin' on and his ol' face turned red as a beet and his eyeballs got as big as saucers. Must have been out of shape. All of a sudden he couldn't go any more, and he just let loose and flopped out. We saw his cowboy hat fly up and I stopped real quick."

Everybody piled out of the pickup and ran forward, Les first, then Dick Payne behind him, then the KOBI-TV cameraman came running up behind the two of them, being careful not to fall over the steep edge.

Foreman's boots were barely under the snout of the pickup.[71]

Moore quickly saw that Foreman was winded but not injured. Moore recounts the exchange: "He was layin' there and I told him, 'You get up from there and you get the hell out of here, now.' I told him, 'You got no right coming here blocking our roads, you communist bastard.' And he said, "I ain't no communist, I'm a registered Republican."

Deputy Bebb stepped in and helped Foreman up. "Well, now, Mr. registered Republican, you got any broken bones?"

Foreman stood and checked himself out. "No."

"Then I have some registered Republican handcuffs for you. Les, you and your boys get back in your pickup and go on to work."

On the trip into Grants Pass, Bebb explained to Foreman that he had probably just rescued him from a sound thrashing. "I expect those boys have had about enough of you folks taking food off their tables."

Foreman was charged with disorderly conduct in Josephine County Court and bailed out by Nancy Morton, his second wife. He tried to file assault charges against Moore, but the booking officer explained that Moore was going about his business in a lawful manner when Foreman had willfully and unlawfully placed himself in Moore's rightful path. You can't break the law and then seek remedy in the courts.

So Foreman used the court house steps to accuse the Plumley crew of attempted murder to the media, claiming Moore had first crashed into him twice and then dragged him under their truck for over a hundred yards like Indiana Jones. That nice dab of theatrical ego-salve has since been repeated by everyone who has told the story, and trusted by other protesters facing big logs trucks: "Just remember what Dave did at Bald Mountain and hang on to the bumper." But it's only charming hokum.

Moore says, "That night Channel 5 in Medford showed it on the news. He wasn't hit and he wasn't being drug, he was a-hoofin' it backwards as fast as his cowboy boots would carry him. And when he couldn't go any more he just flopped and he fell down. He bit off more than he could chew and then turned crybaby."

Foreman was arraigned the next day and went to a jury trial Wednesday, August 24. Foreman acted as his own attorney, arguing that disorderly conduct was necessary to stop the immoral logging road. Foreman explained at length his deep moral commitment to wilderness.

District Attorney Gene Farmer noted that, while Foreman's fine morals were commendable, he didn't really have the right to take the law into his own hands.

They showed the jury television clips of Foreman's incident with the pickup.

Charles "Chant" Thomas, then Frank Silow, both Earth Firsters at the scene, took the stand, one after the other estimating that the pickup had pushed Foreman about 100 yards. Thomas said the vehicle was going 20 to 30 miles an hour. The record shows that neither said anything about Foreman being dragged like Indiana Jones. The video clips showed him upright the whole way. When he fell the pickup stopped.

Foreman called Les Moore to the stand. Foreman stated that Moore had hit him twice with the pickup. Moore shot back, "We didn't hit you. You hit us." The television clips did not show the pickup hitting Foreman.

When asked how fast he was going, Moore said the six-pack was in granny gear, compound low, going about two to three miles per hour, accelerating because he was on an uphill incline to maybe four miles an hour tops.

Foreman asked Moore what he saw when the pickup stopped.

Moore said, "You were laying flat on your back, looking up for a camera to come and take a picture."

The prosecution delivered a one-sentence summation: "We do not have the liberty to break the law." It took the six-member jury less than 15 minutes to find Foreman guilty.

The judge sentenced Foreman to one year of court supervised probation, ordered him not to go within five miles of the Bald Mountain road, and ordered him to pay a $150 fine.

Foreman was taken into custody at the noon recess during his trial for violating a court order for attending a Fourth of July rally organized by Earth First in the Siskiyou National Forest. He was wanted on a warrant issued August 10. He posted $2,525 bail and was released, arraignment on those charges to be held as soon as his disorderly conduct trial was over.[72]

Dave Foreman was just beginning to figure out that if you attack somebody, they're likely to fight back. He had been too busy creating a mass movement to notice.

A lot had happened since the day they cracked Glen Canyon dam. The second Round River Rendezvous on July 4, 1981 in Moab, Utah, attracted triple the first year's attendance, with over 200 affiliates coming from all across the United States. They celebrated Independence Day with declarations that love of the wilderness was true patriotism.

They were beginning to find their identity. One attendee rhapsodized

> [What binds us] is a very deep love of the earth ... The biodiversity of the planet, the air, the water, everything needs to live. It's like a tribe. It's so strong, it's almost like a religion.... It's that sort of feeling, "you feel that way too? I felt so isolated! I thought I was the only wacko out there who wanted to throw myself in front of a bulldozer to protect a tree, and there's others like you!" It's a homecoming; it's really neat to meet your own tribe.[73]

They learned to speak an anti-government jargon of their own, some borrowed from Edward Abbey: "Freddies" is a contemptuous term for Forest Service officers. "Doing a CD" is committing an act of civil disobedience. An "action" is any sort of raid against the establishment. Their mistrust of the FBI was paranoid.

In the growing camaraderie, Dave Foreman clearly linked Earth First with the American founders. He scorned the Reagan administration and its Secretary of the Interior James Watt as destroyers of the land, heirs of Federalist Alexander Hamilton, who wanted the new nation to grow into a large and powerful industrial state. They toasted the contrasting anti-Federalist vision of Thomas Jefferson—a small America of inward-looking yeoman farmers living close to the land—with over 2,500 cans of beer. It was quite a party.

The newsletter published just after the Rendezvous saw Foreman explicitly talking about ecotage for the first time, commenting on a serious action the first night of their Rendezvous in which a Utah Power and Light transmission tower carrying 345,000 volt power lines was toppled seven miles south of their location. Foreman denied any Earth First part in the vandalism, beginning what was to become a typical attention-shifting tactic: blaming it on the victim, "corporate interests themselves," or on "free-lance anti-environmental yahoos."[74]

In the same newsletter Foreman announced an "Ecotricks" contest reminiscent of Environmental Action's "Ecotage" contest of 1970. It was intended to inspire people to defend the earth by any means necessary, but an ecotrick "should not be too fellonious [sic] because we need you out there being active."[75] Earth First now felt confident enough to talk about monkeywrenching in the open. It was an indication that most Earth Firsters believed that the apocalypse was coming soon, that industrial civilization would unravel nature and all would collapse within a few decades, maybe even a few years. Monkeywrenchers had by now become a small secret society of cells within Earth First, never discussed openly even at the Rendezvous.[76]

Then came the recruiting campaign. They kicked off the Road Show on September 9, 1981, a program of speeches by Foreman and songs by Johnny Sagebrush and a film called "The Cracking of Glen Canyon Damn," made by Toby McLeod and Randy Hayes. They took it to forty cities, spreading the gospel, singing eco-songs and peddling T-shirts, cassettes of the music and silent agitators. The message was simple: Americans must "come together nationally to fight the beast of industrial civilization." Here Dave Foreman began to craft the anti-Reagan tirade known as "The Speech" that he gave everywhere. He finally got to snort hellfire and brimstone in the pulpit, to be the preacher he yearned to be as a youth.

In October, an article by Dave Foreman appeared in The Progressive and brought in over three hundred letters of inquiry. For the first time it gave a broader public the message of Earth First: there was a need for a new radical environmental movement that would "fight with uncompromising passion, for Mother Earth" and that industrial civilization was causing a biological apocalypse and must be eliminated.[77]

The Buckaroos had grown into a national movement. By the end of 1981, their founding era was over, signaled by the arrival of ex-newspaperman Pete Dustrud as editor of the newsletter and his shift from the original 8½ by 11 typewritten photocopied product to a tabloid format on newsprint. It opened the newsletter to letters—and articles from others than the leaders. The diversity within Earth First that Foreman talked about began to be real. And it immediately began to have the unanticipated effect of creating factions.

On February 6 and 7, 1982, at a meeting in Eugene, Oregon, members of the Circle of Darkness and other Earth First leaders met and formally decided that their centralized organization would now melt into local contacts only loosely affiliated, a more diffuse cell structure. The organization now became a "movement, not an organization." The meeting also decreed that members would no longer be considered members, but simply Earth Firsters. Anybody could be an Earth Firster, said Foreman, who possessed "the wilderness gene" and acted as "Antibodies against the Humanpox." Now misanthropy, a deep and abiding hatred of humanity, became an explicit part of Earth First rhetoric. It was to become one of the issues that would eventually tear Earth First apart.

The redneck Buckaroo contingent felt themselves to be an elite whose awareness of biological meltdown gave them a special role in saving biodiversity from the imminent apocalypse. They did not see themselves as a necessary part of the post-apocalyptic world. They were misanthropic and pessimistic to the extent of saving enough biodiversity for other species to survive. The phenomenal success of the species *homo sapiens* to them was a cancer, a disease, and need not reach into any future millennium. They advocated a reduction in the total population of earth and many had themselves surgically sterilized. They were the apocalyptics.

But there were those in Earth First who believed that after the apocalypse they and their children would inherit the millennium and set things right. Biologist Reed Noss argued that there is nothing more natural than reproduction, the "overriding concern of our animality." He felt Earth Firsters should reproduce and raise ecologically responsible children. A growing social justice faction within Earth First sided with Noss. Many pledged to create a generation that would defend the earth after the apocalypse in the new millennium. They were the milleniarians.

Political science Professor Martha F. Lee was the first to see this division within Earth First in explicitly religious terms such as "apocalyptics" and "millenarians." The term "millenarian" or "millenialist" comes from the Latin words *mille*, one thousand, and *annus*, year. "It evokes the specter of an imminent apocalypse, and the promise of a thousand year period of glory for the community of believers," Lee wrote in her book, *Earth First! Environmental Apocalypse.* In its Christian form, millenarianism came in two forms, premillennialism and postmillennialism. Postmillennialists believe that Christ will return after the church has established the millennium; premillennialists expect Christ to return to establish the millennium by his own power.

Earth First, of course, has no Christian content, but emerged as a "civil religion" with the characteristics of more traditional millenarian religious movements: the adherents were the chosen people; all subsequent history would be understood as a battle between good and evil, where good was defined as progress toward their goals and evil as reaction; the innocence and purity of the first creation will reappear in a future time; their politics constituted the public good.

Lee contrasted millenarianism with apocalypticism. "Apocalyptics are concerned only with the events and earthly conditions leading up to the apocalypse, the climactic and dramatic event that they believe will soon bring about the end of human history. They are not interested in a millennial future for a chosen race or people; indeed, they may or may not anticipate that human life will continue after the apocalyptic event."[78]

The Buckaroo faction centered on the biocentric worldview, wilderness preservation and monkeywrenching; but the emerging social justice faction focused on transforming society through direct action and civil disobedience. They believed that education and reform are possible; the Buckaroos did not. The social justice faction aimed for a post-apocalyptic millennial community; the Buckaroos simply looked for an imminent apocalypse.

And so Earth First began sifting out into apocalyptics and millennialists.

In the May 1982 issue of the newsletter, a new column called "Dear Ned Ludd" by Dave Foreman replaced one called "Eco-tactics." Now monkeywrenching became an institutionalized tactic of the movement.

Ned Ludd was the icon of early 19th century English workers known as Luddites who resorted to a campaign of breaking machinery to protest unemployment caused by the Industrial Revolution.

For at least three hundred years the weavers in Nottingham, Lancashire and Leeds produced lace and stockings that dominated the English markets and were prominent items in export trade. They were hand made, often in the weaver's home. Today, it would be called a cottage industry. The weavers worked mainly as independent contractors, not as employees of a factory owner, and were accorded high status, even though they were commoners. Apprenticeships, family tradition and community values insured a quality product.

In the first years of the 19th century stocking frames and the early automation of the power loom threatened this long-standing way of life. Because the new equipment was expensive, the weavers could not afford to purchase it themselves and the balance of power shifted away to the factory owners. Simultaneously the Tory government adopted a laissez-faire economic policy. The displaced weavers faced a drastic decrease in income and had to work in the regimented and unpleasant atmosphere of a factory, while the price for their food, drink, and other necessities of life increased. The weavers also complained that the machines made products of shamefully inferior quality.

The new technology was the most powerful tool of the factory owner. A vulnerable tool. Sporadic but well-organized resistance erupted, beginning in the hosiery and lace industries around Nottingham in 1811 and spreading to the wool and cotton mills of Yorkshire and Lancashire.

There are several versions of where the name "Ned Ludd" comes from. It may come from a legendary boy named Ludlam, who, to spite his father, broke a knitting frame. Another version has it that a "feebleminded lad" by the name of Ned Ludd broke two stocking frames at a factory in Nottingham. Of course he meant no harm, and could hardly be punished for his innocent act of clumsiness. Henceforth, when an offending factory owner found one of his expensive pieces of machinery broken, the damage was conveniently attributed to poor Ned Ludd. Other stories present Ned Ludd as a great general or a visionary firebrand.

Whatever the source, the Luddites revived the name. They often appeared at a factory in disguise and stated that they had come upon the orders of "General Ludd," "King Ludd," or "Ned Ludd." Their guerrilla army was a secret army. They controlled the night, they knew the back trails between villages. If threatened by government troops they would simply disappear into the same hills and forests that fostered the legend of Robin Hood. For a while they enjoyed almost universal support of the local people.[79]

They demanded reasonable rates of compensation, acceptable work conditions, and probably quality control. Faced by the intimidating numbers and the surprising discipline of the Luddites, most factory owners complied, at least temporarily. Those that refused found their expensive machines wrecked. At the outset, the Luddites scrupulously avoided violence upon any person.

The non-violent period of Luddism ended at Burton's power loom mill in Lancashire on April 20, 1812. A large body of Luddites, perhaps numbering over a thousand, attacked the mill, mostly with sticks and rocks. The mill was defended by a well armed privately hired group of guards. The guards repulsed the attack, and the Luddites instead burned the owner's house. They were met up with by the military and several were killed. A government crackdown ensued, and many suspected Luddites were convicted, imprisoned, or hanged—14 were hanged in January 1813 in York. Although sporadic outbreaks of violence continued until 1816, the movement soon died out.[80]

Now "Dear Ned Ludd" came to Earth First. As Professor Martha Lee pointed out, "The new name implied that the column would provide advice on tactics that were possibly violent, probably illegal, and usually targeted against corporate property, specifically the implements of environmental destruction."[81] As we saw in Chapter Four, the actual targets turned out to be small owners—very little corporate property was hit.

Although Edward Abbey dedicated *The Monkey Wrench Gang* in memoriam to "Ned *Ludd* or *Lud*" and wrote a scene in which Doc Sarvis used the story of Ned Ludd to inspire the gang to embark upon their adventures, monkeywrenching is not Luddism in its original sense of defending employment, economic gain and social status from new technology. Monkeywrenching does not protest the annihilation of jobs, money or social rank, rather it holds such things in contempt. Monkeywrenching is apocalyptic anti-technology, not even the intellectual neo-Luddism of Kirkpatrick Sale, Jerry Mander, Chellis Glendinning, Langdon Winner and Stephanie Mills. Perhaps it is closer to the primitivist neo-Luddism of Wendell Berry, Jeremy Rifkin, John Zerzan and *End of Nature* author Bill McKibben.

In March, 1982, Foreman announced his plans for a publishing venture, Ned Ludd Books, with the most requested project *Ecodefense: A Handbook on the Militant Defense of the Earth* (later published as *Ecodefense: A Field Guide to Monkeywrenching*). It would include technical information on making explosives, wrecking a bulldozer and destroying an oil rig, as well as suggestions on effectively harassing "villains," and subsequently, going underground, creating a new identity, and minimizing legal charges.[82]

In November 1982, Earth First did its first CD, a civil disobedience protest near New Mexico's Salt Creek Wilderness Area where Yates Petroleum was exercising its mineral rights on an exploration site with a lease set to run out on November 1. Yates cut a right of way across a national wildlife refuge, got to their site and began drilling, which the U.S. Fish and Wildlife Service challenged unsuccessfully. A Sierra Club group from Texas went to protest what they called an "illegal road" and Dave Foreman and Howie Wolke piggybacked on their effort with a separate roadblock. It got them sympathetic coverage from CBS News.

They quickly did a second CD in New Mexico at the Bisti Badlands, an area once proposed for a new wilderness designation, but returned to its original status as a mining site. The "Bisti Mass Trespass" was set to stop the Gateway Mine outside the wilderness area, a project of Sunbelt, a subsidiary of Public Service Company of New Mexico. About fifty Earth Firsters walked into the mine site, burned Interior Secretary James Watt and Sunbelt executive Jerry Geist in effigy, and sang "America the Beautiful," hoping to be arrested for the television news crew they had brought along. Being arrested conferred status. No cops came and the security guards had been ordered to take no action. The Earth Firsters went home disappointed, but claimed victory when James Watt included the Bisti in his next round of wilderness studies.

The two CDs gave them momentum, which was increased in early 1983 by an article on Earth First in Outside magazine emphasizing the movement's "cowboy image." Road Show Two in Oregon and northern California got them nearly three hundred new volunteers. They did an action against the proposed Gasquet-to-Orleans Road in northern California, a connector between the two villages, with the help of recruits from the Road Show's stopover in the college town of Arcata.

It was a bubbling high for Foreman. During this time he, Howie Wolke and Bart Koehler conceived what was to become The Rewilding Project many years later: the Earth First Wilderness Preserve System of 716 million acres, one-third of the United States, containing both existing wilderness and developed areas that were to be depopulated and turned back into wilderness.[83]

While they were formulating this grandiose idea, a guy named Charles Thomas invited Road Show Two to stop awhile at his commune ("intentional community") called Trillium Farm in southern Oregon on the banks of the Little Applegate River. Thomas, who used the aliases of Chant Thomas and Chant Trillium, threw a big party for the Buckaroos and they found that Forest Service employees donated a lot of money.

One thing led to another, and the bunch got to talking about the nearby Kalmiopsis wild area, at 15 million acres, one of the largest preserves in Foreman's planned system. Bald Mountain Road was going into an undeveloped part of the Kalmiopsis, and would be the logical next site for CDs, and that was what brought Dave Foreman to his registered Republican handcuffs on May 12, 1983.

Foreman got even with Doug Plumley by putting his company out of business on that road. On July 1, 1983, an Earth First lawsuit filed in conjunction with Oregon Natural Resources Council produced a temporary restraining order immediately stopping construction of Bald Mountain Road, except for completion of some erosion control work. Thus, by the time of Foreman's August 25 disorderly conduct trial, Plumley had already been defeated.

But fate got even with Dave Foreman. It was at the Bald Mountain campaign that a student from the University of Oregon named Marcy Willow, along with a few other roadblockers, introduced a new element to Earth First: the New Age strain of spiritual linkage with the Earth, a strain that redneck Dave Foreman and the Buckaroos would disparage as "woo-woo."

Woo-woo was Buckaroo slang for the hodge-podge of Native American beliefs, Eastern mysticism, tofu-burger vegetarianism, eco-feminism, bizarre lifestyles and science fiction that absorbed the civil disobedience crowd, who seemed to get "a weird, passive-aggressive sort of high" during confrontations. Cowboy Earth Firsters gagged and snorted "CD junkie!" at that because it was wimpy compared to a proper beer-drinking, meat-eating, monkeywrenching, shit-kicking Buckaroo high.

During the third annual Round River Rendezvous Marcy Willow gave the quintessential woo-woo speech: "[N]o matter how alone you get, as long as there is the Wilderness, there is wild Nature, who is your mother, your child, your lover, ancient, new-born, and the same age as you."

Woo-woo.

Recall that Foreman himself was once fascinated with the woo-woo writer Starhawk (pseudonym of Miriam Simos), a Bay Area Wiccan priestess and author of the 1979 book, *The Spiral Dance: a rebirth of the ancient religion of the great goddess.* He even wrote in Earth First's original statement of principles that "Earth is Goddess and the proper object of human worship."[84] He quickly removed it (see p. 61). Others like Marcy Willow were getting ready to put it back for him, and a lot more.

THURSDAY, NOVEMBER 11, 1983   *Washington, D.C.*
INGRID NEWKIRK HAD JUST RESIGNED from the District of Columbia Animal Disease Control Division. PETA was growing into the largest and wealthiest animals rights group in America and took all her time. Animal rights terrorism surged. PETA made no effort to hide its encouragement. A November 13, 1983 Washington Post article noted that "Newkirk has endorsed—and on occasion served as intermediary for—a clandestine group called the Animal Liberation Front, whose members have stolen research animals from Howard University and a U. S. Navy lab in Bethesda." One of PETA's *Factsheets* stated:

> The Animal Liberation Front's activities comprise an important part of today's animal protection movement just as the Underground Railroad and the French Resistance did in earlier battles for social justice. Without ALF break-ins at the University of Pennsylvania Head Injury Clinic, the City of Hope in Los Angeles, and at many other facilities that had successfully sealed

their atrocities from public scrutiny, many more animals would have suffered....

What you can do:

Offer a permanent home to rescued animals: contact PETA for information. Support PETA's *Activist Defense Fund*, which helps pay legal fees of individuals accused of liberation-related activities. Blow the whistle on facilities where animals are forced to suffer. You may now be working in such a place, or be willing to take a job to keep a particular laboratory or animal supplier under surveillance. All contacts are kept in strict confidence.[85]

MONDAY, JANUARY 27, 1986   *Washington, D.C.*

MIKE ROSELLE, CO-FOUNDER OF EARTH FIRST, was named national campaign coordinator for Greenpeace USA. Roselle's moved to Greenpeace reflected his long-standing preference for civil disobedience that was aimed at changing public opinion. Greenpeace believed in change through education, with the goal of preventing the apocalypse by making industrial civilization more environmentally sensitive. Dave Foreman, although tactful about Roselle's move, saying Earth First had not lost Mike Roselle, but gained Greenpeace, still felt hostility toward big Washington lobbying outfits. Their tactics were in direct opposition to Earth First's. They would do their flashy civil disobedience for the cameras, but in the smoke-filled rooms of Washington they would compromise while wilderness was developed. Foreman felt increasingly that the apocalypse was coming very soon. Mike Roselle did not.

The fracture lines in Earth First were beginning to show. Roselle's influence made the Earth First / Greenpeace combination systematic. For example, in 1990 Susan Pardee of Seattle was both Earth First and Greenpeace, operating out of a shared office.

A new generation was also coming into Earth First. Documentary filmmaker Jessica Abbe, who thought the founders were "brilliant" said that many of their followers were "one step up from street people." Many of the new followers were less educated, came from lower class homes, and brought personal demons with them. Many were grim and deadly serious, without a touch of the founders' humor. Many were throw-away children who attached themselves to the movement because it was the closest thing to family they had ever experienced. Sitting around campfires in a circle where everybody seemed equal and everybody belonged stirred needs in them that had never been nurtured before. Earth First made them part of something bigger than themselves, something that gave them value and meaning and power even though they had nothing to give back but devotion and their hatred of industrial civilization.

In March, monkeywrenchers destroyed the logging equipment of a small Montana firm and left a banner saying, "Earth First!" Dave Foreman chastised the unidentified vandals in the pages of the Earth First Journal as having no clear environmental protection purpose for their act, and complaining that it gave Montana Earth First public relations problems. The Frankenstein effect was setting in. *Ecodefense* was selling quite well. People did what it said. The "Dear Ned Ludd" column was widely read. People did what it said.

Now the double-edged nature of Earth First's "tribal structure" became apparent: without a hierarchy of responsibility, no one could control what the tribe did or deflect blame wrongly placed. Nobody and everybody was a member. Without a formal membership, anybody could be accurately considered part of the movement. Any vandalism that matched the modus operandi written in *Ecodefense* or "Dear Ned Ludd" or published in the Earth First Journal pointed directly to Earth First. A tiny group acting alone with the purest wilderness-saving intention could now commit an act that might conceivably have grave consequences for the entire movement.

In May it happened. A group of Earth First monkeywrenchers cut the power lines into the Palo Verde nuclear plant located twenty-five miles west of Phoenix. The FBI, which had been keeping tabs on Earth First from the beginning, now began an intensive investigation. It would have grave consequences for Dave Foreman.

THURSDAY, NOVEMBER 13, 1986   *New York City*
RODNEY ADAM CORONADO SAID, "THE SEA SHEPHERD TEAM boarded the whaling ships about 4 a.m. Sunday. After determining that no one was on board, the Sea Shepherd team removed the sea valves separating the ships from the ocean, thereby sinking the ships. It was just a coincidence that I was in Iceland at the time."[86]

More than two dozen news agencies listened to the thin 5 foot 11 inch-tall young man, part Mexican, part Yaqui Indian. The government of Iceland had issued a warrant for the arrest of Coronado and a British subject named David Howitt for sinking two Icelandic whaling ships and the wrecking of a whale-meat processing plant, causing $2 million worth of damage. The two had been declared terrorists. The State Department would not comment on whether the 20-year-old Morgan Hill, California resident would be extradited.

Captain Paul Watson, 35, founder of the Sea Shepherd Conservation Society, declared his group responsible for the sabotage and named Coronado as a participant. He said Coronado had gained "historic" stature for his deeds.

At the Coronado family's ranch in Coyote Valley in northern California, the young man's mother, whose name is Sunday, said her son had sent money to the Sea Shepherd group since he was nine years old. He also collected cats, plastered his bedroom walls with posters of whales and of the rock group U2 and persuaded his parents to take in two wild burros captured in Grand Canyon National Park.

The Coronado family had met Watson during a trip to Vancouver, British Columbia, in the fall of 1984. They heard him at a public meeting on aquarium expansion and then met him on the docks. The family left Rodney with Watson for a couple of days and then picked him up for the trip back home.

The weekend after his high school graduation, Coronado went to Santa Monica to join the Sea Shepherds. Watson first sent him to Honolulu to work as gopher for an artist painting a huge whale mural, just to see if he was ready to do the shitwork. He was. Watson took him directly from there to Nova Scotia, then put him in the engine room of the 194-foot converted trawler, *Sea Shepherd II*, for four months, and then to a campaign to stop the authorized hunt of non-threatened pilot whales in the Faeroe Islands, where the meat is a staple for the population. Coronado said he spent five days in jail in the Faeroe Islands after one of the missions was thwarted.

During 1985, when the ship was in London, Coronado got involved with the Hunt Saboteurs, sabbed fox hunts and made many contacts in the animal rights movement. In 1986, he recruited David Howitt to work with him for a month in a whale meat processing plant in Reykjavik. Then one night, as Dean Kuipers later wrote in Rolling Stone, "They tore up the plant's office, leaving logs and computers in ruins, then rushed to the harbor, boarded two huge whaling ships and opened their seacocks, sinking them up to their masts. A guard was sleeping on the third ship, so Howitt and Coronado left it alone, caught a cab to the airport and fled the country."[87]

APRIL 6, 1987   *Davis, California*
THE ANIMAL LIBERATION FRONT BURNED the Animal Diagnostics Laboratory, a veterinary research facility under construction, at the University of California at Davis. Twenty university vehicles were also vandalized. The total damage amounted to $5.1 million. The FBI categorized the raid as domestic terrorism, the first ALF raid so designated. As a direct result the FBI launched an investigation of ALF as a domestic terrorist organization. The FBI continued investigations through September 1990. Only two other ALF incidents have been officially characterized as domestic terrorist acts: 1) the April 1989 arson at the University of Arizona; and 2) the July 1989 theft of animals and destruction of equipment from Dr. John Orem's laboratory at Texas Tech University in Lubbock.[88]

MAY, 1987   *Tucson, Arizona*
THE GREAT AIDS DEBATE in the Earth First Journal decisively split the biocentric faction of Earth First from the social justice faction. An article by the pseudonymous Miss Ann Thropy titled "Population and AIDS" came in the midst of a venomous controversy that had been brewing for some time. Many readers assumed that Dave Foreman wrote it, but the

author was Christopher Manes, a young biocentrist on the editorial staff who would soon write the first book focused exclusively on Earth First, *Green Rage: Radical Environmentalism and The Unmaking of Civilization.*

Miss Ann Thropy's article argued that a disease such as AIDS had the potential to reduce the human population significantly and quickly, thereby benefiting endangered wildlife on every continent. More importantly, as the Black Death had helped bring about the end of feudalism, AIDS might help bring about the end of industrialism. If the population of the United States dropped below 50 million, capital would dry up, governments would lose authority and industrialism would cease to function.

The crass Malthusian tone of such an article in the Earth First Journal outraged the social justice faction. They began to question the legitimacy of Foreman's leadership.

Mike Roselle took action to solidify the social justice faction's position in the September issue of *Earth First!* While avoiding open hostility and making no mention of the social justice/biocentrism conflict, Roselle made an editorial appeal for money from readers to support the Direct Action Fund and the Nomadic Action Group (NAG), a select number of Earth Firsters who traveled the country organizing and running direct action campaigns. [89] The editorial, along with a direct mail piece Roselle sent to all subscribers, brought back over $18,000. Now Roselle had a budget to forward the social justice agenda and increase his power within the movement.

A more select covert cadre of monkeywrenchers who traveled the country committing criminal acts evolved at the same time as a decoupled Nomadic Action Group.

SEPTEMBER 20, 1987   *Vancouver, British Columbia*
RODNEY CORONADO PARTICIPATED in a series of Animal Liberation Front raids on nine fur salons, his first public association with the terrorist group. After gluing the locks of two fur stores, smashing the windows of Grandview Furs and Avenue Furs, throwing red paint inside each store, and spraypainting "Fur is Deadly" and "ALF" on walls, Coronado and other ALF raiders were arrested during an attack on Papas Furs, Grandview Furs and Avenue Furs. Coronado posted $10,000 bail. At the time, Coronado and David Howitt were living on the *Sea Shepherd II* in Vancouver harbor. They piloted the ship down to California, fleeing the jurisdiction and leaving outstanding warrants behind.[90]

Earth First and Rodney Coronado came together on December 5, 1987 when Coronado organized the Hunt Saboteurs in Southern California with Earth Firster Lyn "Lee" Dessaux, recruiting a network of Earth Firsters and local unaffiliated back-to-the-land types. They sabotaged a Bighorn Sheep hunt using air horns, crashing around and yelling to scare away Tule Sheep from trophy hunters.[91]

Coronado soon moved with Dessaux to Santa Cruz, California, and focused on the fur industry. They were creating a cell of Earth First's Nomadic Action Group.

Coronado lived with fellow Hunt Sab Jonathan Paul from 1990 in Santa Cruz. Together they made secret videos of fur farms in an ad hoc group they called Global Investigations.

May 1, 1988   *Ukiah, California*
JUDI BARI BEGAN TO FORGE AN ALLIANCE between the doddering remnant of the Industrial Workers of the World and Earth First. Her idea was that such an alliance could show loggers and environmentalists that they had a common interest in the fall of the corporate industrial monolith that was destroying the environment, that they could create a mass movement to destabilize corporate power and bring about true social change.

Judith Bari was a "pink diaper baby," the daughter of an Italian gemcutter father and a Jewish Ph.D. mathematician, the first woman to be awarded a math doctorate from Johns Hopkins University, both socialists who led a comfortable middle class life on the East Coast. Judi attended the University of Maryland, took part in antiwar protests, smoked dope and became an angry Marxist who toted around Chairman Mao's little red book—the cliché college struggle junkie. She dropped out and took a blue collar job in a postal service mail sorting center, becoming a union shop steward. She got married to a man named Mike Sweeney in 1979 and moved to Santa Rosa, California, where she had two children, took part in antinuclear protests and Sandinista support demonstrations, ever the angry struggle junkie.

In 1987 Bari met a jumpy little guitar-picker named Darryl Cherney, described by Susan Zakin as "a fast-talking Jewish guy from New York who had been a child actor in TV commercials." It was at a benefit concert—Cherney was running for Congress, which was a source of local hilarity during a very short run, but typical of publicity-hound Cherney. He had arrived in 1985 and had run across a small group of '60s dropouts in ponytails called the Environmental Protection and Information Center (EPIC) in Garberville, trading post of the marijuana empire. Cherney's environmental interest led him to a chance meeting with TV reporter-turned activist Greg King, who created "the Headwaters issue" that was to divert Earth First from federal lands protests to direct attacks on private property against Pacific Lumber, the old-line family company that had been taken over by Charles Hurwitz and Maxxam Corporation in a controversial junk bond deal. Pacific Lumber had been very conservative in its cutting policy over the years, leaving it with the largest volume of privately owned old growth redwoods in the world. Now Maxxam was going to rapidly liquidate that redwood to pay for the junk bonds. King wanted Pacific Lumber's lands seized by the state to become a park.

In 1988 Bari and her husband began building a home on Humphrey Lane in Redwood Valley north of Ukiah. The marriage was falling apart and Bari got a carpentry job with a little outfit called California Yurts. Cherney and Bari became an item for a while. Bari felt that wimpy Cherney, who had found himself "scared shitless" of the macho Buckaroos at a Round River Rendezvous, brought out the gentle feminine side of her in-your-face personality. Cherney, in turn, persuaded Bari to join Earth First, although she thought they were a bunch of rowdy sexist assholes. She particularly did not like the strain of misanthropy in Earth First (she would one day write an essay titled, "Why I am not a Misanthrope"). Cherney convinced her that the decentralized organizational style of Earth First meant she could shape her local chapter as she wished, so she got busy.

Courting the Wobblies didn't amount to much beyond founding I.W.W. Local #1 in Ukiah and an article titled, "Fellow Workers, Meet Earth First!, Earth First, Meet the I.W.W.," but it gave Bari the chance to play her fiddle while Cherney plunked his guitar in a few labor songs. They tried to organize some disaffected Louisiana-Pacific workers, and then some Georgia-Pacific workers, miscellaneous truck drivers and construction workers, but mostly Judi Bari established herself as a charismatic leader. She was building her base for later success.

It was another contact near her new home that had a little-recognized but more important impact on the future direction of Earth First.

In Ukiah, in May of 1988, Tim Zell and his wife Martha resumed publication with the eighty-first issue of a little offbeat magazine that had been defunct for ten years: The Green Egg, journal of the Church of All Worlds.

From the unlikely beginning of a tiny group of four students at Missouri's Westminster College in 1962 and modeled on the science fiction novel, *Stranger In A Strange Land* by noted author Robert Heinlein, the Church of All Worlds has grown into an effective focus for Neo-Pagan beliefs and an important influence on radical environmentalism.[92]

The Church of All Worlds was the first Neo-Pagan religion to be incorporated (1968) and obtain federal tax exempt status; it was Tim Zell who coined the term "Neo-Pagan" that drew together the diverse revivals of Wicca—or the Craft, as some prefer—ceremonial magic, Goddess worship and eco-feminism into a single movement.[93]

The Green Egg developed into an electrifying conduit for discussion of the various emerging Neo-Pagan practices, inspiring some and annoying others by its free-wheeling editorial policy of letting anybody throw whatever they wanted into the cauldron and stir it vigorously. Were they reviving genuine old arts or reconstructing illusions about them in strictly modern terms? Let's argue about it. Who really knew anything about what they were doing? Let's argue about it. Wiccans didn't trust ceremonial magicians didn't trust eco-feminists, and so on. The Green

Egg made them aware of each other, prompted a number of conferences and tries at ecumenism, but eventually just wore everybody out. The Green Egg went through 80 lively issues and ceased publication in 1976, upon which many expressed relief that the various practices could now go their separate ways in peace and quiet. But it had given them the boost they needed to become a viable, if marginal, social force. The seeds of the New Age had been sown.

It was Tim Zell who gave the woo-woo crowd what has come to be known as "The Gaea Hypothesis," a theory that the earth is a living organism, something like the scientific account of British scientist James Lovelock, who used the Latinized spelling, Gaia.[94] Lovelock first put the Gaia idea forward at an obscure scientific meeting about the origins of life on Earth held in Princeton, New Jersey, in 1969. It was years later before it gained any recognition.

On the night of September 6, 1970, Zell had a "visionary experience" and delivered it in the form of a sermon to the congregation of the Church of All Worlds a few days later. It was a long tour through biology, cell division, reproduction, and evolutionary theory. Zell saw that all life had evolved from the same original living molecules, so all life was interconnected, part of a single living organism. Zell saw not just that the Earth was a living organism, but it was also a feminine deity he called Terrebia—an awkward coinage of Latin and Greek words meaning "earth" and "living," later replaced by the more elegant term Gaea, the Greek personification of earth as goddess.

Zell's vision was subsequently published as a paper titled *Theagenesis* which made the rounds through the Neo-Pagan community.

Zell claims to have been the first to come up with the Gaea hypothesis in his current promotional material, but the fact is that Lovelock beat him to it by a year. Whether Zell heard of Lovelock's presentation at the 1969 conference and copied it is beside the point; Zell appears to have had a genuine independent insight. At any rate, Zell's version got into the Neo-Pagan and radical environmental movements long before Lovelock's. It stressed the belief that we are all interconnected, and that this interconnection requires us to incorporate ecological principles into our philosophy. Zell also had a few bells and whistles Lovelock did not.

For example, Zell held that the ultimate potential of Gaea was the telepathic unity of consciousness between all parts of the earth's nervous system, between all human beings, and between all living creatures. Such a future had been predicted by the French paleontologist and Catholic priest Pierre Teilhard de Chardin in his 1955 book, *The Phenomenon of Man*. Teilhard called the awakening "the Omega Point." Zell wrote, "Indeed, even though yet unawakened, the embryonic slumbering subconscious mind of Terrebia is experienced intuitively by us all, and has been referred to instinctively by us as Mother Earth, Mother Nature (The Goddess, The Lady)."

Lovelock's work as an atmospheric scientist, on the other hand, was couched in caveats such as "Gaia has remained a hypothesis" and "if Gaia exists," but, even though it left out the telepathic mysticism while proposing that indeed the earth does behave as a single living organism, it lent credence to Zell's religious/science fiction version.[95] Newsweek wrote in 1975 that Lovelock had taken the basic connectedness of all life a step further: "in man, Gaia has the equivalent of a central nervous system. We disturb and eliminate at our peril. Let us make peace with Gaia on her terms and return to peaceful coexistence with our fellow creatures."[96]

The impact of this theory on the development of radical environmentalism can hardly be overstated. The Church of All Worlds advertised itself as "a total, holistic, cultural alternative to the entire fabric of Western Civilization." In time the millenarians in Earth First would use Zell's ideas to make their movement a similar ticket to an alternative universe.

In 1976 Tim and Martha Zell left Missouri and migrated to Northern California, where they became back-to-the-land hippies for ten years, taking on the obligatory ecologically-conscious names: Tim became Otter Zell (sometimes Otter G'Zell) and Martha became Morning Glory Zell. Tim has lately become more stately and goes by the name Oberon Zell.

In April, 1988, while Judi Bari was thinking about hustling the Wobblies, within a few miles of her home Otter Zell updated *Theagenesis* for delivery at the California Institute of Integral Studies' symposium on "Gaia Consciousness: The Goddess and the Living Earth." The next month he, his wife and a new staff revived The Green Egg. It quickly regained its dominance in the field, printing a lot of stuff on meditation, inner healing, herbal remedies, the Goddess, earth rituals, finding your *tamanou*—your power animal—the magic arts of shape shifting and disappearing, on everything you might call woo-woo. Today it is a slick four-color magazine and publishes frequently on what it calls the "earth religions." Mike Roselle had been living in the Bay Area among people of similar views for over a year now. Both Judi Bari and Darryl Cherney would come to write articles for The Green Egg.

We would see the woo-woo influence time and again, for example, in that Live Wild Or Die article we noted on page 94 by the young Earth Firster who recommended leaving everything for a "private" life to avoid detection of illegal monekywrenching. The author also said:

> One thing I would like to touch on briefly is one not many people give much thought or credit to. That thing is magic, specifically, magical protection. Amid your scoffing and teasing I can happily say that there is magical energy to tap into. Most of my protection, outside of all the practical precautions I take, comes from my faith in the magical realm.... I look at it as a kind of chaos energy which engulfs and connects every being and place. Once you learn to see signs of the energy, and begin to connect

with the chaos around you, you can tap into the protective energies. There is no rule of how to do this; everybody must find their own way which suits their particular way of relating. A lot of it has to do with faith and a lot has to do with your connection to other like-minded souls as well as the natural world. Anyway, I'm not writing to convert you non-believers. All I'm saying is just open yourself up to the potentials and the possibilities. What have you got to lose?[97]

Dave Foreman the Buckaroo was not pleased with the Dungeons and Dragons direction his movement was taking. The woo-woo types were weirdos. He had often said things like, "A lot of that Abalone Alliance, vegetarian, non-aggressive, non-ego ball of wax is too New Age for me to swallow."[98] He wasn't much of a Gandhian, either: at a protest where police hurt his wife he lost his temper and attacked the police. But he really didn't like the new people coming into Earth First. He told a Rendezvous crowd of the faithful, "Most radical activists are a dour, holier-than-thou, humorless lot." They needed not only a few laughs, but more importantly "an awareness that we are animals." Emphasizing that point at every speech and Rendezvous, he always led howls, saying "The greatest thing you can do is just howl. Aaaooooooooooooo!" Everybody howled. It became the emblem of Earth Firsters. Foreman said that Earth Firsters "are not devotees of some Teilhardian New Age eco-la-la that says we must transcend our base animal nature and take charge of our evolution in order to become higher moral beings."

Not that he didn't have religious feelings about the earth. He told an interviewer that the environment is "religious in a non-supernatural sense...we have an ethical, reciprocal relationship with the land. We are, for lack of a better term, talking about our souls."[99] In fact, in many interviews he identified monkeywrenching as "a form of worshipping the earth."[100] Monkeywrenching "was very much a sacrament."[101]

He said he was tired of the redneck versus woo-woo debate in his tribe, and if such problems continued, "I will seek my campfire elsewhere."

He was losing touch with the movement he started. Only the old guard kept to the biocentric wilderness-and-monkeywrenching dogma. More and more Earth Firsters *were* devotees of some Teilhardian New Age eco-la-la. Including one of the old guard.

OCTOBER 23, 1987   *Mount Rushmore, South Dakota*
MIKE ROSELLE AND FOUR OTHERS WERE ARRESTED for attempting to unfurl a Greenpeace banner over the carved face of George Washington to draw attention to acid rain. Roselle was sentenced to four months in jail.[102]

While he was in jail, Roselle cracked Earth First as Earth First had cracked Glen Canyon Dam: for some time he had believed Dave Foreman was using part of Earth First's $200,000 annual budget to selectively fund illegal monkeywrenching, siphoning money away from social justice and civil disobedience projects such as the one that had just landed him in jail. How much could it cost to publish a drab newsprint journal, anyway?

Roselle was beginning to have serious problems with Foreman's wilderness-and-monkeywrenching program. His preference for direct action and civil disobedience was sharpening. He realized now that Foreman did not believe that human nature could be changed and believed that humans didn't matter anyway. He realized now that he himself did believe that human nature could be changed and that a new society could be built.

Roselle publicly accused Foreman of using the Earth First journal to pursue his own agenda. He also implied that Foreman was behaving like a dictator. He called his supporters "Foremanistas." Foreman made no reply. The social justice faction smouldered. The crack grew.

DECEMBER 31, 1988 - JANUARY 1, 1989   *Bethesda, Maryland*
THE THREE TEENAGERS LOOKED UP WHEN THE GARAGE DOOR OPENED. "What are you guys doing?" asked Gustavo Machado, peering in. It was three in the morning. New Years Eve had turned into New Years Day.

"Shut the door," said Dov Fischman, leaning over the work table.

The fifteen-year-old came in and did as he was told. He saw Bruno Perrone working with a small cylinder that looked like a piece of metal pipe, assisted by Fischman and Salem Samir Gafsi, three college kids home for the holiday. The three chemistry buffs had been close friends during their years together at Bethesda's Walt Whitman high school. They reunited tonight to finish this secret project.

Machado's older brother Rodrigo had let them in hours earlier. Rodrigo warned them not to get into any trouble; his father was a Brazilian Embassy attaché who could bear no bad publicity.

Then something went wrong with their secret project. The powerful pipe bomb exploded unexpectedly, shattering the garage and killing all four teenagers instantly.[103]

Police discovered that Dov Fischman had a home laboratory with chemicals suitable for pipe bombs: nitrates, peroxides, carbonates and other compounds, plus numerous formulas for making explosives.[104]

A high-school classmate, Alex Ferguson, told police that Fischman's myriad interests included chemistry, especially the explosive potential of various mixtures. "For him it was just curiosity," the young man said. In high school he and Dov had once made a steel containment vessel to test the explosive power of various mixtures. The chemicals were all available by mail order from supply houses and the recipes were easy to get from survivalist magazines.[105]

Fischman had seen Ferguson earlier during the holiday and invited him to join the old gang in a few days, when they would make some "stuff." Ferguson forgot about the invitation and went to a party. He told police, "Dov would never do anything to hurt anybody."[106]

Maryland police and the Bureau of Alcohol, Tobacco and Firearms wrote up the case as an unfortunate accident with four fatalities.

Across the country in a dormitory of the University of California at Berkeley, campus police detective Joseph Chan gathered the personal effects of Bruno Perrone to send to his parents in South America. He found a little black six-ring binder notebook about four inches by six inches. It contained Perrone's thoughts in small cramped handwriting as they developed from his first day at the university in September of 1988. It was horrifying.[107]

Freshman Bruno Perrone began with the observation that people are destroying the earth, so they should all be destroyed—The Apocalypse Creed in full bloom. Then, a few pages later, he realized that killing everyone was impractical. Only the worst need be killed. Who were they? Perrone wrote of seeing a newspaper clipping from last term about animal experiments being carried out at the university: Cleveland Amory, founder of the Fund for Animals, had visited California to stage "Laboratory Animal Liberation Week," marked by protests and arrests at several California universities. Amory singled out U.C. Berkeley ophthalmology Professor Richard Van Sluyters as a culprit. [108]

Van Sluyters, a highly-regarded scientist, performed landmark vision-impairment experiments using kittens, sewing their eyes shut for a brief time to determine what effect the deprivation had on visual development. The research was of great value in treating human blindness, but that was irrelevant to Perrone. He decided that Professor Van Sluyters should be killed. The bomb that went off in that Maryland garage was intended for Van Sluyters.

Now in early January of 1989, campus police detective Chan notified Van Sluyters of the diary's contents while Bureau of Alcohol, Tobacco and Firearms Special Agent Randy Haight was on his way to pick it up. Chan told Van Sluyters, "I'm not supposed to be telling you this, but if I was you I'd want to know." Chan relayed the contents to Van Sluyters just before Haight collected Bruno Perrone's diary, which remains in the custody of the BATF.[109]

Several years later Professor Van Sluyters granted a guarded interview to the San Francisco Chronicle. He did not identify Perrone, only that a bomber had accidentally killed himself while planning to murder him. All he would say of the fallout is, "The police told me, 'Drive a basic-colored car, don't park in the same place and look under your car every day,' I didn't know whether to hide or never go to my office again."[110]

Professor Van Sluyters is still so terrorized by the incident that he refused to be interviewed for this book.

APRIL 21, 1988   *United States*
KAREN PICKETT, A CALIFORNIA EARTH FIRSTER, used money from Mike Roselle's Direct Action Fund to stage the largest protest in the group's

eight-year history, the "National Day of Outrage Against the Forest Ser-vice." This huge anti-federal government protest went nationwide with direct action events in seventy-five locations. Timber-related Forest Ser-vice employees feared for their personal safety. Dave Foreman congratu-lated Pickett on its success in directing public anger at the Forest Service.

The Round River Rendezvous that year was scheduled for Wash-ington State's Okanogan country and the traditional Earth First action was the occupation of a Forest Service office for a day, which resulted in twenty-four arrests (see p. 135).

In the fall of 1988, Mike Roselle and Karen Pickett were married. It didn't last long.

APRIL 3, 1989   *Tucson, Arizona*
THE ANIMAL LIBERATION FRONT BROKE INTO three buildings at the University of Arizona, stole more than a thousand animals being used in medical studies, vandalized equipment worth $300,000 and destroyed two research laboratories, a research center, and an off-campus office by gasoline fires.[111]

Twenty-four hours later, ALF released a videotape of the destruc-tion, filmed by its own crew, to one local TV station, Channel 9. Trans-mitted via satellite from the headquarters of People for the Ethical Treat-ment of Animals (PETA) in Washington, D.C., to Channel 9 [in Tucson, Arizona] the tape was shown on the three major television stations as the lead story on their evening news reports the day after the attack.[112]

The FBI, which had formally classified ALF as a terrorist organi-zation, began to suspect that PETA might have more than a publicist role in ALF raids.[113]

MARCH 20, 1989   *Tucson, Arizona*
EDWARD ABBEY DIED AT HOME OF ESOPHAGEAL VARICES, a throat disease com-mon among alcoholics, at the age of sixty-one. The loss of the great voice behind biocentrism—and monkeywrenching—was a blow to everyone in Earth First, but it hit Dave Foreman hardest. He learned of it on the phone at the Houston airport on his way back from a vacation in Belize. He cried a long time, then wrote a celebration of his friend's life for the cover story of the next issue of Earth First!

> In his death Abbey joined a small company. Perhaps only Henry David Thoreau, John Muir, Aldo Leopold and Rachel Carson have touched so many souls so profoundly. Edward Abbey was a great man because he articulated the passion and wisdom of those of us who love the wild. He was a spokesperson for our generation and for generations to come of those of us who understand where the real world is.[114]

It only provoked the social justice faction. Abbey had believed,

"Whether we live or die is a matter of absolutely no concern whatsoever to the desert."[115] The social justice faction struggled to find their own voice, not echo Abbey's. The factions were obsessed with their own viewpoints.

Not long after Edward Abbey died, Earth First received what Nancy Zierenberg called, "sort of a funny fan letter from Squeaky Fromme," one of the Manson Family girls, addressed from a federal prison where she was serving a life sentence for attempting to assassinate President Gerald R. Ford. At the time Zierenberg was working as the Earth First Journal's merchandise coordinator.[116] It was less odd than Zierenberg realized: Charles Manson constantly warned his followers against the destructive influences that seemed to be everywhere, "in particular, the choking industrial pollution in American cities and the logging and decimation of the irreplaceable redwood forests in northern California," according to author James W. Clarke's *American Assassins: The Darker Side of Politics*.[117] Lynette Alice Fromme, prior to pointing a .45-caliber pistol at Ford, "had been making violent threats for a while and had tried to persuade others to inflict violence on those they felt were responsible for destroying natural resources."[118] No movement is immune from unwelcome personal demons brought by the unbalanced to confuse the issues with unrelated agendas. But the Fromme fan letter presaged a streak of hate mail to Earth First that Susan Zakin characterized as an FBI counterintelligence setup.[119] Zakin's case was weak, but the FBI was indeed watching Earth First.

Nine days after Abbey died, Foreman gave Mark Davis $580, part of which was used to pay for fifty thermite grenades to sabotage high voltage electrical transmission towers and lines at three nuclear facilities: Palo Verde Nuclear Generating Facility of Arizona, the Diablo Canyon Nuclear Generating Facility of California and the Rocky Flats Nuclear Facility of Colorado. Dave Foreman was putting money into illegal monkeywrenching as Roselle suspected.

Mark Davis was a former biker-hippie-yogi Earth Firster who had moved to Prescott, a couple of hours north of Foreman's home in Tucson, where he become involved with eco-feminist Earth Firsters Peg Millett and Ilse Asplund and a botanist named Marc Baker. With varying degrees of complicity, the four had become the Evan Mecham Eco Terrorist International Conspiracy, or EMETIC.

On the night of October 4, 1987, Davis and Millett had used a propane cutting torch to cut the bolts on twelve pylons supporting the main chairlift of the Fairfield Snowbowl ski resort in Northern Arizona's San Francisco Peaks. The next day Davis sent a message from EMETIC to a newspaper in Flagstaff claiming responsibility for toppling the chairlift, citing their opposition of the resort's expansion into "sacred Indian territory."

On September 25, 1988, Davis, Millett, Baker and Asplund partially cut through twenty-nine wooden power poles carrying electricity into the Canyon Uranium mine, causing a power outage. A letter by Davis from EMETIC to the media followed.

On October 20, 1988, Davis and Baker used a propane cutting torch to cut the top pylon supporting the main cable chairlift of the Fairfield Snowbowl ski resort, and another EMETIC letter followed.

The ecoteurs of EMETIC let a cowboy hippie named Ron Frazier in on their exploits—he had helped Davis buy a torch and taught him welding skills and he got a laugh out of seeing how his friend had used the lessons. Some time later an unrelated personal dispute erupted between Davis and Frazier. In anger, Ron Frazier drove to the FBI office in Phoenix to tell them what he knew about Davis. They already knew.

The FBI had been watching Earth First closely ever since that 1986 power line sabotage at Palo Verde. In fact, they had been watching specific events since the 1981 cracking of Glen Canyon Dam. They had reported on the 1983 Bald Mountain road protests in Oregon. They had infiltrated a couple of Round River Rendezvous. They got particularly interested when an Earth Firster with a Ruger Mini-14 fitted with a scope took part in a May 23, 1983 protest at the Glen Canyon Dam birthday celebration where Interior Secretary James Watt was to speak. National Park Service rangers confiscated the rifle and the man disappeared. The FBI knew about the uranium mine hit in advance, but did nothing to stop it. They wanted the man behind Earth First, Dave Foreman. Frazier was accepted, wired with a tape recorder and became a paid FBI informant in January 1988 to get evidence against Foreman.

Frazier's tapes of conversations with the EMETIC four were valuable, but his best work was introducing undercover FBI Special Agent Michael Fain to Peg Millett as "Mike Tait." Millett had a thing for cowboys. "Tait" became the perfect cowboy. Millett loved to dance. Her husband, a forest ranger named Doug Vandergon, didn't dance, so Mike Tait frequently took her out. They got emotionally hot and heavy but not sexually involved. Tait told Millett he was an alcoholic. He had never gone to AA, he said, and struggled to stay dry. Millett's father had been an alcoholic and Tait manipulated that fact with consummate skill. Millett determined to redeem Mike Tait.

On December 8, 1988, Peg Millett invited Tait to join EMETIC in "a plan afoot to get maybe five different nuclear power plants simultaneously where the power going in and the power going out is cut off." Tait was in.[120]

Not quite a month later, Mark Davis asked Ron Frazier for the chemical formula to manufacture thermite, a mixture of powdered aluminum and iron oxide used in high temperature welding and incendiary bombs and sold under the brand name Thermit. Thermite bombs are a favorite of vandals ranging from chemistry geeks attending the Massachusetts Institute of Technology who wedge them between a trolley car's wheels and the rail, ignite them and laugh while it welds the trolley to the tracks, to serious saboteurs who need something guaranteed to start a very hot fire or cut quickly through hard metal.

Two days later, Millett asked Tait for the thermite formula.

Then on January 14, 1989, Mark Davis met with Ron Frazier to discuss thermite production "to hit as many of the nuclear power plants on the west coast and Palo Verde as we can, drop all the lines going in and out of them."

By the end of the month they were talking prices of thermite grenades and ten days later, on February 10, Davis placed an order with Frazier for one-hundred of them. Near the end of February, on the 26th, Mike Tait drove Mark Davis to Tucson to ask Dave Foreman for the money. Foreman would not let Tait in on the discussion, but agreed to come up with the money in about a month.

On March 29, 1989, Dave Foreman gave Mark Davis $580 to be used for equipment and supplies to knock the nukes. Two days later, Davis used the funds to purchase a certified check for $500 from Valley National Bank in Prescott and mailed it as payment for fifty thermite grenades.

On April 25, Mark Davis drove a Volkswagen minibus from Santa Barbara, California, to San Luis Obispo where he reconnoitered the Diablo Canyon Nuclear Generating Station, tailed by six FBI vehicles. A security guard at the nuclear power plant turned Davis away, so he then drove down a side road, where he parked and looked over the power lines. He drove to another location, got out of the minibus, walked up a hill and spent several minutes inspecting a large transmission pole that held lines strung across U.S. 101. Afterward, Davis turned down another side road, where he briefly viewed two other power lines, and then returned to Santa Barbara.[121]

The FBI's case against the Prescott four and Foreman now lacked only one necessary element: Foreman's intent in giving the money for the thermite grenades. They had to get Foreman on tape instructing Tait to sabotage the nuclear facilities. On May 5, 1989, they got it.

Mike Tait came to Foreman's home alone. Foreman took him for a walk in the desert. Tait said he was alarmed that Mark Davis was planning a practice session before hitting the nukes, a raid against the Central Arizona Project (CAP), which supplies water through a canal system to various points in the state. Tait thought it was a bad idea, and Foreman agreed.

"I think it's got to be real targeted and be directed at targets that will have some kind of impact," said Foreman. "Like the nuclear thing, that might help prevent additional plants."

Foreman told Tait to review his book *Ecodefense* in order to get detailed instructions regarding sabotage of power lines. He said the preferred method for high voltage transmission towers and lines would be to unfasten the nuts from the bolts which connect the supports for the towers. At the end, Foreman promised Tait that he would see what he could do to get additional funding to purchase cutting torches and equipment.

Tait came back a week later and Foreman told him to meet with Nancy Zierenberg at the Earth First office, who would give him a check for $100 to be used for sabotage equipment. Foreman siphoning off money for illegal monkeywrenching.

The next day, May 14, Foreman met with Mark Davis and they talked about the significance of the attack on CAP as a practice run for the nukes. The CAP raid, it was decided, would take place. Ilse Asplund and Davis devised an alibi for Davis in the event things went wrong with the later nuclear plant attacks.

On May 30, Davis wrote a draft of a letter acknowledging the responsibility of EMETIC for sabotage of "nuclear energy installations" in the West, and got ready for the practice raid. On the evening of May 31, 1989, Mark Davis, Marc Baker, Peg Millett and Mike Tait stole up to CAP power tower number 40-1 near Salome, Arizona.

They cut the first leg of the tower when suddenly a flare caught them like klieg lights on a movie set. Fifty FBI agents arose from the brush all around them. They got Baker first, then Davis. Millet knew her way around the desert better and got away. As Susan Zakin wrote, "Peg didn't panic. The woo-woo kicked in—all her meditation, the dreams about her power animal, the raccoon. She sank herself into her surroundings, became the rocks, the clean-smelling creosote, the paloverde arcing over the wash." Millett got to a road, hitchhiked back to Prescott—a sixty mile trip—and went to work the next day, but FBI agents arrested her as soon as she got there.

That same morning, FBI agents raided Dave Foreman's home in Tucson and arrested him on conspiracy charges.

JANUARY 14, 1990   *Philadelphia, Pennsylvania*
THE ANIMAL LIBERATION FRONT BROKE INTO Dr. Adrian Morrison's office at the University of Pennsylvania and stole his personal correspondence, files, videotapes, slides and computer disks, and wrote slogans on the walls. His offense? He had testified on Dr. Taub's behalf nine years earlier—the Silver Spring Monkeys had by now been sent to Tulane University in Louisiana—he spoke up for the value of animal research, and—sin of sins—criticized PETA.

As the Philadelphia Inquirer reported of a PETA demonstration at Penn, "Morrison's office was broken into January 14. The militant Animal Liberation Front took credit. Yesterday, PETA released what it said was a 'preliminary examination' of copies of documents taken in the break-in. The review concluded that Morrison had written letters supporting other researchers and that he planned to oppose certain 'animal protective' legislation." [122] There was the nagging question about whether PETA had actually helped plan the raid, and not just reported it.

When *Village Voice* published a story on animal rightist objections to Morrison's research, PETA sent out copies of it to a number of people who lived near Morrison, along with a letter saying, "Please see the enclosed *Village Voice* cover story involving your neighbor Adrian Morrison who lives at [home address]," enabling readers to harass him.[123]

PETA's newsletter bragged: "Tired out from the ALF raid on his Penn lab, Adrian Morrison got a grant to spend a month in Italy this summer visiting fellow cat-electrode implanter Pier Permeggani at the University of Bologna. See how tight funds are? Sure there were sightseeing trips, expensive dinners and vivisection stories swapped, but the real highlight of Morrison's trip was being discovered, exposed and picketed by Animal Amnesty, a PETA contact group. Thanks to the activists, there really *is* no rest for the wicked."[124]

JANUARY, 1990   *Redwood Valley, California*

A HIPPIE KNOWN AS WALKING RAINBOW walked in one day and suggested to Judi Bari that the redwoods needed a mass movement, like the Civil Rights movement of the 1960s. She took the idea to the local environmental center and they formed a committee to organize a series of major rallies. Now Judi Bari was ready to come into her own as a major Earth First leader.

The campaign was first called "Mississippi Summer in the California Redwoods." On March 6, a typewritten flyer went out announcing "'Freedom Riders' needed to save the forest." The text, calling for "Freedom Riders for the Forest" to come from everywhere during the summer of 1990, capitalized as much as possible on 1960s civil rights imagery. It pointed out that two voter initiatives restricting timber cutting would be on the California ballot in the fall. They hoped to bring publicity to them and to "the timber companies' policy of exterminating the redwoods for short-lived profit."

The organizers would "maintain permanent encampments and waves of actions all summer long" to perform "non-violent civil disobedience." They would "provide housing and campsites, guides and support"— some in the form of getting people on the welfare rolls as soon as possible.

Earth First's biggest protest was in the works.

FEBRUARY 8, 1990   *Powell, Tennessee*

AN UNKNOWN ASSAILANT SHOT AND KILLED Dr. Hyram Kitchen, Dean of the Veterinary School of the University of Tennessee. Kitchen, 57, was shot eight times at 6:55 a.m. on Feb. 8, 1990, as he was leaving to have breakfast with a colleague. His body was found 63 feet from his car. He had run toward his home while trying to escape his attacker. Nine shots were fired, eight hit. The final two were fired point blank at the back of Kitchen's head as he sprawled helplessly in the driveway.

A government report stated that, "One month before the incident, a local police department issued an alert through the FBI's National Crime Information Center that various sources, including mail received by the

University of Tennessee indicated that animal rights extremists had threatened to assassinate a veterinary dean within the following twelve months. The Animal Liberation Front's website contains an entry stating, "2/21/90 - Knox County, TN; Report of threat made to assassinate one Dean of a veterinary school every month. Unclaimed." Others who read these reports have since asserted flatly that Dr. Kitchen was assassinated by animal rights activists. He was not.[125]

When I checked these reports with Detective Darryl Johnson of the Knox County Sheriff's Department, who was the investigator of the incident, he informed me of several errors. First, a single threat came after, not before the murder of Dr. Kitchen. Second, it was the University that received the threat, not police. Third, it was a telephone threat, not mailed, and the caller later recanted the threat as a prank. Fourth, the subsequent police alert was intended for internal law enforcement use only and not to accuse any person or group. Fifth, the case is still unsolved, but FBI agents are leaning toward the theory that the killing may have had to do with a possible love triangle. It was positively not an animals rights murder.[126]

APRIL 11, 1990   *Ukiah, California*

THE NORTHERN CALIFORNIA EARTH FIRST CONTINGENT renounced tree spiking. The decision had been made a month earlier, on March 4, at the environmental law conference in Eugene, Oregon where, five years later, I would be invited to expose the Green Cartel. A local mill worker named Gene Lawhorn told the environmentalists that he and his co-workers felt their lives were in danger from tree spiking. If the Judi Baris and Darryl Cherneys of Earth First were serious about building bridges to timber workers with their Wobbly alliance, they should renounce tree spiking.

Bari got up and said, "I agree with Gene."

Lawhorn says that someone went outside to tell Mike Roselle about the exchange. Roselle "had a heart attack"—less than a year earlier he had promised there would be an "unprecedented number of spiked trees" throughout the Pacific Northwest. But he calmed down and finally saw the merits, at least for the duration of Redwood Summer, the name chosen to replace the awkward Mississippi Summer in the California Redwoods. They would "renounce" tree spiking, but not "denounce" it.[127] It was all very tenuous; in fact, Roselle backed out of the agreement a year or so later and subsequently vigorously advocated tree spiking in the Earth First journal. More tree spikings have been done since this "renunciation" than took place before it.

The renunciation would at least help soften the nasty image Earth First got from a critical CBS News 60 Minutes piece broadcast March 4, in which Darryl Cherney said "If I knew I had a fatal disease, I would definitely do something like strap dynamite on myself and take out Grand Canyon Dam. Or maybe the Maxxam Building in Los Angeles after it's closed up for the night," (see p. 22).[128]

None of this pleased Dave Foreman, who, in any event, was preoccupied with his upcoming criminal trial. He had fundamental misgivings about civil disobedience as a form of political protest. Redwood Summer would not be about public land policy, but about state seizure of private land, mere leftist class struggle anti-corporate dogma. Foreman felt Earth First's actions should be directed toward government lands, not privately-held forests. Most of the five-hundred-mile-long strip of coast redwood forests is privately owned, dotted by 108 parks where cutting is off limits. Even Mike Roselle had misgivings, as he later told Susan Zakin: "I would have liked to see us focus on public lands a long time ago, but most of the activists have wanted to really concentrate on redwoods. This was a day of reckoning."[129]

APRIL 22, 1990   *Freedom, California*
EARTH FIRSTERS CALLING THEMSELVES THE EARTH NIGHT ACTION GROUP made two consecutive hits, sawing first through two of wooden power poles and then toppling a steel transmission tower belonging to Pacific Gas & Electric Company, causing a massive failure that cut off electricity at 1:37 a.m. to 100,000 Santa Cruz County residents for 10 to 18 hours. The area was still recovering from the devastation caused by the massive October 1989 earthquake.[130]

Rosina Mazzei of Santa Cruz, a victim of Lou Gehrig's Disease, nearly died when the outage cut off her respirator and her emergency power pack began to fail. Firefighters had to use hand respirators for hours before two registered nurses took over.[131]

The vandals sent a letter to the Bay City New Service and the Associated Press in San Francisco taking responsibility. The FBI investigated the incident as officially recognized domestic terrorism with ties to Earth First.

Mike Roselle's Direct Action Fund and Nomadic Action Group had developed a small corps of reliable activists that included Mike Jakubal, alias Doug Fir, who by 1990 was based out of a remote sixty acres of forested land near Concrete, Washington, in a cabin not much bigger than Ted Kaczynski's, without sewer or running water, but with a small electrical feed. Roving Earth Firsters used it as a staging base for travels and for shelter.

The FBI called Jakubal's core group the "heavies from up North" who could be counted on by Earth Firsters to do any kind of project. The FBI had a list of individuals suspected of making the trip to Santa Cruz as the Earth Night Action Group, but no case could be made and no charges were ever filed.

Among the individuals known by investigators to have visited Jakubal's cabin since 1990 were Erik Bracht, Seattle Earth Firster; Karen Coulter, the significant other of Asanté Riverwind (alias of Charles N. Christensen); Tim Dimock, Seattle Earth Firster; Bill Haskins of the Ecology Center in Missoula, Montana; Elizabeth Loudon, Seattle Greenpeace staff; Jeff McDuff, thought to be an explosives expert; Abe Ringel, unidentified; Argon Steele of Olympia, Washington; Felicia Sue Staub, a Seattle-based administrative director for SANE/FREEZE and Earth Firster; Tony Van Gessel, a tree sitter and one-time roommate of Mitch Friedman; and Greg Winegard, Earth First activist and televised flag burner.

Karen Debraal, a Santa Cruz Earth First contact, endorsed the Earth Night action. Darryl Cherney also endorsed the action. It didn't wash. Even local subversives sympathetic to Earth First deplored the hit as having no point.[132]

Nearly a month earlier, Cherney had circulated an Earth Day poster showing two Earth Firsters hefting monkeywrenches with a bulldozer in the background. "EARTH NIGHT 1990," the poster said. "GO OUT AND DO SOMETHING FOR THE EARTH ... AT NIGHT." Cherney was most likely unaware of the clandestine Nomadic Action Group; although he was highly visible as a spokesman for Earth First, he was not a confidant of either Dave Foreman or Mike Roselle. There was much about Earth First he did not know. Cherney certainly took no part in the power line vandalism; all known NAG participants disdained him as a klutz. But the incident reflected an emerging pattern: Earth First publication recommends illegal action, anonymous perpetrators soon commit recommended crimes.

Cherney was picked up by police on Earth Day 1990 for questioning about an Earth First banner hung from the Golden Gate Bridge.[133]

11:52 A.M. THURSDAY, MAY 24, 1990   *Oakland, California*
SHANNON MARR DROVE HER DATSUN DOWN PARK BOULEVARD, guiding Judi Bari and Darryl Cherney in the white Subaru station wagon behind her to a copy shop. They didn't know their way around Oakland, so Marr was helping them. They had just left David Kemnitzer's house on East 23rd Street, where Bari and Kemnitzer wrote a grant proposal for a non-violence training camp. After making copies, Bari and Cherney would drive together to the college town of Santa Cruz, where a Redwood Summer Roadshow recruiting event had been scheduled.[134]

Marr and Cherney had earlier that morning come from the Berkeley headquarters of Seeds of Peace, a logistical support group for civil disobedience actions at the Nevada Test Site and for the Great Peace March across America in 1987. One of their hot-shot organizers, James McGuinness, could reliably get food and water to 1,500 protesters a day in the most chaotic situations. If somebody drew the crowds, Seeds of Peace could handle them. Redwood Summer looked like it would draw about 3,000 protesters.

Last night Bari and Cherney had driven with folksinger Utah Phillips from a rally in Ukiah to the Seeds of Peace house on California Street in Berkeley. Cherney submitted a two-page proposal to Marr and Kemnitzer outlining how their non-violence training would be structured.[135]

Earth First had approached Kemnitzer asking for Seeds of Peace help in April when he attended a Laytonville meeting. Seeds of Peace was concerned about Earth First's reputation of violence and sabotage. On May 17, negotiations convinced them: Seeds of Peace officially signed on as a co-sponsor of Redwood Summer. Cherney's non-violence training proposal was a necessary followup. Earth First could now write a grant proposal.

Utah Phillips departed about 10 p.m. for his Nevada City home. Cherney crashed for the night at the Berkeley house while Bari followed Kemnitzer to his house in Oakland and stayed there.

When Marr and Cherney arrived in Oakland about 9:30 the next morning, the grant proposal was done. Then Judi got out her violin and Darryl picked up his guitar and they sang a Redwood Summer song for their mentors. Then there were copies of the grant proposal to make and they'd be on their way to Santa Cruz. Bari and Cherney piled a violin case, a couple of guitar cases, and a blue duffel into the Subaru's back seat. Cherney tossed in his camouflage rucksack.

As they drove northbound on Park Boulevard, a couple of minutes after leaving Kemnitzer's house, Marr heard a pop like a large firecracker. She saw Judi's Subaru whiz past her, trailing smoke. It crashed into a pole at the west curb at the corner of MacArthur Boulevard.

Marr pulled over and ran to Bari's car, smoke and sulfurous smell reeking out. Bari said her back hurt and Cherney had facial cuts. Cherney told Marr to take his camouflage bag out of Bari's car because he would need it. Marr put the bag in her car and waited for police and fire crews to respond to the scene.

Allied Ambulance Company Paramedic Sal Taormina treated Bari, who told him, "A bomb went off in the car." Allied Paramedic Brian Buckman treated Cherney who said, "We are political activists with Earth First and they threw a bomb at us."

Judi Bari was seriously injured, her pelvis shattered, her internal organs mangled, her right foot paralyzed from the ankle down. She underwent surgery for her fractured pelvis and was listed in serious but stable condition. She was hospitalized for seven weeks. Darryl Cherney escaped with minor injuries.

The responding Oakland Police Department officer, Sergeant Michael Sitterud, along with FBI Special Agent Frank Doyle Jr. and his FBI crew, examined the blown up Subaru while Bari and Cherney were being taken to the hospital. Oakland Police Sergeant Robert A. Chenault responded to the hospital where he questioned Bari and Cherney.

Doyle found the components of a pipe bomb in the wreckage including a battery, a mechanical watch, electrical wires, pieces of a pipe nipple measuring approximately 2 inches by 12 inches having been capped at both ends and filled with a low explosive filler. Doyle also observed numerous nails bound together by silver duct tape for shrapnel effect. A separate bag of nails found in the Subaru appeared to be identical to those taped around the pipe bomb. The nails were sent to FBI Special Agent David R. Williams of the Explosives Unit for laboratory analysis.

Doyle and Sitterud concluded that the bomb had been on the floorboard behind the driver's seat when it exploded. Doyle specifically stated that his conclusion regarding the location was based on his observation of a large hole in the rear seat floorboard immediately behind the driver's seat and the debris pattern in the roadway and inside the Subaru.

At Highland Hospital, Cherney repeated to Sergeant Chenault what he had told the paramedic and added that it was an assassination attempt. Whatever Cherney told Chenault about the incident itself wasn't consistent with the physical evidence Sitterud found at the scene. That bag of nails was pretty suspicious. And Cherney's insistence that they were political activists with Earth First evoked the same reputation among the Oakland police as it had among the Seeds of Peace.

The FBI briefed the Oakland police on the EMETIC case in Arizona, the recent Santa Cruz power line vandalism and other Earth First-related incidents.

At 2:21 a.m. Friday morning, Sergeant Chenault sought a search warrant for Bari's home, writing in his affidavit, "Affiant believes that Bari and Cherney are members of a violent terrorist group involved in the manufacture and placing of explosive devices. Affiant also believes that Bari and Cherney were transporting an explosive device in their vehicle when the device exploded." They believed the bomb had been manufactured at Bari's home in Redwood Valley. They got their search warrant.

At 5:40 a.m. Chenault and Sitterud with Mendocino County Sheriff's deputies served the warrant and seized an inventory of six items, including a box of finishing nails, two pieces of duct tape, a copy of Foreman's *Ecodefense*, Judi Bari's calendar and a ceramic light socket ceiling unit—significant, because the test circuit of many homemade bombs includes a light socket.

Bari and Cherney were booked Friday for investigation of state explosives laws and held in lieu of $100,000 bail. The District Attorney had until the next Tuesday to file charges if the two were to remain in custody.[136]

Mike Roselle prevailed upon his Greenpeace colleagues to hire a private investigator. They offered to pay the expenses of Sheila O'Donnell of Ace Investigations. O'Donnell went to work with Earth First attorney Susan Jordan.[137]

Four days after the bomb went off in Judi Bari's car, a strange letter arrived at the Santa Rosa Press Democrat's Ukiah News Bureau. It took responsibility for the Judi Bari bombing and for the May 9 bombing of an office at the Louisiana-Pacific mill in Cloverdale, a little town between Santa Rosa and Ukiah on Highway 101.

The Louisiana-Pacific bombing had failed. The bomb had been left on the front porch of the roadside office, where it was supposed to ignite a one-gallon can of oil-gasoline mixture. When the incendiary device exploded at 4:10 a.m., the attached gas can failed to ignite. The damage was minimal. A sign saying, "L-P Screws Millworkers" was left at the scene.

The three page letter explaining all this was signed, "THE LORD'S AVENGER." It is a not very skillful masquerade designed to look like the ravings of a religious fanatic, replete with homiletic cadences, pertinent biblical quotations, randomly capitalized words, obvious misspellings, excessive metaphors and the other usual baggage of the religious zealot.

It was addressed to Mr. [Mike] Geniella, Press Democrat, 215 W. Standley St., Ukiah, CALIF 95482. Whoever wrote it knew exactly which investigative reporter covered the timber beat for the Press Democrat, and the exact street address of his local office—the Press Democrat is a Santa Rosa newspaper, based fifty miles down Highway 101 from Ukiah. It was mailed in a Number 12 envelope postmarked 1:18 PM, 29 MAY 1990, NORTH BAY, CA 949, which tells us it was put in a drop box somewhere between Tiburon and San Rafael. It had one 25-cent denomination Jack London stamp.

The letter begins: "I built with these Hands the bomb that I placed in the car of Judi Bari. Doubt me not for I will tell you the design and materials such as only I will Know." Very coherent and to the point.

The letter's appearance is also neat and professional looking: the biblical quotations are each set apart from the text around them and indented exactly one-and-one-half inches, like the product of a typesetter.

The concerns expressed by the Avenger are a little peculiar:

She [Judi Bari] spoke Satan's Words yet the Lord did not strike her Down. Darkness fell upon my Spirit. My prayers sought Guidance so I would know if the Lord was calling me to Wield His Sword. I could hear no Answer but Satan marched on and caused great Uproar over the land. This Woman Possessed of the Devil set herself on the Honest men of toil who do Gods work to bring Forth the bounty that He has given us to Take. All the forests that grow and all the wild creatures within them are a gift to Man that he shall use freely with God's Blessing to build the Kingdom of God on Earth. They shall be never ending because God will provide.

And God said, Let us make man in our image, after our likeness; and let them have dominion of the fish of the sea, and over the fowl of the air, and over the cattle, and over every creeping thing that creepeth upon the earth.

Genesis 1:26

All of it is God's Gift for us to Take and use so that we can build our Civilization in the Image of the Creator. The Devil is sorely displeased by our Godly Dominion and he sends his demons to sow Confusion and Doubt in our numbers. This possessed demon Judi Bari spread her Poison to tell the Multitude that trees were not God's Gift to Man but that Trees were themselves gods and it was a Sin to cut them. My Spirit ached as the Paganism festered before mine eyes. I felt the Power of the Lord stir within my Heart and I knew I had been Chosen to strike down this Demon.

The perceptive will raise an eyebrow at that greedily capitalized word "Take" tacked on in two unnecessary places. The idea of building "our Civilization in the Image of the Creator" is likewise odd for a religious zealot—for one thing, religious analysts tell me, to do so would be worldliness if not outright idolatry, and for another, *Strong's Exhaustive Concordance of the Bible* shows that neither the word nor the concept of "civilization" appears anywhere in the Bible. Likewise, I could find no pastors of any denomination who preached in defense of civilization in modern terms. These two items smack of an environmentalist's parody on the wise use philosophy rather than fundamentalist religious fervor. Almost everyone who has seen the typescript of the letter, including partisans of Judi Bari, believes the religious tone is a false distraction.

The real puzzler is why a religious zealot claiming to have blown up an environmentalist would use a Jack London stamp. Environmentalists revere Jack London; religious zealots don't know who he was.

Santa Rosa Press Democrat reporter Mike Geniella gave the letter to FBI Special Agent John Raikes, who gave it to Special Agent David R. Williams for laboratory analysis. Raikes asked that the newspaper excise the technical description of the two bombs. Here's what the public didn't see. First, the Louisiana-Pacific Bomb, which, the Avenger tells us, was intended to make people think Judi Bari had planted it because of her long-running feud with L-P:

> The bomb was 1½ inch galvanized pipe with galvanized end caps candle wax on threads. One cap drilled so wires could go to the igniter match heads inside a model paint bottle to be set off by flashlight filament. Epoxy glue in the drill hole. The pipe was Set in a plywood box with a one gallon gas can filled with 70:30 mixture gas & oil. On top of the box was timer-pocket watch with minute hand gone and small hex head screw drilled into the lense (sic). Battery was 9 Volt. A Light switch for safety and light socket for Test lamp. And the bomb was placed and the hour hand touched the screw and the bomb exploded.

In the next paragraph, The Lord's Avenger wrote how the Judi

Bari bomb was constructed:

> It was 2 inch galvanized pipe 11 inches long with black iron end Caps one drilled for wires to go to the matchhead Igniter Bottle. Epoxy glue in the drill hole. Explosive: 3:1 potassium chlorate : aluminum powder. 3 sizes of finishing nails taped on the Outside and pipe taped to a piece of paneling that fit under her seat and the paneling had the pocket watch and also a motion switch of 2 bent wires and a ball. 9 volt battery again and safety switch and Test socket empty. Ticking of the watch was Silenced by a piece of yellow sponge and the whole Device covered with blue towel. The wires were red and black and well Soldered to the Battery. And the Bomb was Hidden and the hour hand Moved.

The Avenger claimed to have put the bomb in Bari's car "whilst she was at the meeting with the loggers," which would have been May 22, two days before she followed Shannon Marr to the copy store in Oakland. The Avenger surmised that the bomb failed to go off at the expected time because of a hangup in the motion sensor or the timer, but when Cherney got in the car with Bari in Oakland, something made it start ticking again and it blew up at 11:53 a.m. on May 24, 1990.

FBI Special Agent David R. Williams, after analyzing the Avenger letter, said that the letter writer "either did build the bombs or knew how they were built."

The FBI internal comparison report shows these facts: both the Louisiana-Pacific bomb and the Judi Bari bomb used a Bull's Eye pocketwatch, manufactured by Westclox, as a time delay mechanism and soldered wire to the watch frame. Both bombs used a pipe nipple and two end caps, filled with the same low explosive mixture. Both bombs used a ceramic lamp base manufactured by the Leviton Corporation and a wall-type switch as a test circuit and safe arm switch. Both bombs used three different colored insulation (red, green and black) 7-strand copper wire. Both bombs used a right hand twist wire connection and/or solder. Both bombs used a Duracell 9 volt battery with a stamped date of January 1993 as a power source. Both bombs used 2-inch wide gray duct tape and ¾-inch-wide black plastic tape.

The differences were matters of accessories: only the Bari bomb had nails wrapped around it for shrapnel effect; only the Bari bomb had a motion sensor. Only the L-P bomb had an incendiary device.

Oakland Police Department officers Sitterud and Chenault believed that Bari or her colleagues built the bombs. They obtained a second search warrant for her home on July 5, almost six weeks after the Subaru blast because they had not searched all of the premises the first time, failing to grasp the nature of the living arrangements Bari had on the site with her former husband. They looked for the bomb components listed above and for typewriter exemplars to check out the Avenger letter. The 73 items on the second seizure inventory included enough to keep their suspicions up: duct tape, yellow sponge foam, pipe with solder, nine types of finishing nails, blue towels, many type exemplars, black electrical tape, and so on.

As the details evolved, the evidence got thinner: none of Bari's type exemplars matched the Avenger letter and police could not single out one of her many friends who could have typed the letter on their machines; the blue towels didn't match; the tape, the solder, and so on didn't match. The nails in the bomb and Bari's car came from the same manufacturing batch, but the batch was so large and distributed to so many retail outlets that it proved nothing about a single suspect. The Oakland Police Department's case evaporated and the Alameda County District Attorney refused to file charges: "Based on the information presented, we will not file charges," Deputy District Attorney Chris Carpenter said. "The evidence . . . is insufficient to secure a conviction."[138]

Who bombed Judi Bari? The case is still unsolved.

Private investigator/journalist David Helvarg and Stephen Talbot made a TV documentary, *Who Bombed Judi Bari?*, that came up with a list of suspects, none very convincing. One was Irv Sutley, who had once taken a photo of Bari posing with his Uzi, supposedly for an album cover: Bari accused him of being a police informer. Then there was ex-linebacker anti-abortionist Bill Staley, thought by some to have written the Lord's Avenger letter—but he probably didn't possess the typographical skill for such a tidy letter. Helvarg and Talbot also included a clip of logger Steve Okerstrom culled from three hours of interview, saying, "If a logger had bombed Judi Bari, she wouldn't be talking about it." Helvarg and Talbot had assured Okerstrom they would allow him to review the film and remove such offhand comments in favor of more substantive remarks, but they broke their promise. The documentary also included Michael Sweeney, Judi Bari's ex-husband, as a suspect, which drove Bari into a rage. Zakin says Bari accused Talbot of ignoring the political implications of her bombing in favor of a sordid domestic violence squabble, and wanted the section cut. Bari fumed that the timber companies and right-wingers of Mendocino County were not on the suspect list. Talbot refused to change his documentary, which had been designed to absolve Bari and Cherney of any guilt in making or knowingly carrying the bomb. San Francisco public television station KQED broadcast it in the spring of 1991, and it did the job.

Today, virtually nobody believes that Bari or Cherney had anything to do with the bomb. After examining the police and FBI reports, I am satisfied that initial law enforcement suspicions were justified, but were based on severely limited background knowledge about Bari and Cherney. Even though the physical evidence still points most strongly to Bari and Cherney and to no other identifiable suspect, I am convinced they had nothing to do with the bomb. In addition, I am convinced that it was not the work of anyone else in a known Earth First chapter, ad hoc decoupling group or the Nomadic Action Group: pipe bombs at the time were outside monkeywrenching modus operandi, although after 1992 they became increasingly common around Eugene, Oregon, used by unknown assailants against a Sony factory, a Hyundai plant and the Associated Oregon Loggers office.[139]

Theories about who really did it abound. Generally, left-wingers blame right-wingers and vice versa. The most far-fetched theories so far are 1) the FBI did it; 2) John Campbell of Pacific Lumber did it; 3) An industry-paid infiltrator disgruntled at both industry and Earth First did it to throw both sides into confusion; 4) Dave Foreman did it; 5) the Nomadic Action Group, the heavies from up north, did it to create martyrs for Redwood Summer; and 6) marijuana growers did it because Bari invited thousands of strangers into their growing area, endangering the crops by possible detection, theft or accidental trampling. In other words, nobody has a clue. Who Bombed Judi Bari has become the trivial pursuit of long rainy North Coast nights.

In 1992, Bari was granted permission by U.S. District Judge Eugene Lynch to file a civil suit in federal court against the FBI and Oakland Police, alleging they had violated her civil rights, specifically that the FBI engaged in conspiracy, false arrest, illegal searches and falsely portrayed Bari and Cherney as responsible for the explosion.[140]

Earth First also convinced Rep. Don Edwards (D-CA) to run a congressional inquiry by the House Judiciary Committee's subcommittee of civil and constitutional rights. Congressman Edwards requested access to the FBI files on the bombing to determine "the nature and extent of the FBI's investigative interest in environmental activists." Edwards received an oral briefing, but in late 1993 the FBI refused to release the documents, effectively ending the probe.[141]

Susan Zakin says that Edwards was convinced to take on his inquiry because Mike Roselle's new girlfriend was Claire Greensfelder, a Democratic party activist who convinced Congressman Ron Dellums (D-CA) to hold a press conference calling for an inquiry into the FBI's handling of the case. Dellums followed up with a letter to Edwards.[142]

TUESDAY, AUGUST 14, 1990   *Tucson, Arizona*
DAVE FOREMAN QUIT EARTH FIRST. He told a reporter, "Essentially, what's happened is there's a lot of class-struggle rhetoric and focusing on evil corporations and that sort of thing, and pulling in a lot of social-justice issues ... whereas I come from the conservation movement and have felt that Earth First! is primarily a wilderness-preservation group, not a class-struggle sort of group."

Mike Roselle, having nursed a long list of grievances against Foreman and seeing the whole Earth First apparatus suddenly falling into his hands, said, "We don't need Foreman in Earth First! if he's going to be an unrepentant right-wing thug."

Foreman said that although he considers himself a conservative, he is not a "right-wing thug." "I'm still a conservative in that I don't like to see unthinking change occur, and also I'm a patriotic American, and I think the American flag symbolizes a great deal of my values. I was very upset, for example, at the Earth First! rendezvous last year in which some nut burned the flag. That doesn't set well with me at all."

Rather than have his energy drained by "internecine warfare," he said, "we figured it was best for everybody if there was sort of a no-fault divorce."[143]

Within days, the staff of the Earth First Journal in Tucson resigned en masse over political disputes and the Journal moved to Missoula, Montana, where it remained for a year or two and then moved to Eugene, Oregon.

The Earth First that Dave Foreman had created ceased to exist. Thereafter, Earth First was a class struggle group. Its agenda was pure millenarianism. Earth First Journal was now run by the social justice faction alone. But the monkeywrenching didn't stop. It got worse.

REDWOOD SUMMER   *Northern California*
MORE THAN 2,000 PROTESTERS PARTICIPATED IN REDWOOD SUMMER without Judi Bari. By Labor Day over 150 had been arrested for criminal trespass. In the interim, they got into scuffles, they made media, but mostly they terrorized the neighbors of Pacific Lumber.

Hundreds of people lived on the boundaries of the private forests that Earth First wanted seized by the government. The hundreds of protesters who occupied Pacific Lumber and other private forest land every day and every night felt no compunction about urinating and defecating near peoples' wells. Seeds of Peace made sure the protesters were fed but not potty trained. After a few weeks, the massive amounts human waste on the ground began to seep into the water table and made the neighbors' well water unsafe. Neighbors went out and took pictures across their property lines of human feces the protesters left randomly scattered on the forest soils. Drinking water had to be trucked in.

The protesters were thieves as well. They stripped the gardens of innocent neighbors, stealing all the food and trampling stalks and vines. Petty burglaries of cash and household items became endemic. Some residents packed up and made other living arrangements for the duration.

One woman who asked not to be identified was so terrified for the safety of her young daughter that she asked her ex-husband to take the girl until things quieted down, prompting a subsequent custody battle.

The welfare rolls of Humboldt and Mendocino Counties bulged with Earth Firsters while their law enforcement budgets were strained nearly to bankruptcy.

JUNE 10, 1991   *Corvallis, Oregon*

THE ANIMAL LIBERATION FRONT OPENED "OPERATION BITE BACK" by setting fire to Oregon State University's mink farm using a timed incendiary device. At the same time, ALF raiders burglarized and damaged associated research offices and spraypainted threats on walls. Later that day anonymous callers contacted the Associated Press in Portland and television stations KATU and KOIN to direct the media to find ALF press releases and videotapes dropped nearby. In the press release, ALF threatened to continue "until the last fur farm is burnt to the ground."[144]

Witnesses recalled seeing a female and a male similar in appearance to Rodney Coronado acting suspiciously in the vicinity of the attack immediately before the blaze.

Five days later, in Edmonds, Washington, ALF firebombed the Northwest Farm Food Cooperative with the same type of incendiary device used in Corvallis. NWFFC provided animal feed to fur breeders throughout the Northwest and had provided financial support to OSU's mink research. ALF issued a press release stating that NWFFC was targeted because of its association with OSU.[145]

Rod Coronado issued a press release on the first of August on behalf of the Coalition Against Fur Farms, using his alias Jim Perez. In it he recounted the OSU and NWFFC ALF actions and stated that they were "crimes of compassion that every animal advocate should support."

On August 12, ALF burglarized and vandalized a fur animal research facility belonging to the Washington State University at Pullman, Washington. ALF issued a press release from the Kinko's copy store in Moscow, Idaho justifying the action. "Until coyotes, and other animals live free from the torturous [*sic*] hand of humankind, no industry or individual is safe from the rising tide of fur animal liberation." The note also threatened scientists engaged in animal research: "Davis Prieur, John Gorham, Fred Gilbert, David Shen, William Foreyt and Mark Robinson, beware. ALF is watching and there is no place to hide." Coronado had admitted to being one of the three individuals who composed and sent these press releases. Coronado was with two women "house sitting" in Pullman during the time of the raid.

After this raid, Rod Coronado was interviewed by Portland, Oregon television station KGW. He identified himself only as someone who was "no stranger to the Animal Liberation Front," acknowledged that he had participated in ALF actions in the past, and did not deny taking part in the WSU ALF action. When asked if he would break the law in the future, Coronado answered, "We have already broken the law, why not do it again."

AUGUST 13, 1991   *Phoenix, Arizona*

DAVE FOREMAN SIGNED HIS PLEA AGREEMENT, guilty of felony conspiracy, for two overt acts: providing money for illegal monkeywrenching, and providing instructions about specific sections of his book *Ecodefense* for illegal monkeywrenching.

It was the unexpected end of a trial that was going sour. Wyoming trial attorney Gerry Spence, the legendary Cowboy Lawyer, had taken Foreman's case, which was joined to the cases of the Prescott four. Spence had never lost. He was the guy who won $1.8 million in damages from Kerr-McGee for the suspicious death of plutonium-plant whistleblower Karen Silkwood. He was joined on the defense team by several others, including a court-appointed attorney named Skip Donau.

The grand jury had delivered three sets of indictments, the original that did not include Ilse Asplund, a First Superseding Indictment that added Asplund and more overt acts, and a Second Superseding Indictment that was devastatingly detailed. They didn't miss a thing. The charges were very specific and hard to answer.

Then, too, the defense had not done well in jury selection. The people impaneled were all real rednecks, not the Dave Foreman-style imitation. Not a sympathetic face in all the twelve, much less any woo-woo types to take pity on Peg Millett. Gerry Spence's usual theatrics didn't work. He bellowed and snorted that Dave Foreman was framed by the FBI and his arrest had been part of a government attempt to squelch the radical environmental movement. The jury didn't seem to think that was a bad thing, the way Spence read them. The entrapment argument was so weak it only convinced friends and supporters of Dave Foreman, bolstered by a single offhand remark Mike Fain had made on tape to FBI colleagues about Foreman being the one they needed to "pop" in order "to send a message." With Foreman siphoning money from Earth First for illegal monkeywrenching, it wasn't a hard message to send. After three months, the prosecution was only half finished presenting its case. The media were ignoring what the defense had hoped to turn into the environmental trial of the century. It didn't look good.

Just before a scheduled one-week recess, Donau suggested that the defendants enter a plea agreement. He thought the jury might acquit Foreman, but the others had no chance, and it wasn't a sure thing for Foreman either. After agonizing over it, all five told Donau to go ahead and send the idea to the prosecution. They bought it.

Foreman pled guilty to felony conspiracy, including all the overt acts described above, with no jail time and sentence deferred for five years— if he obeyed probation rules, he could then withdraw the felony plea for a misdemeanor plea, which he signed at the same time as the felony plea. Mark Davis got six years; Peg Millet got three years. Marc Baker got six months. Ilse Asplund got a one-month sentence.

When it was all over, Foreman and a few friends founded the Cenozoic Society and began to publish a periodical called Wild Earth. Then Foreman occupied himself with the North American Wilderness Recovery

Project, the Wildlands Project, his plan to depopulate and re-wild America, with the help of John Davis, Mitch Friedman, Bill Devall and Reed Noss.

AUGUST 28, 1991   *Sandy, Utah*
THE ANIMAL LIBERATION FRONT DESTROYED an office and spraypainted graffiti on the walls of the Fur Breeders Agriculture Co-op. An incendiary device left behind failed to detonate. No press release followed.

Rod Coronado rented a storage locker in Talent, a little Southern Oregon town on Interstate 5 between Medford and Ashland. In it he kept a typewriter, on which he wrote a report to colleagues and funding sources saying, "LARGEST ... largest fur processors in Montana. After my investigation I discovered that all the fur farmers in Montana used the same company to prepare pelts for auction. The Huggan's Rocky Mountain Fur Company is a building I have been in before. It is all wood, with no alarms and no close proximity to animals. That targeted building contains all the drying racks, and drums used in pelt processing. If we could cause substantial damage to that equipment, we would cause a serious disruption in the pelting season, and also push the Huggan's family (third generation trappers) into a position closer to bankruptcy." Coronado went on to explain that this action could also prevent consumers from buying fur products "for fear of ALF." He also stated that if he could obtain funds, he would mount other attacks "against the fur farm industry this winter."[146]

In fact, Coronado had been there in late 1990, using his alias Jim Perez and posing as a fur buyer. While there, he photocopied a list of addresses for all fur breeders in the Northwest. During that winter of 1990, Coronado moved to Southern Oregon and, with local activists Kimberly Trimiew and Deborah Stout, started the Coalition Against Fur Farms. CAFF shared a Medford, Oregon post office box with the Southern Oregon Hunt Saboteurs. CAFF was to be a decoupling group for ALF, with the goal of destroying all fur farms.

In late 1991, Coronado rented a little cabin for $50 a month across the road from the Trillium Farm on the Little Applegate River, where Earth Firster Chant Thomas had once helped Dave Foreman plan and execute the Bald Mountain road protest. The Trillium community offered to help Coronado with mailings for his work with the Coalition Against Fur Farms.

Coronado's funding sources came through. He was in and out of the little cabin all winter. On December 12, in Hamilton, Montana, ALF burglarized the Huggans Rocky Mountain Fur Farm, but were discovered and fled before doing further damage. That was a disappointment. The next raid had to be done better.

On December 21, ALF burned the privately owned Malecky Mink Ranch in Yamhill, Oregon to the ground. According to telephone records, Rod Coronado called television station KGW stating that he was a member of ALF and reported that the Malecky farm had just been burned. The

ALF accepted responsibility for the Malecky crime in a February 2, 1992 article in Earth First! Journal.

On February 28, 1992, in East Lansing, Michigan ALF burglarized, vandalized and firebombed the Michigan State University offices of Dr. Richard Aulerich. They tore his office to pieces, dumping out files and destroying computer equipment. Then they set the firebomb, a simple arrangement of a Sterno can, light bulb, cube of fire paste, accelerator and lighting material. The blaze also destroyed the adjacent office of Dr. Karen Chou. Two students were inside Anthony Hall when the firebomb detonated.

ALF then went to the mink facilities operated by MSU, pried open the roof to get into the main building, and destroyed the field office with sulfuric acid, melting data logs, feed-preparation machines and gas chambers for killing mink. They then went through the mink shed, released 350 animals, removed experiment cards to destroy work in progress.

ALF then spraypainted "Fur Is Murder" and "Aulerich Tortures Minks" and "ALF" on the walls. Total damage of the raid was set at about $125,000. The lost research had helped define the impact of toxins of wild mink in the Great Lakes in an effort to preserve the health of wild strains.

The week before the raid, Rod Coronado, along with Kimberly Trimiew and Deborah Stout, stayed at the Midland, Michigan home of Deb's parents. Rev. David Stout, a United Methodist minister, and his wife said that the three stayed there for a few days. On the night before the MSU raid, Stout and Coronado checked into a hotel in Ann Arbor.

The day before the raid, Coronado sent a package by Federal Express to Ingrid Newkirk, leader of People for the Ethical Treatment of Animals. Newkirk had arranged—days before the MSU arson attack—with long-time PETA member Maria Blanton of Bethesda, Maryland, to receive the package and then give it to her. Coronado used a fabricated name and address for the sender—Leonard Robideau, 2771 Tecumseh Ave., Toledo, Ohio 92138—and paid with an invalid credit card number to send the package.

A second package was sent immediately after the raid. It contained slides, documents and computer disks stolen from Dr. Aulerich during the MSU raid along with a Hi-8 video of a perpetrator wearing a ski mask taking a mink from the MSU fur farm and holding up the severed head of an otter. This package was also addressed to Maria Blanton from Leonard Robideau. It was sent from a drop box adjacent to the Ann Arbor hotel where Coronado had rented a room. The handwriting on the freight bill was Coronado's. He acknowledges sending the package. But Fed Ex employees intercepted it when they discovered the previous fraud.

A search warrant was executed at the home of Maria Blanton. Records found during the search of Blanton's home showed that Coronado and Alex Pacheco, co-founder of PETA, had planned a burglary at Tulane

University's Primate Research Center in 1990, where the Silver Spring Monkeys had been housed. The records seized included surveillance logs; code names for Coronado, Pacheco and others; burglary tools; two-way radios; night vision goggles; phony identification for Coronado and Pacheco; and animal euthanasia drugs. The raid was never made because the monkeys were sent elsewhere immediately before the raid was scheduled.

The press release publicizing the MSU attack came from People for the Ethical Treatment of Animals. PETA announced that it was acting as a media conduit for the ALF and stated that the ALF took its action in order to end MSU's animal research.

Various grand juries around the country were investigating the Operation Back Bite crimes. When called upon to testify, Jonathan Paul, James "Rik" Scarce, Deb Stout and Kim Trimiew refused, and were jailed to compel their testimony. All were finally released after judges felt that longer incarceration would not change their minds. In July, 1993, the grand jury sitting in the Western District of Michigan indicted Coronado for his role in Operation Back Bite.

Coronado had gone underground more than a year earlier, in April of 1992, vanishing from the Sea Shepherd office where he worked in Los Angeles. He hid out as a Native American, bouncing around from reservation to reservation, steering clear of his animal rights contacts.

In July 1993 he went to the little 222-acre Pascua Yaqui Reservation at 7474 South Camino De Oeste in Tucson, using the alias Martin Rubio—Rubio was his mother's maiden name, known among some Yaqui— and staying with Don Anselmo Valencia, tribal chief and council member. The reservation had been created in 1978 and had only 615 residents. Coronado told the old man he was interested in learning his culture, in coming home. Coronado became helpful in many ways. He was invited to join the Yaqui Coyote dancers, the Coyote Society being entrusted to protect the tribe from outside influences such as alcohol, and to mete out punishments. Coronado was good at working with such afflicted people, trying to straighten them out.

But someone turned him in. Prosecutors will not say who. Coronado's photograph was plastered all over every post office in America on a big Wanted by ATF poster. On September 28, 1994, he was lured to the tribal fire station by a tribal cop claiming there was a wounded hawk to look after. The feds busted him.

On March 3, 1995, Coronado pled guilty to the MSU arson. He admitted assisting others but not starting the fire himself, but it did not avoid the full criminal responsibility for the offense, as aiders and abettors are punished precisely as the one who personally commits a crime. He was sentenced to four years and nine months in prison in August 1995.[147]

The fur raids stopped for a while, but resumed in 1995 and accelerated in 1996.

FEBRUARY, 1994   *Eugene, Oregon*
JUDI BARI WROTE AN ARTICLE TITLED "MONKEYWRENCHING" for the Earth First
Journal. It was a recommendation that Earth First perform the classic
decoupling maneuver separating their above-ground group from underground groups for the sake of gaining legitimacy. Monkeywrenching, she
said, had helped to "isolate and discredit our movement, and drive away
some of our best activists." While strongly emphasizing that "Direct action does not just mean demonstrations. It means action at the point of
production, designed to stop or slow production," Bari warned that "mix-
ing civil disobedience and monkeywrenching is suicidal." She was saying
that she clearly knew that Earth First had been monkeywrenching.

> England Earth First! has been taking some necessary steps
> to separate above ground and clandestine activities. Earth First!,
> the public group, has a non-violence code and does civil disobedi-
> ence blockades. Monkeywrenching is done by Earth Liberation
> Front (ELF). Although Earth First! may sympathize with the activi-
> ties of ELF, they do not engage in them.
>   If we are serious about our movement in the US, we will
> do the same. Earth First! is already an above ground group. We
> have above-ground publications, public events, and a yearly Ren-
> dezvous with open attendance. Civil disobedience and sabotage
> are both powerful tactics in our movement. For the survival of
> both, it's time to leave the night work to the elves in the woods.[148]

APRIL 22, 1996   *San Francisco, California*
MEMBERS OF THE SIERRA CLUB VOTED to support the end of commercial
logging in national forests. The initiative measure was forced onto the
Club ballot by dissidents calling themselves the John Muir Sierrans, including Chad Hanson of Eugene, Oregon, and carried by a 2 to 1 margin
of those who voted.
    In the spring of 1995, Dave Foreman and David Brower had been
elected to the Board of Directors of the Sierra Club. To the surprise of
most Sierra Clubbers, Dave Foreman opposed the logging ban initiative.[149]

1995-1996   *North America*
ANIMAL RIGHTS VANDALS INCREASED THEIR ACTIVITY on several fronts. Their
attacks on restaurants and fast food outlets increased, penetrating security
and damaging McDonalds and Burger King facilities so extensively the
corporate managements will not allow personnel to discuss it. The Animal Liberation Front website, Diary of Actions, 1996, included this sample
of restaurant and food store attacks:

> **3/15/96**-Syracuse, NY; Hickory House BBQ's store front was
> paint bombed, 5 picture windows smashed, all sides of the
> building were covered in A.L.F. slogans, all locks filled with

super glue, 3 large neon signs destroyed. The store remained closed for at least 2 days. -A.L.F.

3/17/96-Cicero, NY; Plainsville Turkey Farms Restaurant - 5 picture windows completely smashed, all locks filled with super glue, all sides of the building spray painted with A.L.F. slogans, front sign paint bombed. This action received intense media coverage. -A.L.F.

6/11/96-Huntington Beach, CA; Two burger restaurants, Bun & Burger and a Jack In The Box were covered with anti-meat messages. Signpost at Bun & Burger were sprayed with "Where's the beef?" "War declared on animal killers," "Meat is Murder," "Next time fire." The 24-hour Jack In The Box at Bolsa Avenue and Edwards Street received the same anti-meat slogans, This is the 3rd hit at this restaurant. -A.L.F.

10/5/96-Redmond, WA; A Kentucky Fried Chicken is spraypainted with 'All meat is murder' and 'Sadist'.-A.L.F.

10/8/96-Bellevue, WA; In the third attack on Honey Bee Hams it had seven large windows smashed, 'Animal Killers' and 'A.L.F.' spraypainted, $8,500 damage.-A.L.F.

10/-/96-Mercer Island, WA; At least 11 actions have occurred since March. McDonald's was spraypainted with ' McDeath', Baskin Robbins spraypainted with 'Dairy = Death' and many windows smashed.-A.L.F.

10/16/96-Ithaca, NY; A McDonald's has it locks glued, other equipment damaged and a huge banner saying 'McDeath: Killing Animals, the Earth and You!' dropped from the roof. - Band of Mercy

10/16/96-Eugene, OR; The following animal abusers had their men's and women's restrooms toilets plugged with sponges; walls covered with slogans; and walls, ceilings, floors and fixtures sprayed with "blood"; 5 McDonalds, 1 Taco Bell, 1 Taco Time, 1 Arby's, 1 Burger King, 1 Carl's Jr, 1 Wendy's.-A.L.F.

10/22-28/96-Vancouver, BC; Several McDonalds billboards spraypainted with 'Go Vegan', 'McDeath' and 'Resist'. -Unclaimed

10/27/96-TX; Blocked the toilet pipes in a McDonalds in rural Texas and wrote slogans in the bathroom including 'Meat is Murder', 'A.L.F.', and 'Tell Your Corporate Masters That the Pipes Were Blocked for the McLibel Two'. -A.L.F.

A MURDEROUS NEW FACTION called the "Justice Department" emerged in October 1993 in the United Kingdom with a wave of parcel bombs, and in Canada in 1996 with razor blades in letters. The Animal Liberation Front's website posted a decoupling notice warning that "the J.D. is not part of the A.L.F. and does not follow the A.L.F.'s guideline of non-violence towards humans." ALF website entries:

1/9/96-BC; 65 envelopes with rat poison covered razor blades, taped inside the opening edge to guide outfitters across B.C. and Alberta. The letter enclosed said "Dear animal killing scum! Hope we sliced your finger wide open and that you now die from the rat poison we smeared on the razor blade. Murdering scum that kill defenseless animals in the thousands every year across B.C., for fun and profit do not deserve to live. We will continue to wage war on animal abusers across the world. Beware scum, better watch out, you might be next! Justice Department strikes again." -Justice Department

3/96-Canada; Fur retailers across Canada sent 87 envelopes containing razor blades allegedly tainted with AIDs infected blood, taped inside the opening edge. -Justice Department

Rod Coronado-style fur farm raids have grown to epidemic proportions.

4/4/96-Victor, NY; Sometime between 9 pm, Wed April 3 and 6 am Thursday morning, A.L.F. activists struck L.W. Bennett & Sons' Fur Farm (Strong Rd, Victor, N.Y. 14564 716-924-2460) and released over 3000 mink. According to the Sheriff's Office, each mink was 'worth' $8, meaning that over $240,000 "worth of pelts" were set free. Some estimates range as high as 1 million dollars.

6/96-Washington State; 80 mink liberated from an unknown fur farm. We received word of this late, with little detail as to location of the farm. -A.L.F.

6/7/96-Sandy, UT; Utah Fur Breeders Agriculture CoOp (a major fur feed supplier) raided and 75 mink used in nutritional research liberated. According to the communiqué, the ALF broke in to the Fur Breeders Agriculture Coop at 8700 South 700 West in the early morning hours of June 7th. Two sheds were half full with mink, and one of them was completely emptied out. -A.L.F.

6/21/96-Riverton, UT; 1,000 mink liberated from fur farm. ALF found dead mink lying in cages, under the cages in piles of feces, with many half eaten by other mink. "More actions are coming. Murderers beware." -A.L.F.

7/4/96-Langley BC; 400 mink were released from Akagami Mink Ranch (26032 16th Ave. Tel: 604-856-4261) in Langley as part of an "Independence Day" operation conducted by the A.L.F. According to the communiqué: "We liberated these mink to save them from a life of torture, enslavement and eventual death. If we would have left them behind, these innocent animals would have been grotesquely killed, through neck-breaking, gassing or anal electrocution." -A.L.F.

7/4/96-Howard Lake, MN; ALF raiders liberated 1000 mink at

Latzig Mink Ranch,as part of an Independence Day action, coinciding with a similar raid on another fur farm in Vancouver, Canada. -A.L.F.

7/4/96-Pleasant View, TN; The A.L.F. visited Mac Ellis Fox Farm in hopes of raiding it for the 3rd time in as many years. The group discovered that the second raid (Nov. 1995) had put the fur farmer out of business. VICTORY! -A.L.F.

8/9/96-HINSDALE, MA; Over 1000 mink were "liberated" from The Carmel Mink Ranch, off Rt. 143. "Most cages were opened....and (we) painted A.L.F. on the shed," said the communiqué. -A.L.F.

8/12/96-Alliance, OH; 2500 mink liberated from Justice Jorney's (president of the Ohio Mink Breeders Assoc.) fur farm. -A.L.F.

9/28/96-Provo, UT; 8000 mink released from Paul Westwood's mink farm, breeding cards destroyed. Huge holes cut in two surrounding fences. "Many animals were left behind and for that we are sorry, but this war is far from over..." from the communiqué. Over $20,000 damages.-A.L.F.

10/5/96-Alliance, OH; Jorney's Mink Ranch hit for the second time in two months, 8000 mink liberated.-A.L.F.

10/5/96-Lyndeborough, NH; Richard Gauthier's fur farm has 35 fox and 10 mink liberated.-A.L.F.

10/11/96-Hinsdale, MA; Carmel Mink Ranch, activists find alarms set to trigger if the cages are opened and still manage to liberate 75 mink.-A.L.F.

10/23/96-Lebanon, OR; Arnold Kroll's mink farm raided, 2000 mink liberated.-A.L.F.

10/24/96-Coalville, UT; Devar Vernon's fur farm raided, 2000 mink and 200 fox liberated.-A.L.F.

10/29/96-Victor, NY; L.W. Bennett & Sons' Fur Farm raided for the second time, three sets of fences cut through and 46 fox released. -A.L.F.

SUMMER, 1996   *Bitterroot Range, Montana*
THE RUCKUS SOCIETY, A SMALL CONSORTIUM OF RADICALS started by Mike Roselle and based in Missoula, held "Action Camp '96," a direct action training seminar ostensibly teaching the 500-odd attendees only nonviolent civil disobedience such as road blockading, media handling and decoupling tactics. As in past training camps of a similar nature, select individuals were quietly taken aside and trained in monkeywrenching techniques.[150]

OCTOBER 30, 1996   *Grangeville, Idaho*
A JURY AWARDED LOGGER DON BLEWETT MORE THAN $1 MILLION in damages from Earth First. After nearly eleven hours of deliberation, a jury of eight women and four men sided with Blewett against twelve Earth First defen-

dants for damages his company suffered during the 1993 protest of the Cove-Mallard area of the Nez Perce National Forest (see page 150).

The jury awarded Blewett $150,000 in compensatory damages and $999,999 in punitive damages.

Defendants included four Earth Firsters from Maine: Billi Jo Barker of Harmony; Rob Borden of Athens; Michael Vernon of Athens; and Dana Wright of Waldboro; two from California: Lawrence Juniper of Bolinas; and Karen Pickett of Canyon; John (Jake) Kreilick of Missoula, Montana; Beatrix Jenness of Montrose, West Virginia; Peggy Sue McRae of Friday Harbor, Washington. Three Earth Firsters lived in Idaho: Peter Leusch of Driggs; Jennifer Prichard of Moscow; Erik Ryberg of McCall.

Robert E. Amon was one of the original defendants, but he declared bankruptcy before the trial and was not included in the jury verdict. Ryberg also declared bankruptcy.[151]

3:30 A.M. OCTOBER 30, 1996   *Oakridge, Oregon*

AN ARSONIST SET FIRE TO THE OAKRIDGE RANGER STATION southeast of Eugene, destroying the complex of U.S. Forest Service buildings.[152] Two days earlier the Detroit Ranger Station was the target of vandals who set a pickup truck on fire and spray-painted anti-logging and anti-Forest Service graffiti on the walls of the building and four other trucks several days before a timber auction. "Earth Liberation Front" was spray painted on a wall.[153] A letter "A" with a wide horizontal bar was painted on the building and trucks—a symbol often used by anarchist groups. An incendiary device—a milk jug filled with a flammable liquid— was found of the roof of the building.

A message titled "ELF Halloween Smash" had appeared in the latest edition of the Earth First Journal, published in Eugene:

"Let the seven nights of the Earth Night allow those who are destroying this planet to be witness to some of the most destructive eco-sabotage and criminal damage ever seen, persuading them to either give up their practices or suffer the consequences!!!"[154]

Earth First spokeswoman Heather Coburn told the Eugene Register-Guard that the Earth Liberation Front is "an anarchist community that gets together and does (things) like that. We're not affiliated with that group. They're a little more radical." It was Judi Bari's decoupling advice from the Earth First Journal of February 1994.

On November 7, the Detroit Ranger Station was evacuated after a threatening note was found in a self-serve information box in front of the office. The note said, "Boom! Boom!"[155]

About the time of the Oregon incidents, Judi Bari was diagnosed with breast cancer which had spread to her liver. She died at her home just outside Willits, California, March 2, 1997.[156] Her passing brought expressions of condolence and respect from supporter and opponent alike, both feeling the absence of a forceful leader.

On Friday, November 1, 1996, the auctioning of five U.S. Forest Service timber sales was guarded by nearly seventy police officers and state troopers in full riot gear as about forty protesters tried to enter but were kept out. Outside, at least one-hundred-fifty angry protesters blocked the pickup of timber company vice president Rob Freres, the winning bidder, from leaving a nearby parking area after the auction, shouting, "Bidder beware!" Motorcycle police roared to the spot to let the driver leave. The auction was moved to Eugene from the Detroit Ranger Station for security reasons.[157]

With time the ALF / Earth Liberation Front ecoterror factions have become more sophisticated in eluding law enforcement. Realizing that new cars do not arouse the suspicions of police, nomadic action groups travel across the nation from one hit to another in brand new rental cars paid for by well-known mainstream environmental groups.

There is an inevitable push toward radicalization. As the ecoterror violence escalates, less radical insiders who object are forced into the position of dissidents and are seen as turncoats to the movement by the more radical. The less radical are threatened and intimidated into silence with reminders that since it is a secret movement, no one will miss them if they should unfortunately disappear. With no one to protect them, they are unlikely ever to come forward and tell what they know.

There is reason to believe that many of the actual perpetrators of unsolved ecoterror crimes come from the ranks of early 1980s adherents and their subsequent clandestine recruits. Some of the later activists are completely unaware of the criminal element among them. None of the radicals mentioned in Chapter Five have likely ever met anyone in the criminal ecoterror element. Some have difficulty believing they even exist. This book should change that.

Most law enforcement officers at the field level have insufficient grip on the true nature of ecoterror and the ecoterrorists. Those who do grasp the problem are thwarted by orders from above. It is politically inexpedient to address the violent agenda to save nature. With the Presidency in mind, Al Gore in particular avoids any recognition that ecoterror exists. Gore has the power and the motive to paralyze any federal investigation. Ironically, prosecutors at the federal grand jury convened to probe the 1996 Forest Service arsons targeted radical pacifist Michael Donnelly but ignored long-time Earth Firsters with blatant ecoterror criminal records. The FBI leadership gave no orders to find nomadic action group members. While the guilty go free, the innocent suffer for it.

The monkeywrenching goes on, worse than ever before.
The ALF crimes go on, worse than ever before.
The anti-industry obstruction goes on, worse than ever before.

Chapter Six Footnotes

[1] Telephone interviews by the author with Mark J. Quinnan, the only one of the Eco-Raiders that can presently be located, November 18, 23, and 30, 1996, and telephone interviews with Pima County Sheriff's Deputy Duane Wilson, now re⁺ired, the lead investigator in the Eco-Raider case, November 19 and 25, 1996, and from clipping files on the Eco-Raiders in the libraries of the Tucson Citizen and the Arizona Daily Star.

[2] Telephone interview with Rick Wilson, former superintendent of Tucson schools and assistant principal at Canyon Del Oro High School. Wilson said that Rowe had later been killed in a traffic accident.

[3] Quoted in *Ecotage*, edited by Sam Love and David Obst, Pocket Books, New York, February 1972, pp. 145-46.

[4] Photo caption only, no related story: "No-Returns Returned," *Arizona Star*, Monday, April 3, 1972, photo by Joe Gold, p. B1.

[5] "'Guerrilla war' hits new homesites," *Tucson Citizen*, Wednesday, February 14, 1973, by Sam Negri, p. A1.

[6] "Raiders are sued by Estes," *Tucson Citizen*, Tuesday, January 8 1974, no byline, p. B2.

[7] "Vandals hit new homes on city's Northwest," *Tucson Citizen*, Friday, March 23, 1973, no byline, p. B1.

[8] The exact cost was never determined, but the $1 million estimate by the end of 1972 was probably not far off the mark according to Bill Estes II of the Estes Company. Telephone interview with Estes November 14, 1996.

[9] "Vandals' toll heavy in housing," *Tucson Citizen*, Wednesday, March 21, 1973, no byline, p. B4.

[10] Quinnan said the Eco-Raiders believed the public reports to be exaggerated. He refused to believe the higher figures my research had uncovered. Telephone interview with Quinnan, November 23, 1996.

[11] "What is the Sound of One Billboard Falling?" *Berkeley Barb*, November 8-14, 1974, by Tom Miller, p. 12.

[12] Miller's story mentioned the New Times spread. The Barb's own story was not published until after the Eco-Raiders were out of jail.

[13] "Vandalism Blamed on 'Ecology Raids,'" *Arizona Star*, Thursday, February 15, 1973, by Ken Burton, p. A1.

[14] This incident was discovered by Sergeant Duane Wilson after he learned the identity of Chris Morrison and ran his records. Sheriff's Department officers had actually stopped different Eco-Raiders several times without realizing who they were. Interview with Duane Wilson, November 19, 1996.

[15] "Vandals plug door keyholes," *Tucson Citizen*, Friday, July 13, 1973, no byline, p. B2.

[16] "Wave of vandalism hits builders," *Tucson Citizen*, Saturday, July 14, 1973, no byline, p. B1.

[17] Quinnan volunteered comments on Walker's "growing extremism" in two interviews, November 18 and 23, 1996.

[18] The law enforcement story was given by former Sergeant Duane Wilson, now retired, in telephone interviews with the author November 19 and 25, 1996,

[19] Quinnan was unaware he had been pursued by police dirt bikes that night, although he recalled the raid on C&D Pipeline Company. Interview with Quinnan, November 23, 1996.

[20] The name is fictitious. Duane Wilson would not reveal the name of a paid informant.

21 "Four Accused of Vandalism," *Arizona Star*, Tuesday, October 2, 1973, no byline, p. B1.

22 "3 Men Enter Guilty Pleas On Eco-Raider Charges," *Arizona Star*, Thursday, October 25, 1973, by Art Arguedas, p. A1.

23 "Estes Suit Charges 5 With Vandalism," *Arizona Star*, Tuesday, January 8, 1974, no byline, no page. See also "'Raiders' are sued by Estes," *Tucson Citizen*, Tuesday, January 8, 1974, no byline, no page, *Citizen* library.

24 Telephone interview with Bill Estes II, November 14, 1996.

25 "Earth First!" *Smart*, September - October 1989, by Susan Zakin, p. 91.

26 Telephone interview with Ken Sleight, November 18, 1996.

27 Telephone interview with Captain Paul Watson, September 26, 1994.

28 Deposition of Alex Pacheco, in the case of Berosini v. People for the Ethical Treatment of Animals, et al., March 9, 1990.

29 "Fox Hunting: Clash of the Classes - English Animal Rights Groups, Preservers of Country Life Face Off," *The Washington Post*, March 1, 1992, by Glenn Frankel, p. A22.

30 David Henshaw, *Animal Warfare*, Fontana Paperbacks, London, 1989, pp. 53-54.

31 *Ibid.*, p. 57.

32 "Crusaders against cruelty: Whether sabotaging hunts, fire-bombing or burgling, anti-vivisectionists are sure right is on their side," *London Independent*, Monday, December 12, 1988, by Nicholas Roe, p. 21.

33 "The Great Silver Spring Monkey Debate," *The Washington Post Magazine*, February 24, 1991, by Peter Carlson, p. w17.

34 "The Silver Spring Monkeys," by Pacheco and Francione, in *In Defense of Animals*, edited by Peter Singer, Blackwell, New York, 1985.

35 Peter Singer, *Animal Liberation: A New Ethics for Our Treatment of Animals*, New York Review, distributed by Random House, New York, 1975.

36 "Activist Ingrid Newkirk fights passionately for the rights of animals, some critics say humans may suffer," *People Weekly*, Oct 22, 1990, Vol. 34, No. 16, by Susan Reed, p. 59. *See also*, "Are animals people too? Close enough for moral discomfort," *The New Republic*, March 12, 1990, by Robert Wright, p. 21.

37 "The Great Silver Spring Monkey Debate," *The Washington Post Magazine*, February 24, 1991, by Peter Carlson, p. w15.

38 The most detailed account of Earth First's origins is in Susan Zakin's *Coyotes and Town Dogs: Earth First! and the Environmental Movement*, Viking Penguin, New York, 1993 (hereafter, Zakin). However, this section also relies on Rik Scarce, *Eco-Warriors: Understanding the Radical Environmental Movement*, Noble Press, Chicago, 1990; "Mr. Monkeywrench," *Harrowsmith*, September / October 1988, by Kenneth Brower, p. 41; and Martha F. Lee, *Earth First! Environmental Apocalypse*, Syracuse University Press, Syracuse, 1995 (hereafter referred to as Lee); and telephone interviews with a former Earth Firster close to Dave Foreman who requested anonymity.

39 One version was so wildly inaccurate it had Foreman, Kohler and Wolke leaving Washington, D.C. and driving west in the VW minibus to form Earth First. "The idea born in that rambling VW trip across the continent became Earth First!" "No Compromise!," *Portland Oregonian* Northwest Magazine Section, Sunday, November 25, 1984, by Katherine Dunn, p. 10.

40 Kenneth Brower, "Mr. Monkeywrench," *Harrowsmith*, p. 43.

[41] Gordon Solberg, *Dry Country News*, cited in *Earth First!* 2, no. 4 (March 20, 1982), p. 3. Cited in Lee, p. 32.

[42] Http://www/unm.edu/~noise.

[43] Dave Foreman memo dated September 1, 1980 cited in Lee, pp.33-34.

[44] Dave Foreman quoted in Lee, p. 35.

[45] Aldo Leopold, *Round River: From the Journals of Aldo Leopold*, ed. Luna Leopold, Oxford University Press, New York, 1953, pp. 158-165. *See also*, Aldo Leopold, *A Sand County Almanac*, Oxford Univerity Press, New York, 1949, p. 188.

[46] "What You Can Do," *Earth First!*, Yule, December 21, 1980, Volumn [*sic*] 1, Number 2, p. 2.

[47] From an interview with George Draffan, in Lee, p. 37.

[48] Christopher Manes, "Green Rage," *Penthouse*, May, 1990, p. 51.

[49] For details of the research, see Edward Taub, "Somatosensory Deafferentation Research with Monkeys: Implications for Rehabilitation Medicine," in *Behavioral Psychology in Rehabilitation Medicine: Clinical Applications*, 1980.

[50] Richard Morgan, *Love and Anger: An Organizing Handbook for Activists in the Struggle for Animal Rights and In Other Progressive Political Movements*, second edition, Westport, Connecticut, Animal Rights Network, 1981.

[51] "The Great Silver Spring Monkey Debate," *The Washington Post Magazine*, February 24, 1991, by Peter Carlson, p. 15.

[52] Draft of "Correction and Clarification," reached by the Multi-Door Dispute Resolution Division of the District of Columbia Superior Court, David R. Anderson, Esquire, mediator, in the case of Alex Pacheco v. Katie McCabe, Civil Action 90-0A01627, February 15, 1990.

[53] *Science*, December 11, 1981.

[54] Paid advertisement by PETA, *Washington Post*, September 20, 1981.

[55] Taub v. State, 296 Md. 439, 463 A.2d 819 (Md. 1983).

[56] Quoted in Howard Goodman, "Medical Ethics and Animals, *Inside*, Fall, 1987, p. 98.

[57] James M. Jasper and Dorothy Nelkin, *The Animal Rights Crusade: The Growth of a Moral Protest*, The Free Press, New York, 1992, p. 47.

[58] Ingrid Newkirk, *Free the Animals! The untold story of the U.S. Animal Liberation Front and its Founder, Valerie*, Noble Press, Chicago. 1992, 372 pages.

[59] Susan Reed and Sue Carswell, "Animal passion," *People Weekly*, January 18, 1993, vol. 39 no. 2, p. 34.

[60] "Activists Subpoenaed In Animal-Lab Thefts," *Philadelphia Inquirer*, Wednesday, October 3, 1984, by Martha Woodall, p. B1.

[61] "Animal Rights Activists Stage Sit-In at NIH to Protest Experiments," *The Washington Post*, July 16, 1985, by Mark Katches and Eve Zibart, p. D3.

[62] Jennie Dusheck, "Protesters prompt halt in animal research," *Science News*, July 27, 1985. See also Barbara J. Culliton, "HHS halts animal experiment," *Science*, August 2, 1985.

[63] "The Rise of Medical Vigilantes," *The Washington Post*, September 8, 1987, by Abigail Trafford, p. z15.

[64] "University of Pennsylvania Head Injury Laboratory," in *Newsletter, 1985 - The Year In Review*, Animal Legal Defense Fund Newsletter No. 1, 1986, p. 2. See also, Jennie Dusheck, "Protesters prompt halt in animal research,"

Science News, July 27, 1985, p. 53.

[65] Telephone interview with Douglas Plumley, December 16, 1996.

[66] Memo to Directors of Forest Protective Associations, James B. Corlett, Oregon Forest Protection Association, 1326 American Bank Building, Portland, Oregon 97205, July 9, 1980.

[67] "Bald Mountain Road Work Continues," *Grants Pass Courier*, April 27, 1983, by Paul Fattig, p. A1.

[68] Telephone interview with Johnny O'Connor, December 16, 1996.

[69] Telephone interview with Dick Payne, December 17, 1996.

[70] Telephone interview with Les Moore, December 17, 1996.

[71] An Earth Firster on the site told the Grants Pass Courier that Foreman stayed in front of the truck for about 100 yards. "Dave tripped and the truck went part-way over him. Then the crew jumped out and gathered around him," said the unidentified witness. "Earth First! Leader Jailed," *Grants Pass Courier*, Thursday, May 12, 1983, by Paul Fattig, p A1.

[72] "Activist Makes Point, Before Being Found Guilty," *Grants Pass Courier*, Thursday, August 25, 1983, by Paul Fattig, p. B1.

[73] Helen Wilson quoted in Lee, p. 51.

[74] Dave Foreman, "The Reichstag Fire—1981," *Earth First! Newsletter*, vol. 1, no. 7, (Lughnasad [*sic*] / August 1, 1981, pp. 7-8.

[75] "Earth First! Announces 'Ecotricks' Contest," *Earth First! Newsletter*, vol 1, no. 7, (Lughnasad [*sic*] / August 1, 1981, p. 8.

[76] Lee, p. 55 and note 63, p. 167.

[77] Dave Foreman, "Earth First!," *The Progressive*, vol. 45, no. 10, October 1981, pp. 39-42.

[78] Lee, pp. 18-19.

[79] Much of this section is paraphrased from the web site of "The Ballad of Ned Ludd," Techno-Folk Opera by Corinne Becknell and Marty Lucas, the liveliest account of the Luddites ever. URL: http://town.hall.org/places/ludd_land/index.html.

[80] See Dinwiddy, J.R., *From Luddism to the First Reform Bill* (1987); Liversidge, Douglas, *The Luddites: Machine-Breakers of the Early Nineteenth Century* (1972); and Thomis, Malcolm I., *The Luddites: Machine-Breaking in Regency England* (1970).

[81] Lee, p. 65.

[82] Dave Foreman, "Ludd Readers," *Earth First! Newsletter*, vol. 2, no. 4 (Eostar Ritual / March 20, 1982, p. 11.

[83] Dave Foreman, Howie Wolke and Bart Koehler, "The Earth First! Wilderness Preserve System," *Earth First!* vol. 3, no. 5, Litha / June 21, 1983, p. 9.

[84] Memorandum regarding Earth First Statement of Principles and Membership Brochure, Sept. 1, 1980, by Dave Foreman, p. 1. Cited in Lee, p. 39.

[85] "The Animal Liberation Front: Army of the Kind," *PETA Factsheet*, Miscellaneous #5, no date, Washington, D. C.

[86] "'Saboteur' Describes Sinking of 2 Whalers, Does Not Confess; Says Activist 'Team' Scuttled Vessels," *Arizona Republic*, Saturday, November 15, 1986, by Knight-Ridder, p. A1.

[87] Dean Kuipers, "The Tracks of the Coyote," *Rolling Stone*, June 1, 1995, p. 54.

[88] *Report to Congress on Animal Enterprise Terrorism*, August 1993, p. 14.

[89] Interviews with county law enforcement officers nationwide revealed a pat-

tern of monkeywrenching crimes that appeared in the wake of traveling Earth Firsters, some of which are listed in Chapter Four. Ben Hull and Carla Jones, law enforcement agents of the U.S. Forest Service, stated in interviews it was their opinion that the pattern was real.

90 Government's Sentencing Memorandum, *United States v. Rodney Adam Coronado*, July 31, 1995, p. 6. *See also*, Dean Kuipers, "The Tracks of the Coyote," *Rolling Stone*, June 1, 1995, p. 54.

91 "Eco-warriors group targeted key research," *Detroit News*, March 13, 1995, by Paige St. John, p. B4.

92 Robert A. Heinlein, *Stranger in a Strange Land*, G. P. Putnam's Sons, New York, 1961.

93 Margot Adler, *Drawing Down the Moon: Witches, Druids, Goddess-Worshippers, and Other Pagans in America Today*, Beacon Press, Boston, 1979, second edition 1986.

94 J. E. Lovelock, "Gaia as seen through the atmosphere," *Atmospheric Environment*, no. 6, p. 579, 1972.

95 James E. Lovelock, *Gaia: A new look at life on Earth*, Oxford University Press, New York, 1979, 157 pages.

96 *Newsweek*, March 10, 1975, p. 49.

97 "What's It Gonna Take? *Live Wild Or Die* #4, anonymous, 1994. There are numerous books in the occult literature on vanishing, for example, see *Invisibility: Mastering the Art of Vanishing. A guide to hiding yourself from sight using techniques culled from alchemy, rosicrucianism, medieval magic and advanced yogic practices*, by Steve Richards, The Aquarian Press, Wellingborough, Northhamptonshire, England, April, 1982, 160 pages.

98 "Environmental radicalism backed - Local response favorable," *Portland Oregonian*, January 23, 1983, by John Hayes, p. C2.

99 Lee, p. 56.

100 "The Environmental Guerrillas," *Boston Globe Magazine*, March 27, 1988, by Jim Robbins.

101 Lee, p. 56.

102 "Arrests cut short acid rain protest," *Rapid City Journal*, Friday, October 23, 1987, by Hugh O'Gara, p. 1.

103 "4 Youths Die in Md. Explosion; Powerful Blast Rips Doors Off Bethesda Garage," *The Washington Post*, January 1, 1989, by Paul Duggan and Lisa Leff, p. A1.

104 "Police Look for Clues In Blast That Killed 4; Youths May Have Been Involved in 'Prank,'" *The Washington Post*, January 2, 1989, by Fern Shen, p. B1.

105 "Home Blasts Fascinated Md. Student, Friend Says; Mail-Order Books Were Used in Experiments," *The Washington Post*, January 4, 1989, by Paul Duggan, p. B1.

106 "Chemical Pipe Bomb Suspected in 4 Deaths; Exact Type of Blast May Never Be Known," *The Washington Post*, January 5, 1989, by Paul Duggan, p. D1.

107 Telephone interview with Detective Joseph Chan, currently of Alameda County, California, District Attorney's investigations office, March 1, 1996.

108 "Cleveland Amory Joins Animal Demonstrations," *San Francisco Chronicle*, Wednesday, April 20, 1988, by Martin Halstuk, p. B7.

109 Freedom of Information request to University of California Police Captain Pat Carroll, dated March 1, 1996. BATF Case No. 89-0048.

110 "Labs a Target of Rights Activists - Animal Researchers Feel Hunted," *San Francisco Chronicle*, Friday, October 1, 1993, by Janet Wells, p. A1.

111 "Battling the Animal Liberation Front," by Assistant Chief Harry R. Hueston II, University of Arizona Police Department, in *The Police Chief*, September 1990, p. 52.

112 *Ibid.*

113 *Terrorism in the United States, 1989*, Terrorist Research and Analytical Center, Counterterrorism Section, Criminal Investigative Division, U.S. Department of Justice, Federal Bureau of Investigation, Washington, D.C., December 31, 1989.

114 Dave Foreman, "Goodbye Ed," *Earth First!* vol. 9, no. 4, Eostar / March 21, 1989, p. 19.

115 Edward Abbey, *Desert Solitaire: A Season in the Wilderness*, McGraw-Hill, New York, 1968, p. 255.

116 Zakin, p. 371.

117 James W. Clarke, *American Assassins: The Darker Side of Politics*, Princeton University Presss, Princeton, New Jersey, 1982, p. 148.

118 Clara Livsey, M.D., *The Manson Women: A "Family" Portrait*, Ricahrd Marek Publishers, New York, 1980, p. 57

119 Zakin, pp. 371-372.

120 Second Superseding Indictment, *United States v. Davis, et al.*, No. CR-89-192-PHX

121 "Agent Says FBI Tailed Defendant To Nuke Plant," *Phoenix Gazette*, Friday July 26, 1991, by Anthony Sommer, p. B13.

122 "2 Sides Meet At Animal-Research Rally," *Philadelphia Inquirer*, Sunday February 4, 1990, by Jerry W. Byrd, p. B2.

123 Letter on PETA letterhead dated May 8, 1990, addressed to a resident of the same street as Morrison and signed by "Ann Cynoweth, Researcher."

124 "PETA Smells a Lot of Bologna," *PETA News*, September / October 1990, p. 26

125 "Slaying of Veterinary Dean Stirs Animal Rights Controversy," *Lexington Herald-Leader*, Wednesday February 28, 1990, by The Associated Press, p. A4.

126 "Slaying of Dean Remains a Mystery," *Memphis Commercial Appeal*, Tuseday February 5, 1991, by The Associated Press, p. D12.

127 "Environmental Group Says It Won't Spike Trees," *San Francisco Chronicle*, Wednesday April 11, 1990, by Elliot Diringer, p. A24.

128 Darryl Cherney, quoted in *Sixty Minutes Transcripts*, vol. 22, no. 24, March 4, 1990, p. 3.

129 Zakin, p. 386.

130 "Earth Day power outage - Power poles cut; cops investigate," *Santa Cruz County Sentinel*, April 23, 1990, by Steve Perez, p. A1.

131 "Outages cut woman's lifeline," *Santa Cruz County Sentinel*, April 23, 1990, by Maria Guara, p. A1.

132 "Group claims responsibility - Letters say 'sabotage' directed at PG&E," *Santa Cruz County Sentinel*, April 25, 1990, by John Robinson, p. A1.

133 Zakin, p. 387.

134 Affidavit for Search Warrant, Municipal Court of the Oakland-Piedmont Judicial District, Sergeant Robert A. Chenault, Oakland Police Department, May, 25, 1990.

[135] "Bomb charge absurd, says activists' friend," *Santa Rosa Press Democrat*, May 27, 1990, by Tobias Young, p. A5.

[136] "Bomb hurts timber activists," *Santa Rosa Press Democrat*, May 25, 1990, by Blews W. Rose, Mike Geniella, and Alvaro Delgado, p. A1.

[137] "Private investigator probing bomb blast," *Sacramento Union*, Tuesday, May 29, 1990, by The Associated Press, B3.

[138] "D.A. Won't File Charges In Bombing Of Earth Firsters," *San Francisco Chronicle*, Wednesday July 18, 1990, by Martin Halstuk, A1.

[139] Internal security reports show that both fake pipe bombs and real ones have been placed in or near the Eugene, Oregon Hyundai site during 1996.

[140] "San Francisco Car-Bomb Victims Allowed To Sue FBI," *San Francisco Chronicle*, Friday, September 4, 1992, no byline, p. A26.

[141] "FBI Won't Open Car Bombing Files - Environmentalists, Congressman Protest Agency's Refusal," *San Francisco Examiner*, Friday, October 1, 1993, by Eric Brazil, A5.

[142] Zakin, p. 394.

[143] "Earth First! Co-Founder Quits - Is Unhappy With Group's New Focus," *Arizona Republic*, Wednesday August 15, 1990, by Sam Negri, p. B1.

[144] "Animal-Rights Groups' Motive Questioned," *Portland Oregonian*, Wednesday June 19, 1991, by Ty Weisdorfer, p. E07.

[145] "Animal-Rights Group Claims It Started Big Edmonds Fire," *Seattle Times*, Sunday, June 16, 1991, by William Gough and Dave Birkland, p. A1.

[146] *United States v. Rodney Adam Coronado*, Government's Sentencing Memorandum, Case No. 1:93-CR-116.

[147] "Animal Rights Case Suspect Is Ordered To Pay Restitution," *Portland Oregonian*, Thursday, August 17, 1995, by staff, wire and correspondent reports, p. D5.

[148] "Monkeywrenching" *Earth First*, February 1994, by Judi Bari, p. 8.

[149] "Victory for Sierra Club Dissidents," *San Francisco Chronicle*, April 23, 1996, by Alex Barnum, p. A1.

[150] "We Few, We Happy Few, We Band of Fledgling Monkeywrenchers, Learning to Speak in Sound Bites," *Outside*, October 1996, v. xxi, no. 10., by Tad Friend, p. 48.

[151] "Highland Enterprises bulldozes Earth First!; Jury awards Blewett more than $1 million in suit over Cove-Mallard protest in 1993," *Lewiston Tribune*, Thursday, October 31, 1996, by Kathy Hedberg, p. 1A.

[152] "Fire destroys ranger station," *Eugene Register-Guard*, Thursday, October 31, 1996, by Lance Robertson, p. A1.

[153] "Vandals target Detroit ranger station," *Eugene Register-Guard*, Tuesday, October 29,1996, by Lance Robertson, p. 1C.

[154] "Federal property becomes a target," *Portland Oregonian*, Thursday, October 31, 1996, by Dana Tims and Bryan Denson, p. 1A

[155] "Scare clears out ranger station," *Eugene Register-Guard*, Friday, November 11, 1996, by Lance Robertson, p. 1C.

[156] "Judi Bari has breast cancer," *Santa Rosa Press Democrat*, November 2, 1966, Mary Callahan, p. 1. "Judi Bari, environmentalist, Earth First activist" (obituary), *Seattle Times*, March 3, 1997, p. B8.

[157] "Sale brings a show of force," *Eugene Register-Guard*, Saturday, November 2, 1996, by Lance Robertson, p. 1A.

Chapter Seven
REASONS

In mid-1996, The New York Times Magazine ran a story on Esther Dyson, internet guru, writer, futurist, and "the most influential woman in all the computer world." While telling of her mother, who taught mathematics at the University of California at Berkeley, she commented that it was at the same time Ted Kaczynski taught there. "For all I know, my brother and I ran into him when we played tag in the math department elevators."

Dyson said she was "fascinated by the Unabomber." She remarked, "No. 1, he's a maniac. No. 2, he's asking valid questions: is technology bad?"

She rated his manifesto as "an example of a freelance writer who wasn't very good, but then, his writings are what got him caught. Interestingly, he could have put his manifesto on the Internet without going to the New York Times or The Washington Post."

Esther Dyson mused, "I keep thinking that if he were even remotely plugged in, he could have been spouting all his stuff on the Net and that might have kept him from getting all bottled up inside."[1]

Esther Dyson could be any of us. Like Dyson, we ponder whether Technology Is Bad while we enjoy its fruits and suffer its dilemmas. In that regard, we all live in the world of the Unabomber. Like Dyson, we haven't a clue what goes on in the minds of those who slip beyond mere pondering to apocalyptic fatalism, to cultivate a fanatical hatred of technological civilization and then act to destroy it. In that regard, we simply can't grasp the world of the Unabomber.

283

We just can't believe there is a violent agenda to save nature. But there is.

We just can't believe there are people planting bombs, destroying equipment and obstructing workers to save nature. But there are.

We just can't believe ecoterror exists. But it does.

Why?

I have considered that question for many years. The explanations I find most compelling are also the most disturbing, for they do not provide a comfortable platform from which the advocates of industrial civilization may look down upon those who hate it. We may well call for more adequate laws to protect us from the haters, but we will find scant justification to hate the haters. We are more likely to find a mirror in the dark.

Why did the environmental movement spawn ecoterrorism? The answers lie deep in our culture and our own minds. Just as movements of social change do not suddenly appear full blown for no reason, so violent fanaticism does not suddenly appear without cause. Like all movements of social change, environmentalism itself was a reaction to wrongs that needed to be set right. It was rooted in 19th Century American industrial history and popular culture and took shape in a series of social and political upheavals beginning in the 1960s.

The older conservation movements from Teddy Roosevelt's era, both the utilitarian "gospel of efficiency" of bureaucrat Gifford Pinchot, who sought "the greatest good for the greatest number over the long run," and the preservationism of naturalist John Muir, who sought federal protection for wilderness and wildlife, were reactions to perceived abuses of nature. Pinchot, as a trained forester and the first Chief of the U.S. Forest Service in 1905, sought to perpetuate forest commodities by ending the cut-and-run tactics of private timber entrepreneurs in favor of government ownership of forests and the imposition of regulations to insure reforestation and prevent depletion of soils, watersheds, grazing and timber supplies. Muir, as a co-founder of the Sierra Club in 1891, sought to prevent the incursion of homesteading and civilization into wild places in favor of government ownership of wilderness and the imposition of regulations to insure the growth of a federal parks and nature preserve system to encompass as large an area of wild lands as possible.

The anti-pollution movement that grew from public outcry against the "killer smogs" in Donora, Pennsylvania in 1949; the nuclear fallout scare of the mid-50s with its headlines "Strontium-90 in Babies' Milk"; the pesticide alarm dramatized in 1962 by Rachel Carson's *Silent Spring*; the plight of our garbage-laden oceans revealed in a 1966 National Geographic television special, *World of Jacques Cousteau*; oil-soaked birds from a seafloor wellhead blowout in the Santa Barbara Channel in January 1969 followed in June by the Cuyahoga River igniting from an oil and kerosene slick in a hundred-foot fireball that destroyed two railroad bridges; and Barry Commoner's catchphrase, "the environmental crisis," in his 1971

The Closing Circle, all cultivated the sense of an imminent global disaster and the possible end of all life on earth. The dread radicalized a growing number of people and transformed the earlier conservation movement into the environmental movement, bringing in whole new constituencies and goals. It is noteworthy that the terms "environmentalist" and "environmentalism" in their modern sense did not exist until the late 1960s; previously there were only "conservationists."[2]

At about the same time, a series of political protests erupted that moved America toward extending the concept of rights—the civil rights movement, the women's movement, the gay rights movement and other social and political conflicts. Those movements were linked to grave suspicions about society's major institutions: that which resisted demands for extended rights was necessarily the oppressor. Soon the idea of rights for nature found support.

A precursor of the idea of rights for nature appeared in Aldo Leopold's 1949 *A Sand County Almanac* in the form of the land ethic. The central idea was that land is not a commodity that humans may own, it is a trust for which we can only act as stewards. "In short, a land ethic changes the role of *Homo sapiens* from conqueror of the land-community to plain member and citizen of it," Leopold wrote. "It implies respect for his fellow-members, and also respect for the community as such."[3]

The nub of the land ethic is fundamentally an economic concept affecting ownership and income, and a political idea of title and control. "Land," Leopold complained, "is still property."

However, Leopold read history as an "ethical sequence," a continual extension of ethics that was actually "a process of ecological evolution." His extension of ethics to land, connecting the natural world to questions of right and wrong, struck a deep emotional chord, evoking love, guilt and highminded morality, setting the stage for explicit rights:

"The land-relation is still strictly economic, entailing privileges but not obligations."[4]

"It is inconceivable to me that an ethical relation to land can exist without love, respect, and admiration for land, and a high regard for its value. By value, I of course mean something far broader than mere economic value; I mean value in the philosophic sense."[5]

"Your true modern is separated from the land by many middlemen, and by innumerable physical gadgets. He has no vital relation to it; to him it is the space between cities on which crops grow."[6]

"A thing is right when it tends to preserve the integrity, stability, and beauty of the biotic community. It is wrong when it tends otherwise."[7]

"Individual thinkers since the days of Ezekiel and Isaiah have asserted that the despoliation of land is not only inexpedient but wrong. Society, however, has not yet affirmed their belief. I regard the present conservation movement as the embryo of such an affirmation."[8]

The explicit premise that natural objects should have rights came as a consequence of a Sierra Club lawsuit, *Sierra Club v. Morton*, that went to the U.S. Supreme Court in 1972. The Sierra Club had sued to revoke the U.S. Forest Service permit granted to Walt Disney Enterprises, Inc., for construction of a $35 million ski and recreation complex in Mineral King Valley adjoining California's Sequoia National Park. The Sierra Club lost, but the dissenting opinion of Justice William O. Douglas gave them a strategic victory: "Contemporary public concern for protecting nature's ecological equilibrium should lead to the conferral of standing upon environmental objects to sue for their own preservation. See Stone, Should Trees Have Standing? Toward Legal Rights for Natural Objects, 45 S. Cal. L. Rev. 450 (1972). This suit would therefore be more properly labeled as *Mineral King v. Morton*."[9]

The essay Justice Douglas cited was University of Southern California Law Professor Christopher D. Stone's law review article, *Should Trees Have Standing? Toward Legal Rights for Natural Objects*, which argued that natural objects should be able to be plaintiffs for their own injuries in a court of law. Stone prepared it specifically for publication in a Symposium issue of the Southern California Law Review in order to influence the Supreme Court case, which it did: In addition to Justice Douglas's dissenting opinion, Justices Blackmun and Brennan endorsed the idea of legal standing for natural objects. Justice Blackmun even called attention to the deep reason why change was needed by quoting the famous poem of John Donne beginning, "No man is an island," taken as a metaphor for the ecological notion that the world is a seamless web.[10]

Stone's essay was later published as a paperback including the high court's opinions in the case and an introduction by noted biologist Garrett Hardin, who, like Justice Blackmun, underscored the power of emotion and poetry to change policy. Hardin wrote: "'Poets,' said John Keats, 'are the unacknowledged legislators of the world.' ... Surely it is time now to make explicit the implications of the poets' insights and rebuild the written law 'nearer to the heart's desire.'" Hardin confused John Keats with Percy Bysshe Shelley, who was the actual author of that famous last line from *In Defense of Poetry*, but the idea is clear: Eloquent pleas to aesthetics, ethics and right and wrong will radicalize the people and they will change the law. A succession of American laws has given implicit rights to nature, including the Wilderness Act of 1964, the Endangered Species Act of 1972, and various animal cruelty acts at the state level.

In 1973, Arne Naess, a Norwegian philosopher, solidified the idea of extending rights to nature and provided a rationale for further radicalization when he first suggested that two distinct environmental movements were forming. The first movement, said Naess, was the large, popular, mainstream cluster of groups presided over by professionals—

bureaucratic and shallow in that it merely sought reforms of pollution and resource depletion. The other was small, personal, and deep in that it envisioned a fundamental change in the way human cultures related to the natural world: Deep Ecology, the liberation of nature from human exploitation.[11]

Naess, born in 1912, was a professor of philosophy at the University of Oslo from 1939 to 1970, when he resigned to become a radical environmental activist. He defined a quasi-religious outlook on nature that he called "ecosophy" [ecological philosophy], an elaborate system embodying his vast expertise in empirical semantics, the philosophy of science, Gandhi's theory of nonviolence, and Spinoza's theory of freedom and ethics.[12]

Shorn of technicalities, the essence of Naess's Deep Ecology philosophy is a complete rejection of Western humanistic civilization and all anthropocentrism. Anthropocentrism, argued Naess, includes only humans in its list of moral subjects, and it must be replaced with biocentrism, which includes the entire ecosystem in its moral framework.

Bill Devall and George Sessions elaborated Naess's ideas into a set of basic principles that gained wide currency among radical environmentalists:

1. The well-being and flourishing of human and nonhuman Life on Earth have value in themselves (synonyms: intrinsic value, inherent value). These values are independent of the usefulness of the non-human world for human purposes.

2. Richness and diversity of life forms contribute to the realization of these values and are also values in themselves.

3. Humans have no right to reduce this richness and diversity except to satisfy *vital* needs.

4. The flourishing of human life and cultures is compatible with a substantial decrease of the human population. The flourishing of nonhuman life requires such a decrease.

5. Present human interference with the nonhuman world is excessive, and the situation is rapidly worsening.

6. Policies must therefore be changed. These policies affect basic economic, technological, and ideological structures. The resulting state of affairs will be deeply different from the present.

7. The ideological change is mainly that of appreciating *life quality* (dwelling in situations of inherent value) rather than adhering to an increasingly higher standard of living. There will be a profound awareness of the difference between big and great.

8. Those who subscribe to the foregoing points have an obligation directly or indirectly to try to implement the necessary changes.[13]

Deep Ecology was a turning point in that it implicitly asserted that Technology Is Bad and advocated the dismantling of industrial civilization. Radical environmental groups emerged to "implement the necessary changes" by direct action, first with Greenpeace, which grew out of the 1970 "Don't Make A Wave Committee" in Vancouver, British Columbia, a group of Quakers who protested U.S. nuclear weapons testing because the Sierra Club would not. The committee rented a boat and went to the test site at Amchitka, Alaska, spending their travel time reading a book of Indian legends. According to *The Greenpeace Story* by Michael Brown and John May, they adopted one prophetic passage:

> [T]here would come a time, predicted an old Cree woman named Eyes of Fire, when the earth would be ravaged of its resources, the sea blackened, the streams poisoned, the deer dropping dead in their tracks. Just before it was too late, the Indian would regain his spirit and teach the white man reverence for the earth, banding together with him to become Warriors of the Rainbow.[14]

Greenpeacers thereafter became the Warriors of the Rainbow and named their boat Rainbow Warrior. The metaphor of the Warrior became their persona: they believed that humankind's action in the environment was leading to an imminent apocalypse and they could help stop it. It was a declaration of war on industrial society in the name of a millenarian community, nuclear-free and ecologically sensitive. Although Greenpeace grew into a multimillion-dollar international lobbying and fund-raising network that sought credibility among lawmakers, it opened the path to organized radical environmentalism.

Captain Paul Watson's Sea Shepherd Conservation Society broke away from Greenpeace and criticized their former colleagues for abandoning their radical roots. Others emerged: Earth First, the Animal Liberation Front, People for the Ethical Treatment of Animals and dozens of other increasingly radical groups such as the Animal Rights Militia and the Earth Liberation Front.

In 1975, Australian philosopher Peter Singer's book *Animal Liberation: A New Ethics for Our Treatment of Animals*, explored the concept of rights for animals. Singer argued that because animals can feel pleasure and pain, they deserve moral consideration and he demanded drastic reduction in their use. He argued that to assume that humans are superior to other species is "speciesism"—an injustice parallel to racism and sexism. Singer grounded his philosophy in utilitarianism: animals might still be used by humans, but only with consideration of their feelings. In short, animals are worthy of moral consideration.

Almost every animal rights activist either owns or has read *Animal Liberation*. Almost every animal rights activist uses "speciesism" as a catchword. And almost every animal rights activist has come to see Singer as too moderate. He sealed his own eclipse by positioning animal rights as a moral crusade. In questions of morality, there can be no compromise, and positions can only harden. In a crusade, the extension of rights becomes a permanent revolution, ever pushing toward an ever-flying goal.

Tom Regan entered the animal rights movement with an essay titled "The Moral Basis of Vegetarianism," and later hardened to the position that animals have inherent worth as living organisms and should never be used as resources. His 1983 book, *The Case for Animal Rights*, abandoned utilitarianism for absolutism. No benefits to humans can justify using animals for medical or scientific research or substance testing. The rights of animals are absolute.

Vegetarians became natural allies to animal rightists. The modern vegetarian trend emerged in the 1960s with the growing interest in Eastern philosophy and alternative lifestyles. Vegetarian journals show sympathy for the animal rights cause, particularly on the issue of factory farming. Animal rightists are commonly vegan: strict vegetarians who avoid not only meat but also any products from animals, including leather and fur. Their strong sympathy for animals evokes the deep feeling that "Meat is Murder." Their feelings of disgust for those who eat meat readily slip into feelings of hatred for those who eat meat. Their campaign against meat as a form of cannibalism separates them from the rest of society, which gives no symbolic meaning to eating meat.

The Animal Liberation Front used violence against property while disavowing violence against people. However, the moral crusade inevitably hardened positions and escalated actions: more extreme factions such as the "Justice Department" emerged to use death threats, bombings and booby trapped letters against those they hated.

In *Eco-Warriors*, James "Rik" Scarce included the animal rights factions in the radical environmental movement. He used "radical" in both senses of "going to the root of things"—fighting for the root of human existence, "the lifegiver Earth"—and "extreme" in their doctrines and tactics. Radical environmental groups, Scarce said, share basic characteristics that distinguish them from more moderate environmental groups:

> 1. They fight for nature through direct action and the destruction of property.
> 2. Their goal is preservation of biological diversity.
> 3. They act without direction from an organizational hierarchy.
> 4. They are poor.
> 5. They have little hope of actually ending the actions they protest.

6. They believe they are in a war and must "rise, fight back against the onslaught of technomania sweeping every corner of the world...from the high seas to the highest mountain that holds an ounce of silver or gold."

7. They believe that the earth's capacity to withstand industrial civilization is almost at an end.[15]

Christopher Manes rejects Scarce's inclusion of the animal rights faction in truly radical environmental movements because they extend ethical and moral standing only to animals and not to nonsentient entities such as plants, forests, rivers and mountains.

Manes insisted that all species are equal, a philosophy known as biocentric equality, and shared by many radical environmentalists. Manes was a member of Earth First's apocalyptic faction. When it split from the millenarian faction in 1990, it changed the face of radical environmentalism and created a serious threat to society. As Professor Martha Lee wrote:

> That split was frustrating for the movement's founders, but it also caused great problems for American law enforcement agencies. While Earth First!ers had been difficult to track while they remained a decentralized but united movement, their activities were more difficult to predict during the movement's periods of instability. Those problems only increased after Earth First!'s final split. The "new" millenarian Earth First!ers remain fairly visible, but their faith in education and social change render them less dangerous to the state than their predecessors. The apocalyptic biodiversity faction, however, poses more of a problem. Its adherents left the movement to pursue their goals independently; they still hope for an imminent apocalypse, and they still believe that their function is to preserve as much wilderness as possible before that event, using whatever tactics they deem necessary. They no longer belong to an identifiable movement, however, and thus are more difficult to track than the "new" Earth First!ers. The belief system of these individuals is also much more extreme: it gives no special status to human life.[16]

In fact, it is not terribly difficult to track the apocalyptic ecoterror factions. They typically obtain publication for their extreme rhetoric in the Earth First Journal, commonly a short time before criminal attacks similar to those advocated in the published rhetoric. Earth Firsters clearly know who provides the extreme rhetoric and knowingly use decoupling tactics to protect themselves from prosecution. However, sympathizers with radical environmentalists cannot see or admit the decoupling process, and they justify radical actions with philosophical arguments.

Bron Taylor, associate professor of religion and social ethics at the University of Wisconsin in Oshkosh, who wrote "The Religion and Politics of Earth First!" argues that radical environmentalism, Earth First in particular, contains both religious and political themes. Even though most Earth Firsters reject organized religion, they all adhere to "a radical 'ecological consciousness' that intuitively, affectively, and deeply experiences a sense of the sacredness and interconnectedness of all life."[17]

Taylor wrote in a 1996 newspaper essay:

> For the past five years, I have explored the diverse subcultures of radical environmentalism. This research convinces me that applying the terrorist label to radical environmentalism is inaccurate.
>
> Though not all radical environmentalists think alike, most would agree on three broad claims. They believe that the natural world is inherently valuable, apart from its usefulness to human beings. Indeed, the Earth and all life is sacred. This essentially religious perception provides a powerful restraint on violence because humans and nonhumans alike are seen as deserving of respect because all life participates in a sacral landscape.
>
> They also claim, as do many scientists, that humans are causing an unprecedented extinction crisis. Radical environmentalists believe that industrialism, consumerism and the domination of life by corporations intent on extending market capitalism into all planetary corners contribute to the global decline of biodiversity and the widespread desecration of land. These activists clearly deserve the label "radical" because they envision and hope for the destruction (or at least retreat) of industrial life ways. They generally believe that overturning industrialism is a prerequuisite to ecological sanity and to the reharmonization of life on earth. But few among them think this will occur as a result of their activism, and to my knowledge, none sees terrorism as a solution. Rather, if we do not change our ways, they believe, nature will take its course; great suffering will flow, including more species extinctions, perhaps even our own, and eventually an ecological equilibrium will be restored.
>
> Radical environmentalists do not see electoral politics as a way to bridge the gap between what is (the present extinction crisis) and what ought to be (the flourishing of all life forms). Democracy is seen as broken or as never having existed in the first place, and elections as dominated by corporate elites. Consequently, many laws are illegitimate, and illegal tactics, both civil disobedience and "monkey-wrenching" (movement parlance for destroying equipment used to damage the environment) may be morally permissible or even obligatory.[18]

To anyone who has encountered a range of radical environmentalists, Taylor's account is naïve. Some radicals accept violence more than
others. There is a scale of acceptance of violence. The social justice
millenarians generally lie on the pacifist end of that scale, but the most
devoted deep ecology ideologists and the criminal element lie at the other,
grim, humorless and hate-ridden. They obsessively hate loggers, miners,
ranchers, farmers, fishers, or any other resource industry workers. They
are truly consumed by hate, as Dave Foreman indicated when he quit Earth
First (see p. 60). The ecoterror factions feel hate and fatalism, which
justifies any desperate act and on occasion degenerates into self-destruc-
tive behavior, consuming their colleagues who object to violence with
threats and intimidation.

Ecoterrorism was a natural outgrowth of our changing society.
Changing values and increasing political skills in the years after World
War II allowed environmentalism to come into being. Political scientist
Ronald Inglehart found that "The values of Western publics have been
shifting from an overwhelming emphasis on material well-being and physical security toward greater emphasis on the quality of life." The generation that grew up in the Great Depression of the 1930s was obsessed with
the basics of food, clothing and shelter. The generation that grew up during World War II was obsessed with physical security and survival in the
face of total war. But then the productive forces released during the 1950s
brought affluence and security on an unprecedented scale.

During the 1960s, four important "system level" changes took
place, according to Inglehart:

1. Economic and technological development brought satisfaction
of basic sustenance needs to an increasingly large proportion of the population.

2. Distinctive cohort experiences gave the younger generation a
new outlook on life shaped primarily by the absence of total war, despite
the traumas of the Korean and Vietnam conflicts.

3. Rising levels of education, especially a higher proportion of the
population obtaining a college education, changed society's values.

4. Expansion of mass communication and increasing geographic
mobility of our society also changed values.

These system-level changes brought about individual-level
changes. Economic development and the affluence that came with the
absence of total war freed our minds from basic needs and led to an increasing emphasis on our individual needs for a sense of belonging, self-
esteem and self-realization.

Higher education and mass communication gave the 1960s generation increasing skills to cope with politics on a national scale. Education also changed personal values. Inglehart discovered that college life
makes students more liberal, more tolerant and more likely to challenge

authority. Mass communications such as television and computer networks introduce dissonant signals into our homes and show alternate lifestyles and competing mindstyles. Both higher education and mass communications made it more difficult for parents to transmit their personal values to their children in unaltered form.

In their turn, these individual-level changes brought about further system-level changes. Political issues changed during the 1970s. "Lifestyle" issues became increasingly salient. The social bases of political conflict changed: elite-directed political mobilization and class conflict gave way to elite-challenging issue-oriented special interest conflicts. The civil rights movement and anti-Vietnam war protests made elite-challenging groups permanent features of the American political landscape.

Inglehart found the best explanation for the direction society took in psychologist Abraham Maslow's "needs hierarchy." Based on many years of clinical experience, Maslow discovered the observable fact that as people become able to satisfy their basic needs for food, clothing, shelter and physical security, those basic needs no longer motivate as powerfully and a new set of needs arise. Thus it was predictable that as the privations of the Depression and the threat of World War II receded and people were able to satisfy their basic needs, they would begin to feel new needs arising.

These new higher level needs are non-material and arise in a more or less regular order. They progress from the need for love, for a sense of belonging, for self-esteem, and for "self-actualization"—to be all that one can be. During the 1960s and '70s many were able to rise up this ladder of personal well-being.

Maslow discovered that when these higher level needs are themselves gratified, they too no longer motivate as strongly and a new and final highest-level set of needs arise: The knowledge needs and the aesthetic needs—the need to know, to understand the universe we live in, and the need to live in beautiful surroundings, to create beautiful things, to live a beautiful life.

Maslow was struck by the power of the aesthetic need: "I have attempted to study this phenomenon on a clinical-personological basis with selected individuals, and have at least convinced myself that in *some* individuals there is a truly basic aesthetic need. They get sick (in special ways) from ugliness, and are cured by beautiful surroundings; they *crave* actively, and their cravings can be satisfied *only* by beauty. It is seen almost universally in healthy children. Some evidence of such an impulse is found in every culture and in every age as far back as the caveman."[19]

In the 1960s the science of ecology entered our culture to fill an emerging knowledge need, because it seemed to explain everything in nature, and later the science was popularized as environmentalism and elevated to an aesthetic—a key to beauty, to a promise of ultimate gratification, to a life of perfect harmony, all the things that Aldo Leopold wrote about. During the 1970s and '80s, a substantial fraction of the total U.S.

population had risen to the knowledge needs and the aesthetic needs, and they swelled the ranks of the environmental movement, which offered gratification.

But Maslow discovered in his long career as a clinical psychologist that there are unexpected consequences of growing all the way to the top of the needs hierarchy. People at the highest levels, he found, may go two ways: growth to loftier levels of human nature, or toward a blindly destructive pathology. The latter at these highest levels begin to feel an "independence of and a certain disdain for the old satisfiers and goal objects, with a new dependence on satisfiers and goal objects that hitherto had been overlooked, not wanted, or only causally wanted." Old gratifiers "become boring, or even repulsive." New ungratified needs are overestimated. Lower basic needs already gratified are underestimated or even devalued. In a strikingly prophetic passage, Maslow warned:

> In a word, we tend to take for granted the blessings we already have, especially if we don't have to work or struggle for them. The food, the security, the love, the admiration, the freedom that have always been there, that have never been lacking or yearned for tends not only to be unnoticed but also even to be devalued or mocked or destroyed. This phenomenon of failing to count one's blessings is, of course, not realistic and can therefore be considered to be a form of pathology. In most instances it is cured very easily, simply by experiencing the appropriate deprivation or lack, e.g., pain, hunger, poverty, loneliness, rejection, injustice, etc.
>
> This relatively neglected phenomenon of post-gratification forgetting and devaluation is, in my opinion, of very great potential importance and power.[20]

Maslow provided us with clues to explain the behavior of environmentalists in general, who come from middle and upper middle class origins, who never had to struggle for "the food, the security, the love, the admiration, the freedom that have always been there," who devalue and mock and destroy the loggers and miners and ranchers and farmers and fishers who invisibly supply their now-despised basic needs. You could call it The Spoiled Brat Syndrome. It explains how Aldo Leopold could envision a land ethic that calls economic needs "mere" and disregards the consequences on others of ending property rights in land.

Maslow also helps us understand the individuals who have a truly basic aesthetic need, those who get sick in special ways in ugliness and actively crave and can only be cured by being in beautiful surroundings. We immediately call to mind Edward Abbey and Dave Foreman and some other biocentric apocalyptics. But we do not see the social justice millenarians quite so dependent upon beauty; they are more in the knowledge needs, the intellectual needs, at the ideological level. The millenarians are less radical than the apocalyptics, perhaps because they have not made that final step away from teeming humanity into the more abstract realm of beauty. The apocalyptics are misanthropes. The millenarians are not.

But why did the growing affluence of post-World War II America and the rise of large populations up the needs hierarchy generate the environmental movement and turn towards nature as icon and idol? Why not some other direction? For that answer we must turn to the work of two influential early anthropologists, Arthur O. Lovejoy and George Boas. Their researches convinced them that primitivism lies at the root of most human behavior. Primitivism, as they define it, is "the belief of men living in a relatively highly evolved and complex cultural condition that a life far simpler and less sophisticated in some or in all respects is a more desirable life." Primitivism reflects the assumption "that correctness in opinion and excellence in individual conduct or in the constitution of society consists in conformity to some standard or norm expressed by the term 'nature' or its derivatives."[21]

That defining essence applies to all environmentalisms, deep, shallow or what-not, but is particularly strong in the biocentric apocalyptic faction and the ecoterror faction. Even apocalyptic Christopher Manes agreed on this point:

> With uncharacteristic insight, Ron Arnold writes in *At the Eye of the Storm* that "eco-terrorists are not preservers of the status quo, or even 'New Luddites' anxious about technology stealing their jobs, but rather deeply primitivist activists opposed to industrial civilization itself." Except for the unflattering use of the ecoterrorist epithet, the statement is an essentially correct description of how most radical environmentalists feel toward industrialism.[22]

Lovejoy and Boas studied two kinds of primitivism that are pertinent to environmentalism: Chronological primitivism and cultural primitivism.

Chronological primitivism is a kind of philosophy of history answering the question: When is the best of times, the past, the present, or the future? Chronological primitivists answer: The past. The Theory of Decline supposes that the highest degree of excellence or happiness in man's existence came at the beginning of history. The early Greeks, for example, held that a primal Golden Age was the best of times:

> First of all the deathless gods who dwell on Olympus made a golden race of mortal men who lived in the time of Cronos when he was reigning in heaven. And they lived like gods without sor-

row of heart, remote and free from toil and grief: miserable age rested not on them; but with legs and arms never failing they made merry with feasting beyond the reach of all evils. When they died, it was as though they were overcome with sleep, and they had all good things; for the fruitful earth unforced bare them fruit abundantly and without stint. They dwelt in ease and peace upon the lands with many good things, rich in flocks and loved by the blessed gods.[23]

That description of The Golden Age by Hesiod, circa 700 B. C., embodies the inducements offered by all utopias; its outlines are clearly recognizable in modern ecotopias. Live simply and Nature will take care of all your needs—like it used to be. Back to Nature is best, back in time. Whether back to the Golden Age of the Greeks or back to the Late Paleolithic, it's always *back*. Though they may differ on the details, most environmentalists feel that earlier times were better times. They idealize pristine America prior to European settlement. They idealize ancient societies, as we see in the revival of the Goddess, Druidic practices, Celtic mythology and so forth. They are backward-looking chronological primitivists.

Cultural primitivism, on the other hand, is the discontent of the civilized with civilization. To people living in any phase of cultural development it is always possible to conceive of some simpler one by pointing to contemporary tribal peoples. Cultural primitivism has had enduring roots in human psychology ever since the civilizing process began. As Lovejoy and Boas wrote,

> It is a not improbable conjecture that the feeling that humanity was becoming overcivilized, that life was getting too complicated and over-refined, dates from the time when the cave-man first became such. It can hardly be supposed—if the cave-men were at all like their descendants—that none among them discoursed with contempt upon the cowardly effeminacy of living under shelter or upon the exasperating inconvenience of constantly returning for food and sleep to the same place instead of being free to roam at large in the wide-open spaces.[24]

The earliest coherent narrative on earth, a tale of a Sumerian hero from the Third Millennium BCE, *The Epic of Gilgamesh*, is suffused in such primitivism. It gives us a glimpse of the first monkeywrencher, animal rights activist and ecoterrorist in its opening sequences. The tale begins with Gilgamesh, King of Uruk, oppressing his people, who pray to the gods to send them deliverance. Aruru, the Mother, the Great Lady, hears their prayers and creates the wild man Enkidu as a diversionary tactic, so that the two would "square off one against the other, that Uruk may have peace." Enikdu is first encountered by The Stalker, a farmer and hunter living in a rural area between Uruk and the wilderness. Here is a recent translation of the crucial passage from the Akkadian version composed by an exorcist-priest, Sîn-leqi-unninnï, in the Middle Babylonian Period, between 1600-1300 BCE:

When Aruru heard this, she formed an image of Anu in her heart.
Aruru washed her hands, pinched off clay and threw it into the wilderness:
In the wilderness she made Enkidu the fighter; she gave birth in darkness and silence to one like the war god Ninurta.
His whole body was covered thickly with hair, his head covered with hair like a woman's;
the locks of his hair grew abundantly, like those of the grain god Nisaba.
He knew neither people nor homeland; he was clothed in the clothing of Sumuquan the cattle god.
He fed with the gazelles on grass;
with the wild animals he drank at waterholes; with hurrying animals his heart grew light in the waters.

The Stalker, man-and-hunter,
met him at the watering place
one day—a second, a third—at the watering place.
Seeing him, the Stalker's face went still.
He, Enkidu, and his beasts had intruded on the Stalker's place.
Worried, troubled, quiet,
the Stalker's heart rushed; his face grew dark.
Woe entered his heart.
His face was like that of one who travels a long road.

The Stalker shaped his mouth and spoke, saying to his father

"Father, there is a man who has come from the hills.
In all the land he is the most powerful; power belongs to him.
Like a shooting star of the god Anu, he has awesome strength.
He ranges endlessly over the hills,
endlessly feeds on grass with the animals,
endlessly sets his feet in the direction of the watering place.
For terror I cannot go near him.
He fills up the pits I dig;
he tears out the traps I set;
he allows the beasts to slip through my hands, the hurrying creatures of the abandon;
in the wilderness he does not let me work."[25]

It could be a recent newspaper story.

Though they may differ on the details, most environmentalists feel that the simpler life of which they dream has been somewhere, at some time, actually lived by human beings. In primitive tribal societies environmentalists see living replicas of the character and life they wish to emulate. The simpler, better life *must* have existed, and present-day primitive cultures are the proof that such a life is possible for everyone in the primitive future. Environmentalists look not only to the past for better times, but also to extant preliterate cultures for better lives. Environmentalists are cultural primitivists.

The key to primitivism is looking to "nature" as the measure of all things. Only nature is good. Man can be good only by following nature. Lovejoy and Boas found at least seven varieties of "the state of nature" that shape primitivist belief, themes that pervade Western literature from ancient times:

1. The original condition of things, and especially the state of man as nature first made him, whatever this condition may be supposed to have been, is best. As Rousseau put it, "Everything is good when it leaves the hands of the Creator; everything degenerates in the hands of man."

2. That condition of human life is best which is most free from the intrusion of technology, where none or only the simplest practical arts are known. Emerson and Thoreau are American exemplars of this creed.

3. Natural societies should have no private property, particularly private property in land. Most radical environmentalists oppose private property rights in land.

4. Natural societies should have the simplest marital states, such as the community of wives and children; in its extreme form, sexual promiscuity, including incest. Such sexual insurgency has a long history: The Greek philosopher Diogenes is reputed to have advocated the community of wives, "considering marriage to consist in nothing but the union of the man persuading with the woman consenting. And for this reason he also thought that children should be held in common."[26]

Diogenes also asserted that incest is not against nature: "Oedipus discovered that he had had intercourse with his mother and had had children by her; whereupon—when he should, perhaps, have concealed this, or else have made it lawful for the Thebans—he first of all announced it to everybody, and then reproached himself and moaned loudly that he was father and brother to the same children, and husband and son to the same woman. But cocks do not see anything wrong in such unions, nor do dogs or asses, nor yet the Persians, who are considered the best people of Asia."[27]

5. Vegetarianism is best, not on hygienic grounds but as an expression of the feeling that bloodshed in all its forms is sinful, that man in an ideal state should—and once did—live at peace with the animals as well as with his own kind, as Enkidu did in the Epic of Gilgamesh. Gilgamesh tells us that Enkidu lived among the animals as one of them until seduced by a prostitute sent to lure him to the city of Uruk where his great strength could relieve the populace of the tyrant Gilgamesh. After having sex with the prostitute, Enkidu could no longer communicate with the animals. They fled from him. He was polluted because sex with his own kind had made him distinctively human. He went to the city where he abandoned his vegetarian ways and his "natural" life, a metaphor for the transition each of us makes from uncivilized infant to civilized adult. Note the implicit psychological equation: sex equals eating meat equals pollution equals human.

6. Society is best without organized political government, or without any except the "natural" government of the family, clan or tribe—anarchism, in the nonpejorative technical sense of the word.

7. Natural ethics operates when man is in unity with himself, controlled by "natural" impulses, without deliberate and self-conscious moral effort, the constraint of rules, or the sense of sin.

We can see the roots of Deep Ecology clearly in primitivism, as well as the behavior of radical environmentalists, including animal rights activists. But primitivism is not the exclusive domain of radicals. Primitivism dwells in all of us. We each wear civilization more or less uneasily. Change is the most difficult aspect of existence and technology has sped the rate of change beyond the endurance of many. It is not difficult to feel what the radical environmentalist feels.

But it is very difficult to empathize with the political fanaticism of the ecoterrorist who bombs and vandalizes and obstructs others to save nature. Researchers from historians to criminologists to psychoanalysts have studied fanaticism since early in the 20th century. In a 1983 study, *Fanaticism: A Historical and Psychoanalytical Study*, two historians, Gérard de Puymège and Miklos Molnar of the University of Lausanne, and a professor of psychiatry, André Haynal of the University of Geneva, discover useful insights.

Even though the modern concept of "fanaticism" does not reach even as far back as the time of Shakespeare, its ancient Latin roots cast light on our contemporary understanding.

In Rome, inspired soothsayers interpreting omens were called *fanatici*, as were the priests of the oriental mystery cult to the goddess Ma Bellona who in their delirium struck themselves with swords and hatchets, causing their blood to gush forth. The word, initially without pejorative connotations, derives from *fanum*, the temple where the oracles were pronounced, and has the same root as *vates*, meaning prophet. The *fanum* is the place of prophecy....

Receiving inspiration from the other world, the *fanaticus* expressed himself with extravagance, like one demented, in ecstatic and sometimes violent contorsions to the point of self-mutilation. [28]

The followers of the fanatic crowded around the temple. But entry into the temple was forbidden to the noninitiate—the profane (from the Latin *profanus*, literally *pro*, "in front of," plus *fanum*, "the temple") must stay outside. The profane are dangerous to the religion as long as they remain unconverted; their every act is *profanation*, disregard of the sacred through ignorance. They desecrate everything they touch. This is how many radical environmentalists regard those who are not converted to their earth religion.

Something similar came from ancient Judea among the Zealots: the new convert, unlike the initiated *zealot*, must remain on the square in front of the temple during services until the circumcision which marks his total integration. During this ceremony the convert is given a new name for a new life, he becomes a *neophyte*, a "new child" (Greek, *neo-phuton*) in the "family." The sect of the Zealots bequeathed to psychiatry the concept of "zealotry" (from the Greek *zelotupos*, meaning "jealous"), designating both maniacal jealousy and argumentative faith.

Radical environmentalists likewise take on new names at their conversion, naming themselves after animals, natural objects or symbolic actions—Catfish, Riverwind, Digger, Elk Herd—and join the tribe, which gives some their first real sense of family. As Dave Foreman (Digger) said, "We created a community...and you need that...[but] you don't have that in your family anymore, and you don't have that in your neighborhood anymore.... To a lot of people in Earth First!, the tribal belonging became the main thing."[29] It was Maslow's needs hierarchy at work, the need for a sense of belonging.

The key point about fanaticism is that fanatics break with tradition for an idea or an ideal which becomes in their mind an absolute, worth sacrificing themselves and others for. Fanatics designate villains against whom they can unleash their rage without guilt. They feel contempt or indifference for everything other than the object of their passion. They have an unshakable certainty in the rightness of their ideas. They project aggressiveness on the presumed enemy, who is seen as a persecutor. They justify the transgression of morality ("we are monkeywrenching for the good of earth"). They are consumed by hatred, which is concealed behind the facade of the "just person."

André Haynal took the view that fanaticism "can only be understood through the depths and subterranean currents of our psyche." Fanaticism, he concluded, is a megalomaniacal condition. Pointing to Sigmund Freud's work, Haynal noted that megalomania has a structure traceable to child psychology and the "fiction of omnipotence" that must give way to reality as the child grows. And "the little primitive creature," as Freud called the infant, must turn into a civilized human being in the space of a few short years. The child must pass through an immensely long stretch of human cultural development in a highly abbreviated form. Sometimes it doesn't work. Fanatics, Haynal wrote, search for "a pathological omnipotence which most often masks feelings of impotence and despair, the inability to accept one's own limits and submit oneself to the rules of civilization as interiorized in the superego."[30]

Most strikingly, Haynal wrote that fanaticism can only be understood against the background of the psychoanalytical conception of civilization, the tension between instincts and culture. Freud was one of only a handful of researchers who directly investigated the question of why certain individuals want to abandon or destroy civilization. Freud's conclusions are thought-provoking.

> But how ungrateful, how short-sighted after all, to strive for the abolition of civilization! What would then remain would be a state of nature, and that would be far harder to bear. It is true that nature would not demand any restrictions of instinct from us, she would let us do as we liked; but she has her own particularly effective method of restricting us. She destroys us—coldly, cruelly, relentlessly, as it seems to us, and possibly through the very things that occasioned our satisfaction. It was precisely because of these dangers with which nature threatens us that we came together and created civilization, which is also, among other things, intended to make our communal life possible. For the principal task of civilization, its actual *raison d'être*, is to defend us against nature.[31]

It is this precise point that the radical environmental cannot, must not accept. Rejection of civilization, rejection of defending ourselves *against* nature is the core of their worldview. The idea of defending ourselves *against* nature is outdated, unecological, they argue. Civilization is the culprit. Nature is benign. Human beings are the culprit. Anthropocentrism is the culprit. We must learn to live *with* nature, not *against* it. Biocentrism is the solution. Policy must be made considering earth first, even if that hurts humanity. After all, earth and its ecology is the basis of everything, including our human existence.

They may be telling us more about themselves than about the world.

Freud considered civilization as "a process in service of Eros," of life, whose purpose is to combine individuals, families, races into one great unity, "the unity of mankind." But man's aggressive instinct, "the hostility of each against all and of all against each," which Freud felt to be a "self-subsisting instinctual disposition in man," opposed the program of civilization. He asserted, "This aggressive instinct is the derivative and the main representative of the death instinct which we have found alongside of Eros and which shares world-dominion with it."

And now, I think, the meaning of the evolution of civilization is no longer obscure to us. It must present the struggle between Eros and Death, between the instinct of life and the instinct of destruction, as it works itself out in the human species. This struggle is what all life essentially consists of, and the evolution of civilization may therefore be simply described as the struggle for life of the human species.[32]

If Freud was right, an aggressive instinct lurks in each of us that thwarts "the struggle for life of the human species." Perhaps we will find the roots of ecoterrorism in all of us, controlled and disarmed.

Presented with a popular environmental movement, charismatic leaders, favorable media attention and foundation-funded condemnations of the opposition, is it any surprise that some radical environmentalists would turn their aggressive instinct to destroying the civilization they had been taught was the source of all evil?

The wise use movement has arisen to question environmentalism and defend civilization. It says we should not give in to hate.

What if civilization is not the destructive evil we think it is?

What if civilization is "a process in service of Eros?"

What then?

Chapter Seven Footnotes

1 "The Cyber-Maxims of Esther Dyson," *New York Times Magazine*, Sunday, July 7, 1996, by Claudia Dreifus, p. 19.

2 The original 1961 edition of *Webster's Third New International Dictionary*, the authoritative dictionary of the American language at the time, defined "environmentalism" as "a theory that views only environment rather than heredity as the important factor in the development of the individual or a group—compare HEREDITARIANISM." The 1971 edition has an entry for environmentalist ("one concerned with the quality of the human environment, *esp* a specialist in human ecology) but none for environmentalism. Merriam-Webster, Inc. files show its earliest record of "environmentalism" in the *Annual Report to the Stockholders* by the President of ITT in 1970: "...our broad obligation to society at large through many programs as well as social action, urban affairs, environmentalism and consumerism." Telephone interview with editor E. W. Gilman of Merriam-Webster, Inc., Springfield, Massachusetts, November 15, 1996.

3 Aldo Leopold, *A Sand County Almanac*, p. 240.

4 *Ibid.*, p. 238.

5 *Ibid.*, p. 261.

6 *Ibid.*, p. 261.

7 *Ibid.*, p. 262.

8 *Ibid.*, p. 239.

9 The text of the Opinions in Sierra Club v. Morton is in Christopher D. Stone, *Should Trees Have Standing: Toward Legal Rights for Natural Objects*, William Kaufmann, Inc., Los Altos, California, 1974.

10 John Donne, Devotions XVII: "No man is an Iland, intire of itselfe; every man is a peece of the Continent, a part of the maine; if a Clod bee washed away by the Sea, Europe is the lesse, as well as if a Promontorie were, as well as if a Mannor of thy friends or of thine owne were; any man's death diminishes me, because I am involved in Mankinde; And therefore never send to know for whom the bell tolls; it tolls for thee."

11 Arne Naess, "The Shallow and the Deep, Long-Range Ecology Movement. A Summary," *Inquiry* 16 (1973): pp. 95-100.

12 Arne Naess, *Gandhi and Group Conflict: An Exploration of Satyagraha—Theoretical Background*, Oslo, 1974; "Spinoza and Ecology," in S. Hessing, ed., *Speculum Spinozarnum 1677-1977*, London, 1978; "Through Spinoza to Mahayana Buddhism, or through Mahayana Buddhism to Spinoza?" in J. Wetlesen, ed., *Spinoza's Philosophy of Man; Proceedings of the Scandinavian Spinoza Symposium 1977*, Oslo, 1978; "Self-realization in Mixed Communities of Humans, Bears, Sheep and Wolves," *Inquiry* 22, 1979, pp. 231-241.

13 Bill Devall and George Sessions, *Deep Ecology: Living as if Nature Mattered*, Peregrine Smith Books, Salt Lake City, 1985, particularly Chapter 6, "Some Sources of the Deep Ecology Perspective," p. 70.

14 Michael Brown and John May, *The Greenpeace Story*, Dorling Kindersley, London, p. 9.

15 Rik Scarce, *Eco-Warriors*, pp. 4-13

16 Lee, p. 149.

[17] Bron Taylor, "The Religion and Politics of Earth First!," *Ecologist* vol. 21, no. 6, November/December 1991, p. 259.

[18] "Ecologist to Unabomber? Culture: So-called radical environmentalists may fight rough but aren't terrorists," *Los Angeles Times*, Friday, May 17, 1996, by Bron Taylor; p. 9.

[19] Abraham H. Maslow, *Motivation and Personality*, Second Edition, Harper & Row, New York, 1970, p. 51.

[20] *Ibid.*, p. 61.

[21] Arthur O. Lovejoy and George Boas, *Primitivism and Related Ideas in Antiquity*, Octagon Books, New York, 1973, first published by The Johns Hopkins Press, 1935, p. 7, p. 103.

[22] Christopher Manes, *Green Rage: Radical Environmentalism and the Unmaking of Civilization*, Little, Brown and Company, Boston, 1990, pp. 225-226.

[23] Hesiod, "Works and Days," in *Hesiod: The Homeric Hymns and Homerica*, English translation by Hugh G. Evelyn-White, Loeb Edition, published in the United States by Harvard University Press, Cambridge, 1914, p. 11.

[24] Lovejoy and Boas, p. 7.

[25] *Gilgamesh*, translated from the Sîn-leqi-unninnï version by John Gardner and John Maier, Vintage Books, New York, 1984, p. 68, p. 73.

[26] Diogenes Laertius, VI, 72. See the Loeb Edition, *Diogenes Laertius: Lives of Eminent Philosophers*, translated by R. D. Hicks, Vol. II, published in America by the Harvard University Press, Cambridge, 1925, p. 75.

[27] Dio Chrysostom, *Discourses*, X, 29-30. Edited by J. von Arnim, 1893.

[28] André Haynal, Miklos Molnar and Gérard de Puymège, *Fanaticism: A Historical and Psychoanalytical Study*, Schocken Books, New York, 1983, p. 17.

[29] Interview by Martha Lee with Dave Foreman, cited in Lee, p. 36.

[30] Haynal, Molnar and de Puymège, p. 38.

[31] Sigmund Freud, *The Future of an Illusion* (1927), W.W. Norton & Company, New York, translated by James Strachey, 1961, pp. 18-19.

[32] Sigmund Freud, *Civilization and its Discontents* (1930), W.W. Norton & Company, New York, translated by James Strachey, 1961, p. 69.

BIBLIOGRAPHY

Books

Edward Abbey, *Desert Solitaire: A Season in the Wilderness*, McGraw-Hill, New York, 1968.

Edward Abbey, *The Monkey Wrench Gang*, J.B. Lippincott, New York, 1972.

Margot Adler, *Drawing Down the Moon: Witches, Druids, Goddess-Worshippers, and Other Pagans in America Today*, Beacon Press, Boston, 1979, second edition 1986.

Ron Arnold, *At the Eye of the Storm: James Watt and the Environmentalists*, Regnery Gateway, Chicago, 1982.

Ron Arnold, *Ecology Wars: Environmentalism As If People Mattered*, Free Enterprise Press, Bellevue, Washington, 1987.

Murray Bookchin, *Remaking Society: Pathways to a Green Future*, South End Press, 1990.

Michael Brown and John May, *The Greenpeace Story*, Dorling Kindersley, London.

James W. Clarke, *American Assassins: The Darker Side of Politics*, Princeton University Presss, Princeton, New Jersey, 1982.

Alexander Cockburn and James Ridgeway, *Political Ecology: An Activist's Reader on Energy, Land, Food, Technology, Health, and the Economics and Politics of Social Change*, Times Books, New York, 1979.

Alexander Cockburn and Ken Silverstein, *Washington Babylon*, Verso, New York, 1996.

Barry Commoner, *Making Peace With The Planet*, Pantheon, New York, 1990.

Garrett De Bell, editor, *The Environmental Handbook*, Ballantine Books, New York, 1970.

Bill Devall and George Sessions, *Deep Ecology: Living as if Nature Mattered*, Peregrine Smith Books, Salt Lake City, 1985.

Dio Chrysostom, *Discourses*, X. Edited by J. von Arnim, 1893.

Diogenes Laertius, VI, Loeb Edition, *Diogenes Laertius: Lives of Eminent Philosophers*, translated by R. D. Hicks, Vol. II, published in America by the Harvard University Press, Cambridge, 1925.

Environmental Action staff, editors, *Ecotage!*, Pocket Books, New York, 1971.

Dave Foreman and Bill Haywood (pseudonym), editors, *Ecodefense: A Field Guide to Monkeywrenching*, Ned Ludd Books, Tucson, Arizona, 1985.

Dave Foreman, *Confessions of an Eco-Warrior*, Harmony, New York, 1991.

Dave Foreman and Murray Bookchin, *Defending the Earth*, South End Press, Boston, 1990.

The Foundation Center, *Grants for Environmental Protection and Animal Welfare, 1991-1992*, New York, 1992.

The Foundation Center; *National Guide to Funding for the Environment & Animal Welfare*, New York, 1992.

Michael W. Fox, *Returning to Eden: Animal Rights and Human Responsibility*, Viking, New York, 1980.

Sigmund Freud, *The Future of an Illusion* (1927), W.W. Norton & Company, New York, translated by James Strachey, 1961.

Sigmund Freud, *Civilization and its Discontents* (1930), W.W. Norton & Company, New York, translated by James Strachey, 1961.

Gilgamesh, translated from the Sîn-leqi-unninnï version by John Gardner and John Maier, Vintage Books, New York, 1984

Marija Alseikaite Gimbutas, *The Language of the Goddess: Unearthing the Hidden Symbols of Western Civilization*, San Francisco, Harper & Row, 1989.

Alan Gottlieb, editor, *The Wise Use Agenda*, Free Enterprise Press, Belleuve, Washington 1988.

David T. Hardy, Esq., *America's New Extremists: What You Need to Know About the Animal Rights Movement*, Washington Legal Foundation, Washington, D. C., 1990.

André Haynal, Miklos Molnar and Gérard de Puymège, *Fanaticism: A Historical and Psychoanalytical Study*, Schoken Books, New York, 1983.

Samuel P. Hays, *Conservation and the Gospel of Efficiency: The Progressive Conservation Movement 1890-1920*, Harvard University Press, Cambridge, 1959.

Robert A. Heinlein, *Stranger in a Strange Land*, G. P. Putnam's Sons, New York, 1961.

David Helvarg, *The War Against the Greens*, Sierra Club, San Francisco, 1994.

David Henshaw, *Animal Warfare*, Fontana Paperbacks, London, 1989.

Doug Henwood, *The State of the U.S.A. Atlas: The Changing Face of American Life in Maps and Pictures*, Simon and Schuster, New York, 1994.

Doug Henwood, *Wall Street*, Verso, New York, 1997.

Hesiod, "Works and Days," in *Hesiod: The Homeric Hymns and Homerica*, English translation by Hugh G. Evelyn-White, Loeb Edition, published in the United States by Harvard University Press, Cambridge, 1914.

Bruce Hoffman, *The Contrasting Ethical Foundations of Terrorism in the 1980s*, The Rand Corporation, Santa Monica, California, 1988.

Robert Hunter, *Warriors of the Rainbow: A Chronicle of the Greenpeace Movement*, Holt, Rinehart and Winston, New York, 1979.

James M. Jasper and Dorothy Nelkin, *The Animal Rights Crusade: The Growth of a Moral Protest*, The Free Press, New York, 1992.

W. Alton Jones Foundation, "The wise use movement," by John Peterson Meyers and Debra Callahan, Charlottesville, Virginia, February 6, 1992.

Aldo Leopold, *A Sand County Almanac*, Oxford University Press, Oxford, 1949.

Martha F. Lee, *Earth First! Environmental Apocalypse*, Syracuse University Press, Syracuse, 1995.

Clara Livsey, M.D., *The Manson Women: A "Family" Portrait*, Richard Marek Publishers, New York, 1980.

BIBLIOGRAPHY

Arthur O. Lovejoy and George Boas, *Primitivism and Related Ideas in Antiquity*, Octagon Books, New York, 1973, first published by The Johns Hopkins Press, 1935.

James E. Lovelock, *Gaia: A new look at life on Earth*, Oxford University Press, New York, 1979.

MacWilliams Cosgrove Snider, "The wise use movement: Strategic Analysis and Fifty State Review," Clearinghouse on Environmental Advocacy and Research, Washington, D.C., March 1993.

Christopher Manes, *Green Rage: Radical Environmentalism and the Unmaking of Civilization*, Little, Brown and Company, Boston, 1990.

Kathleen Marquardt, *AnimalScam: The Beastly Abuse of Human Rights*, with Herbert M. Levine and Mark Larochelle, Regnery Gateway, Washington, D.C., 1993.

Robert Marshall, *The People's Forests*, H. Smith & R. Haas, New York, 1933.

Abraham H. Maslow, *Motivation and Personality*, Second Edition, New York, Harper & Row, 1970.

John McPhee, *Encounters with the Archdruid*, Farrar Strauss and Giroux, New York, 1971.

Donella H. Meadows, Dennis L. Meadows, Jørgen Randers, and William W. Behrens III, *The Limits to Growth*, Universe Books, New York, 1972.

Carolyn Merchant, *The Death of Nature: Women, Ecology, and the Scientific Revolution*, Harper & Row, San Francisco, 1980.

John G. Mitchell and Constance L. Stallings, editors, *Ecotactics: The Sierra Club Handbook for Environment Activists*, Pocket Books, New York, 1970.

Richard Morgan, *Love and Anger: An Organizing Handbook for Activists in the Struggle for Animal Rights and In Other Progressive Political Movements*, second edition, Westport, Connecticut, Animal Rights Network, 1981.

Arne Naess, *Gandhi and Group Conflict: An Exploration of Satyagraha—Theoretical Background*, Oslo, 1974.

Roderick Frazier Nash, *The Rights of Nature*, University of Wisconsin Press, Madison, 1989.

Roderick Frazier Nash, *Wilderness and the American Mind*, Yale University Press, New Haven, 1967, revised edition 1973.

Ingrid Newkirk, *Free the Animals! The untold story of the U.S. Animal Liberation Front and its Founder, Valerie*, Noble Press, Chicago, 1992.

Marvin Olasky, *Patterns of Corporate Philanthropy: Public Affairs Giving and the Forbes 100*, Capital Research Center, Washington, D.C., 1987.

Judith Plant, editor, *Healing the Wounds: The Promise of Ecofeminism*, New Society, Philadelphia, 1989.

Tony Poveda, *Lawlessness and Reform: The FBI in Transition*, Brooks/Cole, Pacific Grove, California, 1990.

Tom Regan, *The Case for Animal Rights*, University of California Press, Berkeley, 1983.

Andrew Rowell, *Green Backlash: The Subversion of the Environment Movement*, Routledge, London, 1996.

Rik Scarce, *Eco-Warriors: Understanding the Radical Environmental Movement*, The Noble Press, Inc., Chicago, 1990.

Screaming Wolf [pseudonym attributed to Sidney and Tanya Singer], *A Declaration of War: Killing People to Save Animals and the Environment*, Patrick Henry Press, Grass Valley, California.

Peter Singer, *Animal Liberation: A New Ethics for Our Treatment of Animals*, New York Review, distributed by Random House, 1975.

Brent L. Smith, *Terrorism in America: Pipe Bombs and Pipe Dreams*, State University of New York Press, Albany, 1994.

Starhawk (pseudonym of Miriam Simos), *The Spiral Dance, A Rebirth of the Ancient Religion of the Great Goddess*, Harper & Row, San Francisco, 1979.

Jacqueline Vaughn Switzer, *Green Backlash: The History and Politics of Environmental Opposition in the U.S.*, Lynne Rienner Publishers, Boulder, 1997.

Carol Van Strum, *A Bitter Fog*, Sierra Club, San Francisco, 1983.

Captain Paul Watson, *Earth Force! An Earth Warrior's Guide to Strategy*, Chaco Press, Los Angeles, 1993, Foreword by Dave Foreman.

Paul Watson as told to Warren Rogers, *Sea Shepherd: My Fight for Whales and Seals*, W. W. Norton & Company, New York, 1982.

Susan Zakin's *Coyotes and Town Dogs: Earth First! and the Environmental Movement*, Viking Penguin, New York, 1993.

Articles and Studies

Ron Arnold, "EcoTerrorism," *Reason*, February 1983, vol. 14, no. 10.

Doug Bandow, *Ecoterrorism: The Dangerous Fringe of the Environmental Movement*, The Heritage Foundation Backgrounder, Washington, D.C., April 1990.

James Barnes, "Barry Clausen: Flim Flam Man or Private Dick?" *Earth First Journal*, Beltane (May-June), 1996.

Kenneth Brower, "Mr. Monkeywrench," *Harrowsmith*, Vol. III, No. 17, September-October, 1988.

Alexander Cockburn, "Earth First!, the Press and the Unabomber," *The Nation*, May 6, 1996.

Claudia Dreifus, "The Cyber-Maxims of Esther Dyson," *New York Times Magazine*, Sunday, July 7, 1996.

Kimberly D. Elsbach and Robert I. Sutton, "Acquiring organizational legitimacy through illegitimate actions: a marriage of institutional and impression management theories," *Academy of Management Journal*, October 1992 vol. 35 no. 4.

John Elson, "Murderer's Manifesto: Threatening more attacks, Unabomber issues a screed against technology," *Time*, July 10, 1995 Volume 146, No. 2.

Dave Foreman, "Earth First!," *The Progressive*, vol. 45, no. 10, October 1981.

Jonathan Franklin, "First They Kill Your Dog," *Muckracker: Journal of the Center for Investigative Reporting* , Fall 1992.

David Helvarg, "The anti-enviro connection (paramilitary groups and anti-environmentalists)," *The Nation*, May 22, 1995 v260 n20.

David Helvarg, "Anti-enviros are getting uglier: the war on Greens," *The Nation*, Nov 28, 1994 v259 n18.

Assistant Chief Harry R. Hueston II, "Battling the Animal Liberation Front," University of Arizona Police Department, in *The Police Chief*, September 1990.

Joe Kane, "Mother nature's army; guerrilla warfare comes to the American forest," *Esquire*, Feb. 1987, vol. 107.

Dean Kuipers, "Eco warriors," (Interview with Mike Roselle), *Playboy*, vol. 40, no. 4, April 1993.

Dean Kuipers, "The Tracks of the Coyote," *Rolling Stone*, June 1, 1995

Aldo Leopold, *Round River: From the Journals of Aldo Leopold*, ed. Luna Leopold, Oxford University Press, New York, 1953.

J. E. Lovelock, "Gaia as seen through the atmosphere," *Atmospheric Environment*, no. 6, p. 579, 1972.

Christopher Manes, "Green Rage," *Penthouse*, May, 1990.

Arne Naess, "The Shallow and the Deep, Long-Range Ecology Movement. A Summary," *Inquiry* 16 (1973).

Susan Reed, "Activist Ingrid Newkirk fights passionately for the rights of animals, some critics say humans may suffer," *People Weekly*, Oct 22, 1990, Vol. 34, No. 16.

Susan Reed and Sue Carswell, "Animal passion," *People Weekly*, January 18, 1993, vol. 39 no. 2

Carmelo Ruiz-Marrero, "The International PR Machine: Environmentalism á la Burson-Marsteller," *Earth First Journal*, vol. 14, no. 3, Brigid, February-March 1994.

Ken Silverstein and Alexander Cockburn, "The Collapse of the Mainstream Greens," *CounterPunch*, Vol 1, No. 17, October 1, 1994.

Leslie Spencer, "Fighting Back," *Forbes*, July 19, 1993.

Christopher D. Stone, *Should Trees Have Standing: Toward Legal Rights for Natural Objects*, William Kaufmann, Inc., Los Altos, California, 1974.

Bron Taylor, "The Religion and Politics of Earth First!," *Ecologist* vol. 21, no. 6, November/December 1991.

Robert Wright, "Are animals people too? Close enough for moral discomfort," *The New Republic*, March 12, 1990.

Susan Zakin, "Earth First!" *Smart*, September - October 1989.

Government Reports and Criminal Court Documents

Affidavit of Special Agent Donald J. Sachtleben, FBI, to U.S. District Judge Charles C. Lovell, United States District Court, Helena Division, District of Montana, Criminal Complaint in the case of *United States of America v. Theodore John Kaczynski*, filed 96 APR 4 AM 10 58.

General Accounting Office, "Illegal and Unauthorized Activities On Public Lands—A Problem With Serious Implications," Washington, D.C., 1982, CED-82-48.

"Government's Presentencing Memorandum," in the case *United States of America v. Rodney Adam Coronado*, No. 1:93-CR-116, United States District Court for the Western District of Michigan, Southern Division, by United States Attorney Michael H. Dettmer.

Report to Congress on the Extent and Effects of Domestic and International Terrorism on Animal Enterprises, U.S. Department of Justice, U.S. Government Printing Office, Washington, D.C., August, 1993.

Terrorism in the United States, 1989, Terrorist Research and Analytical Center, Counterterrorism Section, Criminal Investigative Division, U. S. Department of Justice, Federal Bureau of Investigation, Washington, D. C., December 31, 1989.

Terrorism in the United States: 1994, FBI Terrorist Research and Analytical Center, Washington, D.C., U.S. Department of Justice, 1995.

INDEX

INDEX